Synthetic Biology
HANDBOOK

Darren N. Nesbeth
University College London, UK

CRC Press
Taylor & Francis Group
Boca Raton London New York

CRC Press is an imprint of the
Taylor & Francis Group, an **informa** business

CRC Press
Taylor & Francis Group
6000 Broken Sound Parkway NW, Suite 300
Boca Raton, FL 33487-2742

First issued in paperback 2019

© 2016 by Taylor & Francis Group, LLC
CRC Press is an imprint of Taylor & Francis Group, an Informa business

No claim to original U.S. Government works

ISBN-13: 978-1-4665-6847-1 (hbk)
ISBN-13: 978-0-367-86772-0 (pbk)

Library of Congress Cataloging-in-Publication Data

Names: Nesbeth, Darren N., editor.
Title: Synthetic biology handbook / editor, Darren N. Nesbeth.
Description: Boca Raton : CRC Press/Taylor & Francis, 2016. | Includes
bibliographical references and index.
Identifiers: LCCN 2016003967 | ISBN 9781466568471 (hardback : alk. paper)
Subjects: | MESH: Synthetic Biology--methods | Synthetic Biology--standards |
Bioengineering--methods | Genetic Engineering
Classification: LCC QH438.7 | NLM QT 36 | DDC 572.8/6--dc23
LC record available at http://lccn.loc.gov/2016003967

Visit the Taylor & Francis Web site at
http://www.taylorandfrancis.com

and the CRC Press Web site at
http://www.crcpress.com

Contents

Preface

Writing a synthetic biology textbook is beset with many of the same challenges faced by the field in the wider scientific world. What is the 'definitive' definition? What are the core topics that define the field? What terms and phrases have the community accepted? What terms and phrases have the contributing authors accepted? Ultimately these issues must give way to providing an account of the research, even though debate may continue. Initially, this book was envisaged as an olive branch to the more cynical or unaccepting members of the fundamental life science community. But as the sheer scale of elegant and interesting work piled up for inclusion in each chapter, it became clear that a comprehensive textbook would also be valuable for the diehard biohackers, academics, entrepreneurs and captains of synthetic biology industry who have rolled their sleeves up and got to work founding companies, making research breakthroughs and democratizing biology over the last decade. I would like to thank all the contributing authors for their hard work in producing these chapters and the publishers for their patience in the face of my spaghettified deadlines.

Introduction

Scepticism is a common response from researchers in the established life sciences when first introduced to the aims and methods of synthetic biology. For some, this view is eventually replaced with acknowledgement of synthetic biology as an exciting and potentially revolutionary new field of scientific and engineering endeavour. But for many researchers, industrialists and citizen scientists, gaining an accurate sense of the scope and methods of synthetic biology can be a challenge. This is particularly true when faced with joyous media announcements at one extreme and opaque primary research articles at the other. Assessment of any new field involves enquiry beyond the hype and the headlines to scrutinise the underlying science. The purpose of this book is to provide a source of accessible information for interested observers who seek to decide for themselves what aspects of synthetic biology are most valuable and also to offer an overview for researchers wishing to enter the field.

Birth of a Discipline

Synthetic biology is more than a revitalisation of molecular biology, although parallels do exist between the births of each field. Research centres opened in the 1960s such as the European Molecular Biology Organisation (EMBO) were not founded by biologists or biochemists but by physicists. Synthetic biology arguably emerged from computer scientists' desire to apply the principles of engineering and computer science to biology. In this sense, both fields arose from non-biologists taking a new perspective on biology.

From the 1960s onward, molecular biology was driven by collaboration across multidisciplinary teams of biochemists, mathematicians, chemists, physicists, computer scientists and engineers. Powerful recombinant DNA techniques developed in biochemistry and microbiology combined with developments in crystallography and computing. Molecular biology became the principle route for 'forward genetics', identification of genes involved in a given phenotype, and 'reverse genetics', characterisation of multiple phenotypes arising from a given gene. The tools of molecular biology have been sufficiently useful for increasing our understanding of the biological world that the field remains a fulcrum of philosophical debate concerning concepts of reductionism and complexity. By the 1990s, standard molecular biology techniques had become an indispensible life science tool, but also a very technically demanding skill to acquire.

The rallying call of synthetic biology activities, such as the International Genetically Engineered Machines (iGEM) competition, has also energised multidisciplinary approaches to biology. This is in major part due to the streamlining of recombinant DNA methods to enable greater accessibility for people with no prior background in molecular biology. More importantly, the availability of affordable, sizeable, user-defined fragments of DNA has in many ways spawned a new 'synthetic DNA' era, distinct from the previous 30 years for which recombinant DNA was the cheapest and most convenient option for assembling experimental DNA sequences. The established quantitative biology research communities have also provided much of the technological framework and expertise that supports and drives a great deal of synthetic biology research. If synthetic biology can follow a similar success path to that taken by molecular biology since the 1960s, there is certainly a fascinating journey ahead.

Building Biology

The most straightforward task of synthetic biology is to build new things. This is in contrast to the goal of fundamental research to discover and understand natural biological phenomena. A shortlist of model organisms has emerged over time that meets many of the needs of discovery bioscience, including workhorses such as bacteriophage (libraries, promoters), *Caenorhabditis elegans* (genes, neurons), *Drosophila melanogaster* (genes, development), *Saccharomyces cerevisiae* (protein interactions, fermentation), *Escherichia coli* (plasmids), cells from Henrietta Lacks (tissue culture), mammalian viruses (promoters, gene transduction) and *Xenopus laevis* (genes, development, neurons). These organisms have been selected based on their usability and their track record of application in research is vast.

Model organisms illustrate the power of having the right research tool to help to answer a particular question. As such, it is unsurprising that synthetic biologists seek to go beyond these legacy organisms and engineer *de novo* biological tools to build new biologies. The first step towards this goal is to establish highly modified genomes and genetic codes. As these efforts reveal the rules governing design of novel genomes and genetic codes so the ability to invent new methods for programming complex, biological materials becomes a more realistic proposition. Organisms conceived and designed purely for the purpose of enhanced tractability can be regarded as the logical progression for which the above model organisms are the conceptual progenitors.

Close Relatives

What delineates synthetic biology from other disciplines, such as biotechnology, systems biology and metabolic engineering, that can appear at times to be confusingly similar? Scientific disciplines can be defined by their characteristic methods to a varying degree. Metabolic engineering could be defined as the application of biochemical, mathematical and chemical approaches to co-opt metabolic pathways within living cells, for the production of desired chemical entities. Biotechnology is a broader set of methods and a broader set of desirable products. Systems biology seeks to gain insight from relationships within large data sets rather than individual experiments at the laboratory bench.

For closely related fields, the critical delineators may not be found in techniques or experimental approaches but in overarching philosophical goals. For example, a field such as exobiology encompasses the goal of studying extra-terrestrial environments with respect to their capability to support forms of life. The tacit expectation of exobiology is that extra-terrestrial life exists now or has done so in the past. Of course, most exobiologists will accept the formalism that their efforts may result in a proof that extra-terrestrial biological phenomena do not exist and have never existed within the portion of the universe tractable to human investigation. However, it is doubtful that even the most taciturn of exobiologists actually labours under this expectation. By contrast, astrobiology, although it employs largely the same techniques of exobiology, has a distinct overarching goal, namely to understand how life arose on Earth. In this way, astrobiology and exobiology encompass similar activities but are defined by their distinct overarching goals.

In common with metabolic engineering, systems biology and biotechnology, synthetic biology can employ a broad range of experimental approaches, including bioinformatics, mathematical modelling, chemistry, computer science, chemical engineering and essentially all of the practices and approaches of the life sciences. However, the overarching philosophical aim of synthetic biology is to prove that the molecular understanding of biology can be exploited by and defined within archetypal engineering frameworks. The extent to which a given research activity advances, or is indifferent to, this goal is a good indicator as to how near or how far it is from synthetic biology.

Of course all modern science shares the tacit assumption that biology obeys the known rules of physics with respect to concerns of matter and energy. However, rather than being a mere background assumption, demonstrating that biological matter can be predictably designed and utilised as an engineering material lies at the heart of synthetic biology. Even if used for fundamental research, synthetic biology will always be an applied discipline as the construction of each new biological tool ideally makes biology subsequently more amenable to engineering approaches.

Big Ambitions

Does sufficient insight currently exist to enable biology to be fully transformed into an engineering discipline? The clear answer at the time of publication of this book is no. No is also the answer to whether there is yet sufficient understanding of consciousness, cancer, development, the immune system, weather, gravity and turbulence. Research activities continue in all these fields because all scientific endeavours are initiated while the path ahead is uncertain. The publication of evolutionary theory preceded mechanistic understanding of the heritable material it described. As we fly to the Moon before Mars, so we design plasmids before we design genomes. So synthetic biology seeks to use engineering approaches to generate designed biological complexity, while fundamental science does indeed still wrestle with natural biological complexity.

However, it can pay to be cautious when deciding which challenges are daunting and which are doable. Researchers who devised the Paracelsus challenge, to change one protein into another while retaining half of the original protein's sequence, thought they had devised an impossible bioinformatics task for the ages back in 1994. The challenge was met in 1997, prompting the organisers to sound a note of caution for future grand biological challenges: offer T-shirts rather than large cash sums as incentive.

Towards Standardisation

A distinct feature of synthetic biology is the streamlining and standardisation of DNA assembly procedures, which is examined at length in Section I of this book. This standardisation inevitably involves setting limits on the number of unique restriction sites used for cutting and ligating DNA, a key example being the BioBrick™ format. This constitutes a change to the traditional approach, which has persisted for some 30 years, of furnishing plasmids and related cloning vehicles with as many unique restrictions sites as are practical within a 'multiple cloning site'. Molecular biology has thrived since the early 1980s without restrictive standards, so what is the benefit of imposing rules now? Synthetic biology seeks to harness the network effects of large numbers of adopters to expand and maintain extensive open source libraries of freely or cheaply available plasmids. Indeed, most DNA library platforms involve the use of unique restriction sites for ease of manipulation or to define fragment size. Simple, streamlined cloning strategies tend also to be cheap and accessible in terms of lowering skill barriers and opening participation beyond specialists. As we shall see in Section I, the BioBrick™ format in particular is designed to enable repetitive, recursive rounds of cloning for every individual cloning step, an approach ideally suited to robotic automation and computer aided design.

Terminology – Illustrative or Irritating?

Most life scientists find it easy to appreciate that an engineerable biology is one that may be more readily harnessed for mankind's benefit. A more troublesome proposition may be the nomenclature that emerges when addressing biology as an engineered system rather than a complex phenomenon. Why use terms such as 'chassis', 'part' and 'device', in preference to 'cell', 'genetic element' or 'gene'? The most important answer is that it does not matter, as long as good research is taking place.

However, all scientific names and terms gloss over the complexity of the underlying phenomena they are used to identify to a degree. So even though nomenclature is not the most important facet of synthetic biology, it can serve a useful and important purpose. First, synthetic biology is multidisciplinary and using terms likely to be familiar to researchers from fields beyond chemistry and biotechnology, such as control engineering and physics, helps to foster this. Second, when the aim is to use defined components to fabricate new biologies, words and concepts common to design and engineering can help to focus efforts and identify challenges. For instance, using the term 'chassis' rather than 'cell' in one sense serves only to highlight the huge distance between the straightforward relationship an automotive 'chassis' has with the engine it encases compared with the as yet intractably complex relationship between a DNA genome and its surrounding cytosol and organelles. However, it is precisely the use of this term that can help to focus researchers' efforts on mapping the ways a cell *can* be considered as a chassis, with respect to the intended function of synthetic gene circuits, and those ways in which it currently *cannot*.

Future Biomanufacturing

Applying rigorous engineering methods to derive economic value from biological material is a key synthetic biology aim that also has a lengthy track record in the fields of chemical and biochemical engineering. Since the mass production of penicillin in the 1950s, chemical engineering has been characterised by combining fundamental science and cutting-edge engineering to provide new industrial capabilities. Crude, complex, low-value biological material is transformed into pure, defined and valuable material by the action of a sequence of large-scale preparative procedures. The sequence of these steps, or 'unit operations', defines a process. Upstream processing (USP) defines the initial unit operations required to grow and handle sufficient quantities of biological material for final materialisation of a product. Downstream processing (DSP) typically involves separation of biological materials by centrifugation, filtration and chromatography to progress material from a crude mixture to a final, useable state.

Most DSP steps can be accurately modelled, and therefore designed, based on the application of fundamental physicochemical principles and statistical approaches to data analysis. USP is far different and presents challenges much more familiar to life scientists. In USP, recombinant DNA techniques are frequently used to optimise factors affecting transgene expression in host cells, such as gene dosage, codon usage and chromosomal location. Ancillary steps can also be taken, such as down-regulating genes encoding host cell proteins, down-regulating genes whose associated phenotypes interfere with DSP and co-expressing genes that enhance DSP.

In this way, the capabilities and limitations of USP are defined by biological challenges, such as predictable genome design, protein folding and cell growth. Compared with DSP, the biological phenomena that delimit USP are less amenable to purely physicochemical and analytical approaches but are more effectively addressed by the efforts of synthetic biology and related fields.

Processing of physically robust and biologically complex materials, such as lignocellulosic biomass, or highly fragile materials, such as stem cells, poses very distinct challenges. Addressing these challenges with 'bioprocessing' led to the emergence of the field of biochemical engineering as a specialisation of chemical engineering. Surprisingly, the aims and methods of chemical and biochemical engineering are often unknown to fundamental life science researchers despite their critical role in translating basic research into the medicines and the valuable fine chemicals that underpin economic development.

The philosophical aim of synthetic biology fits perfectly within the nexus between USP and DSP. Indeed, synthetic biology has the potential to expand the proportion of bioprocessing that can be described in physicochemical terms beyond traditional DSP operation to include USP. If successful, synthetic biology could place entire bioprocesses firmly within the bounds of rigorous engineering frameworks and, in so doing, realise the benefits of predictable design, performance and scalability for some of the world's most prohibitively expensive medicines and therapies.

The industrial potential of synthetic biology goes beyond greater integration of conventional biotechnology and chemical engineering. Pursuing the goals of synthetic biology can help to accelerate the process by which basic research insights and fledgling technologies become real-world products and medicines. Arguably the most pressing example of the disconnection between elegant molecular bioscience and real-world impact is in the field of gene therapy. Currently, most gene therapies are developed in a bespoke manner with respect to the format of the delivery vehicle, the active therapeutic agent and the platform strategy for production. Each and every bespoke approach requires separate regulatory acceptance, stretching timelines from discovery to mass-produced medical solution. Standardisation and tractability could bring huge benefits to gene therapy if successfully incorporated into production platforms. Multi-platform interoperability, universal measurements and predictable design of biomolecular tools could all accelerate development of good manufacturing practices for gene therapy applications and are all current targets of much synthetic biology research.

This Book

This book attempts to provide an overview of current areas of focus in synthetic biology. As with any new field, progress is fast moving and inevitably there are developments and topics that are not discussed here. However, the following chapters provide the background necessary to build an accurate picture of synthetic biology research areas and how they fit within the overarching goals of the discipline.

Section I comprises three chapters and presents synthetic biologists' efforts to standardise some of the classic tools and approaches of molecular bioscience and to devise new standards. This section also goes beyond the laboratory to explore the wider societal context and potential impacts of synthetic biology and how this nascent field can distinguish itself from traditional industrial-academic models of research by authentically engaging with and empowering general publics.

Section II deals with synthetic biology's exploitation of so-called legacy systems, organisms that evolved naturally and have been chosen, due to their ease of use and other natural characteristics, as tools for design and construction of new products and capabilities.

Finally, Section III examines arguably the most technically demanding goal of synthetic biology, that of designing and building *de novo* cells and genetic codes. Software, robotics and automation are critical to a synthetic biology vision of the (near) future in which bespoke organisms can be designed *in silico* by non-expert users and automatically materialised in a process comparable with how millions of people design and materialise documents using computers, software and printers every day. As such, this section also includes a chapter on computational approaches to designing genes and gene networks.

Editor

Darren N. Nesbeth earned his BSc in molecular biology from University College London (UCL) and PhD in molecular cell biology from Imperial College, London, United Kingdom. He did postdoctoral work on the scale-up of gene therapy at King's College London, fundamental cell biology at Imperial College and industrially applied cell engineering at UCL. At the UCL Department of Biochemical Engineering, he develops synthetic biology solutions to industrial challenges in the manufacturing of biopharmaceuticals and fine chemicals.

Contributors

Chris P. Barnes
Department of Cell and Developmental
  Biology
University College London
London, United Kingdom

Philipp Boeing
Department of Cell and Developmental
  Biology
University College London
London, United Kingdom

Yanika Borg
Department of Biochemical Engineering
University College London
London, United Kingdom

René Daer
Ira A. Fulton School of Biological and
  Health Systems Engineering
Arizona State University
Tempe, Arizona

Marcus K. Dymond
Division of Chemistry, School of
  Pharmacy and Biological Sciences
University of Brighton
Brighton, United Kingdom

Karmella Haynes
Ira A. Fulton School of Biological and
  Health Systems Engineering
Arizona State University
Tempe, Arizona

Pier Luigi Luisi
Science Department
University of Roma Tre
Rome, Italy

Darren N. Nesbeth
Department of Biochemical Engineering
University College London
London, United Kingdom

Heinz Neumann
Institute for Microbiology and Genetics
Georg-August University Göttingen
Göttingen, Germany

Tanel Ozdemir
Department of Cell and Developmental
  Biology
University College London
London, United Kingdom

Nicola J. Patron
The Sainsbury Laboratory
Norwich Research Park
Norwich, United Kingdom

Desmond Schofield
Department of Biochemical Engineering
University College London
London, United Kingdom

Pasquale Stano
Science Department
University of Roma Tre
Rome, Italy

Yo Suzuki
J. Craig Venter Institute
Synthetic Biology and Bioenergy Group
La Jolla, California

Alexander Templar
Department of Biochemical Engineering
University College London
London, United Kingdom

Philip D. Weyman
J. Craig Venter Institute
Synthetic Biology and Bioenergy Group
La Jolla, California

Section I

Standardising Biology

1

Synthetic Biology: Culture and Bioethical Considerations

Marcus K. Dymond

CONTENTS

1.1 Introduction

Synthetic biology made its way into popular culture after the global publicity surrounding J. Craig Venter's 2010 announcement of the world's 'first cell under the control of a chemically synthesised genome'.[1] In the many news reports that followed, the potential for synthetic biology to solve the world's most pressing problems, such as pollution, the energy crisis, global warming and disease, was presented alongside the risks posed by synthetic biology. As awareness grows among the public, this mixture of downside risk and upside potential, in combination with political and economic interests, has made synthetic biology a hot topic with ethicists, scientists, politicians and the press. The reason for this intense social interest in synthetic biology is due to the fact that synthetic biology is in many respects an advanced form of genetic engineering and therefore many vestiges of the genetic modification (GM) debate are carried over into the discussion on synthetic biology.[2] This is one of several prior associations that synthetic biology has inherited. Another, often negative, association is made between synthetic biology and the exaggerated myth of science fiction. One might even go as far as to suggest that in popular culture the science fiction genre is part of the language of synthetic biology, since it would seem that any discussion on synthetic biology sooner or later makes reference to science fiction as a metaphor for risk, ethics or social responsibility. In addition, the ethical issues of synthetic biology are associated with the impacts that past technologies have had on the planet and its inhabitants.

A tradition of engaging the public in the genuine bioethical debates around synthetic biology has emerged among the pioneers of the field. In the hope of fostering this tradition in researchers new to the field, this chapter presents a snapshot of the cultural and political reaction to synthetic biology followed by a discussion of issues related to risk, societal responsibility and governance.

1.2 Perceptions and Portrayals of Synthetic Biology in Popular Culture and Politics

References to synthetic biology in the scientific literature occur sparsely prior to 2003.[3] Two scientists, Waclaw Szybalski and Stephane Leduc, are associated with the concept. The first proposal of a synthetic phase emerging from molecular biology is attributed[4,5] to the geneticist Waclaw Szybalski[6,7] who in the 1970s posed the rhetorical question, 'what might follow the descriptive phase of molecular biology?' One possibility Szybalski foresaw was a synthetic biology where whole genomes might be synthesised.

> This would be a field with unlimited expansion potential.[6]
>
> **Szybalski**

Whilst Szybalski's comments allow us to understand the synthetic biology of Venter and his contemporaries, there is an earlier reference to synthetic biology in the scientific literature. French biologist Stephane Leduc details his search to explain life (and synthesise life) from physical and chemical processes in his 1912 work titled 'La Biologie Synthétique'.[8] Leduc's experiments were in an area of synthetic biology that we would probably now call the *de novo* synthesis of life, and Leduc, like many other scientists throughout history, sought to answer one of the most fascinating scientific questions, namely what is life?

Historically, scientists who have pursued answers to the question 'What is life?' have in their own lifetimes raised ethical concerns among the society of their time, largely because it has brought science into conflict with other societal groups with their own views of what life is or what is sacred within nature. Synthetic biology is the most recent technology to enter into this discussion and predictably older ethical debates have resurfaced in response to it, with a dialogue rich in historical reference. An example of this recurring debate is discussed by Yeh and Lim[9] through the contrast of the historical debates that surrounded synthetic organic chemistry in the early 19th century and current debate on synthetic biology. At a time when vitalism dominated biological theory, to the extent that biological molecules (organic) were thought impossible to synthesise outside of living organisms, Wöhler's[10] chemical synthesis of urea was one of the first steps in demystifying biological chemistry.[11] The parallels between the ethical debate surrounding the birth of synthetic organic chemistry and the current ethical debate surrounding synthetic biology are interesting. For example, at the time Wöhler's synthesis of urea gave rise to concerns that synthetic humans might soon follow.[9] In modern times, the image of synthetic chemists creating life in a round-bottomed flask is comical. However, the synthetic biological counterpoint of designer babies[12] does raise some of the same ethical debates about playing God and creating artificial life now as were previously raised in the past.

As noted previously, science fiction literature and movies are frequently referenced metaphorically in debates over the ethics and social responsibility of synthetic biology. Much of the reason this is possible, even if in reality it is implausible, is due to previous generations of social commentators exploring these ethical themes in response to the potential perils that the new technology of their time posed. Classic examples are George Orwell's *Nineteen Eighty-Four*[13] and Aldous Huxley's *Brave New World*.[14] Both novels explore the loss of individual freedoms that future worlds might bring. This is an ethical theme directly relevant to synthetic biology as will be discussed, where these two novels differ is in their opinion of the technologies that are the threat. The biological technology described in *Brave New World* makes it a frequently used risk metaphor in synthetic biology.

There are many other novels that warn of the perils of technology, for example H.G. Wells' *The World Set Free* (1914)[15] foretells the atomic age. In the novel, Wells extends the atomic theory of his time to create an optimistic picture where the escalation of centuries of atomic war ultimately results in a peaceful utopia, with a population rich in artists. Pursuing a more biological theme with *The Island of Dr Moreau*, H.G. Wells[16] presents the tale of the vivisectionist (Dr Moreau) who flees London in disgrace to an island where he surrounds himself with chimeric creatures, such as *The Leopard Man*, grafted by his own surgical skills. Written at time when vivisection was advancing physiological knowledge at a rapid rate, Wells discusses the ethics of animal experimentation. The novel's themes are complex, but one of the overriding messages is to ask whether the potential consequences of vivisection are worth the potential gains, a question that is asked at a time when the unknown aspects of nature and the natural world were considered big threats to humanity.

Whilst the novels described earlier, and others like them, do raise ethical themes that are discussed within synthetic biology, such as apocalypse and the morality of meddling with nature (and the consequences thereof), it is worthwhile to point out that the technologies are not typically comparable and thus the risk issues are not always directly relevant to synthetic biology. To the press and to politicians, this lack of relevancy does not dilute the potency of the science fiction metaphor. A perfect example of this is Mary Shelley's early 19th century novel *Frankenstein*.[17] The technology described by Shelley (a mixture of galvanism and surgery) has little to do with the molecular techniques of synthetic biology, but as will be discussed the Frankenstein metaphor is the most potent of all.

Building on 19th century literary traditions, 20th century cinematographers have given every kind of fictional chimera form, function and conscience in the emerging genre of the science fiction film. One of the common themes of these stories is the tale of a scientist who becomes the unwitting destabiliser of the 'natural order of things' – and a catalyst for releasing all the forms of chaos envisioned in the author's mind. The moral, it would appear, is that scientists should not attempt to investigate what they do not (and cannot hope to) understand, a moral that is completely at odds with the aspirations of the research scientist. In the 21st century, numerous science fiction films have plot lines which derive directly from the perils of science and technology associated with synthetic biology. Recent examples include *Splice*,[18] where two genetic engineers illicitly introduce human genes into animals and struggle to control the resulting life forms, or the *Resident Evil*[19] series where the uncontrolled release of a genetically engineered virus devastates the fictional Raccoon City. Each of the large number of novels and films in this genre remind the audience that science is a journey into the frontiers of knowledge, where, dimly illuminated by hypothesis, there lurks dystopia and apocalypse.

What these fictional stories tend to ignore is the research governance and risk mitigation strategies of life-scientists who have also tried to learn the lessons of past mistakes, mistakes that occurred when the physical sciences led to industrialisation. In the early part of the 20th century, the role of biology, as a science, appeared to be limited to understanding the natural world. Creativity, in the synthetic sense, was the promise of the physical sciences and over the first half of the century, driven by two world wars and imperial ambitions; polymers, electronics, antibiotics, atomic energy, pesticides and fertilisers improved the quality of life for the inhabitants of the now developed world. In the latter half of the 20th century, environmental awareness of the cost of these technologies increased and the impact of humankind in global warming, pollution, the poverty gap and global health problems was recognised. By the early 21st century, genetic engineering and cloning technologies had been developed and biology, it seemed, could be creative too but with this creativity there came a new set of risks.

The reaction of scientists to these risks is evidence of a community learning from the mistakes of history. In the 1970s, a voluntary moratorium on DNA cloning technologies was undertaken, by researchers active in the field, until such a time when the risks were better understood.[20–22] On the one hand, it was feared that DNA technologies might cause the manmade equivalent of *The Andromeda Strain*, a 1971 cinematic blockbuster penned by Michael Crichton where an alien virus wipes out large swathes of humanity.[23] On the other hand, it was speculated the loss of individual freedoms envisioned by Aldous Huxley in his novel *Brave New World*[14] might be closer than previously debated. The ethics of the recombinant DNA debate are discussed retrospectively by Robb[24] and many of these issues have resurfaced with respect to the concerns raised by synthetic biology.

1.2.1 Impact of Synthia

In the wake of Craig Venter's synthetic genome announcement Barack Obama mobilised the Presidential Commission for the Study of Bioethical Issues to look into synthetic biology. Politics is of course concerned with economic and social stability; whilst the wholesale acceptance of synthetic biology has ethical considerations, its economic potential is unfathomably valuable to any government. Luminary thinkers, notably the physicist Freeman Dyson, foresee the domestication of biotechnology as largely inevitable, as discussed by Dyson in his influential essay 'Our Biotech Future'[25] and regularly cited within the canon of synthetic biology literature. Taking computer technology as an example, Dyson explains how, despite being originally envisioned as enormous centralised facilities, computers have had their greatest impact on our world through decentralisation and domestication. The home computer was a key step in this process, which has given us a wave of smartphones and touch pads that are blank canvases for human personalisation. In Dyson's vision of the future, decentralisation of biotechnology might spread to the point where children are embracing genomic programming for fun. It remains to be seen if society, as a whole, will accept this do-it-yourself (DIY) biology so readily[2] but a less than orderly queue of small and medium-sized enterprises (SMEs), multinational corporations and research institutions stand in line to reap the potential rewards.

One of Dyson's acknowledged influences is the work of Carl Woese, microbial taxonomist, who in 2004 called for 'a new biology for a new century'.[26] Perhaps synthetic biology is the answer to that call and the fulfilment of Dyson's prognostication. One thing is clear: the growth of synthetic biology into domestic markets brings the promise of huge economic rewards in tandem with risks that are as yet unquantified. The Canadian Action Group on Erosion, Technology and Concentration (ETC) is one of many organisations raising public awareness about risk issues in synthetic biology.[27] It was also among the 111 organisations that have called for a moratorium on synthetic biology research until the risks are better understood.[28]

Whilst it is true that the risks of synthetic biology are not completely understood, it is also true that the representations of these risks are often sensationalised by allusion to science fiction. This makes it very difficult to have a well-informed public debate on synthetic biology. For example, consider the headlines of 2010 and the press allusion to fictional classics. In the afterglow of Venter's synthetic genome announcement, the *Daily Mail Online*[29] led with the following headline:

> Scientist accused of playing God after creating artificial life by making designer microbe from scratch – but could it wipe out humanity?

Following up on this article, the *Daily Mail* science editor, Michael Hanlon, poses the question[30] 'Has he (Venter) created a monster?' before alluding to the 2007 Francis Lawrence film *I Am Legend* based on the Richard Matheson novel[31] of the same name:

> There are fears the research could be abused and lead to millions being wiped out by a plague like in the Will Smith film *I Am Legend*.

In the novel, Robert Neville is the last uninfected survivor of a bacterial pandemic that can infect both living and dead hosts. Symptomatic of the infection is the transformation of humans into vampire-like beasts. In the filmic version, Smith plays Robert Neville, a military virologist, again the lone survivor of a pandemic, but this time the virus is a human intervention gone wrong.

Other newspapers were quick to report the breakthrough by Venter and his team. News International's UK newspaper *The Sun* leads its article on the 'first synthetic cell' with the subdued[32] 'Scientist creates first man-made cell' before following up with:

> A US biologist has stepped into the shoes of Baron Frankenstein by breathing life into a bacterium using genes assembled in the laboratory.

1.2.2 Frankenstein Parallax

The Baron Frankenstein referred to is, of course, none other than the literary creation of Mary Shelley in her landmark 1818 novel *Frankenstein*.[17] Inspired by the experiments of scientists in her lifetime, Shelley penned a reworking of the classical tale of Prometheus, the man who stole the secret of fire from the gods, only to be punished eternally by Zeus. Shelley's modern Prometheus, Baron Victor Frankenstein, is a maverick young scientist carrying out experiments in reanimation. Fittingly in 1910, another scientific maverick, Thomas Edison, gave Shelley's tale its first filmic reworking (see Figure 1.1), thus beginning a popular culture fascination with the monster present to this day.

It is difficult to quantify the influence of Shelley and the length of the shadow that Baron Frankenstein's nameless monster has cast over popular culture. Jon Turney does an admirable job in his *Frankenstein's Footsteps*.[33] In Shelley's novel, the central theme is a scientist taking science into the realm of creation (of life). The central premise is that the act of creation of life is something that only God, the giver and taker of life, can bestow.

FIGURE 1.1
Stills from Thomas Edison's 1910 film adaptation of Mary Shelley's *Frankenstein*. Shelley's novel tells the story of Baron Victor Frankenstein's ungodly dabbling in creating life.

Complicit with this stance is the idea of God as omnipotent, all-knowing and all-seeing and humankind, as his creation, is in possession of limited knowledge and foresight. The moral of the story is in a sense simple, that is, when humans are arrogant enough to presume to be gods (by seeking to create life), God abandons them. Victor Frankenstein's self-loathing, a developing theme of the novel, expresses his sense of abandonment by God thoroughly.

Turney[33] demonstrates just how frequently the word Frankenstein has been used to label the work of 20th century scientists who operate at the edge of social acceptance. Retrospectively, many of the accused now stand as scientific heroes. In the 21st century, the Frankenstein terminology has a much wider connotation than 'playing God'. It has been extended to playing with nature, where nature remains sacred regardless of whether it is the product of intelligent design or evolution. This transition is, of course, understandable given the rise of the environmental movement and the growing concern of the impact and sustainability of humanity on nature. By the 1990s, Frankenfoods became a recognisable metaphor for genetically modified (GMed) foods, in part driven by the lobby against the production of GMed foods.[34] As noted, GMed technology was the forerunner of synthetic biology and as a result the Frankenstein label is never far from sight in synthetic biology debate.

Few people are more aware of the constant comparison of synthetic biology to fiction than those researchers who have thrust themselves forward to promote their endeavours. Harvard geneticist George Church and science writer Ed Regis take on the battle to explain synthetic biology as a positive force for societal change in their excellent book *Regenesis: How Synthetic Biology Will Reinvent Nature and Ourselves*.[35] The book seeks to reassure the reader that synthetic biology will not bring the science fiction nightmare depicted in films and to explain in simple terms what synthetic biology is, what the real success stories are and what it might provide in healthcare advances, energy generation and conservation. Church and Regis conclude by looking at how these scientific triumphs will change the future – Our Biotech Future, to quote Freeman Dyson.[25]

Over the last decade, the pioneers of synthetic biology have laid down a strong public engagement movement in the field. Much of the motivation for public engagement has come from the lessons learned in the nanotechnology and GMed food debates.[36] Prior to Venter's announcement of the 'first synthetic cell' in 2010, the prominent synthetic biologists Drew Endy and Christina Smolke were interviewed by *Esquire* magazine.[37] The article, entitled 'How to make life', presents both the practice and practitioners of synthetic biology in an exclusively positive light, highlighting the potential for synthetic biology to help cure cancer, clean up pollution and detect land mines.

1.2.3 Synthetic Biology Olympiad: International Genetically Engineered Machines

A major impetus driving positive perceptions of synthetic biology is the annual International Genetically Engineered Machines (iGEM) competition.[38–40] iGEM has, since its start in 2003, grown from a small competitive undergraduate course to a global event (see Figure 1.2) attracting many thousands of competitors. Campos (2012)[41] documents the genesis of the iGEM competition in detail for the interested reader.

The iGEM premise is straightforward. Teams of (mainly undergraduate) researchers design genetically modified cells to carry out tasks that range from the simple, such as making *E. coli* smell of bananas, to the more complex integration of bacteria with electronic technology. Competitors are awarded medals (bronze, silver or gold) based on a series of defined criteria.[38] One of the criteria for gold is to propose novel approaches to issues such as safety,[42] security, ethics or ownership, sharing or innovation – all so-called 'human practices'. In short, the iGEM competition encourages its competitors, some of whom will go on to become research leaders of the future, to engage with the risk and societal responsibility issues that synthetic biology raises. A previous winner of the 2009 grand prize, University of Cambridge, tackled this human practice aspect by collaborating with UK-based artist and designer Daisy Ginsberg. Ginsberg's work, often the product of collaboration with scientists, articulates some of ethical and cultural dialogues that come with new technology as part of the Synthetic Aesthetics[43] approach. This is best articulated by Ginsberg herself:

> I am a designer/artist exploring emerging technologies and their social, cultural and ethical implications. Having spent nearly 6 years researching synthetic biology, my interest is in seeking new roles for – and approaches to – design that could help us to develop more sustainable, ethical modes of innovation, by using design as a tool for critical thinking and as a mediator between diverse disciplines. Rather than using design just to solve problems, can we use it to ask better questions? Otherwise, promised disruptive technologies risk disrupting nothing.[44]

FIGURE 1.2
Finalists at the 2012 iGEM competition (iGEM Foundation and Justin Knight). (Reproduced with permission under the Creative Commons Licence.)

The 2009 winning project *E. chromi* modified bacteria (*E. coli*) to be sensors of gastrointestinal disease, which are then intended to be digested by humans. Disease diagnosis occurs through a colour chart and a faecal colour examination, a process embodied by the *E. chromi* Scatalog sculpture of Daisy Ginsberg and James King, shown in Figure 1.3. The Scatalog garnered online debate of such fervour and technical detail that onlookers could be forgiven for thinking they were witnessing the roll-out of real-world technology, rather than a piece of sculpture intended to stimulate debate around human interaction with possible technologies arising from synthetic biology.

Whilst all steps made by synthetic biologists to engage the public in the field should be commended, a sobering reminder of the magnitude of the task of rebranding synthetic biology in popular culture can be seen in newspaper readership statistics. Suffice to say these statistics demonstrate that the tabloids are a gateway into the minds of the populus. In 2010, average newspaper circulation for a typical month[45] shows that the two papers quoted earlier (*The Daily Mail* and *The Sun*) alone sold over five million copies, a number well in excess of newspapers committed to less sensationalist reporting. For example, *The Independent*'s reporting[46] of the 'first synthetic cell' was more balanced and contained no allusion to sci-fi disasters. With a circulation of less than 200,000 copies over the same period,[45] it can hardly be considered a dominant force in shaping the population's reaction to, and impression of, synthetic biology. Taking a broader view, two publications from the Woodrow Wilson Centre, as a part of its Synthetic Biology Project, examine press coverage of synthetic biology in 2003–2008[47] and 2008–2011[48] across America and Europe. The research shows that across the continents the reporting of synthetic biology has increased and that generally a balanced view predominates – although it appears that the word Frankenstein is still associated with the field.[48,49]

FIGURE 1.3
E. chromi: The Scatalog. Alexandra Daisy Ginsberg and James King, with the University of Cambridge iGEM Team 2009. (Reproduced with permission.)

1.3 Synthetic Biology in Academic Literature

The earliest references to synthetic biology as a scientific concept appear at the start of the 20th century.[8] However, its discussion in more familiar terms stems from the 1970s where Waclaw Szybalski made several insights into the synthetic future of molecular biology.[6,7] Regular publications in the field begin to appear post-2003.[3] In their article 'Synthetic Biology: Mapping the Scientific Landscape' Oldham et al. (2012)[3] adopt a scientometrics approach to appraise the scientific literature. One of the key findings of this study is the diversification of synthetic biology with the growth of the field. The net result appears to be a redefinition of what synthetic biology is as new researchers try to align their own research interests with the latest funding opportunities. Whereas traditionally synthetic biology was defined by its approach of employing engineering principles for constructing genetic systems, a process that seems inherently targeted towards *in vivo* application,[50] usage of the term in the scientific literature has diversified to encompass terminology such as cell-free synthetic biology[51] and *in vitro* synthetic biology,[52] as shown in Figure 1.4.

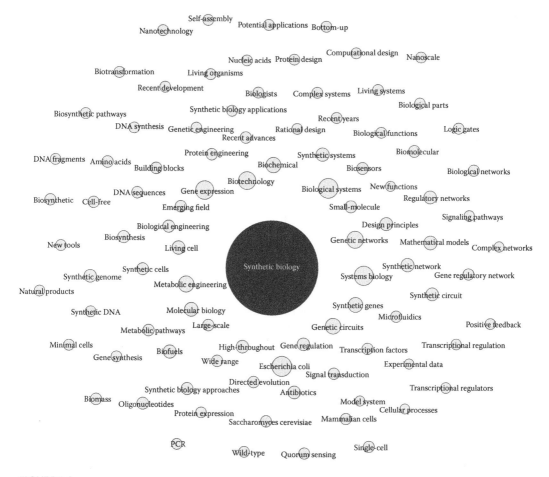

FIGURE 1.4

The landscape of synthetic biology literature. (Reproduced from Oldham et al., *PLoS ONE*, 7(4), e34368, 2012. With permission under the Creative Commons Licence.)

Terms like systems biology, protein expression, microfluidics, nanoscale and self-assembly are also closely associated with the field.[3] Synthetic biology research topics include genetic re-engineering[53] of bacterial,[54] yeast[55] and mammalian cell lines,[56] the genome syntheses of Venter,[1,57] the manufacture of synthetic ribosomes,[58] protein modification,[59] genetic code expansion,[60,61] synthetic organelles[62] comprised of nucleic acid–lipid[63,64] or protein–lipid assemblies[65] and *de novo* synthesis of life.[66] What appears to unite the field is the synthesis of the molecular machinery of life, or life processes. Given that genetics has been described as 'discovering the alphabet of life',[34,67] one might describe synthetic biology as writing with the language of life. It is this vagueness in the definition of synthetic biology research[3] that can create much confusion over what bioethical considerations are relevant to, and new to, the field.

1.4 Societal Risks and Responsibilities Posed by Synthetic Biology

A number of publications in the field of bioethics[68] provide a rigorous bioethical framework which can be used to explore how synthetic biology fits into the existing set of bioethical concerns.[69–73] In view of potentially new bioethical threats the Nuffield Council on Bioethics report on emerging biotechnologies gives a comprehensive overview of the risks, choice and governance viewpoints.[74]

Broadly speaking the ethical concerns surrounding synthetic biology fall into the following categories:

1. Playing God and creating artificial life
2. Uncontrolled release and ecosystem disruption
3. Biosecurity, bioterrorism and biowarfare
4. Global justice and intellectual property
5. Biopunks, biohackers and DIY science

Bioethical consideration in synthetic biology is typically articulated by posing questions such as the following:

> Since there are many previous examples where the application of technological advances have resulted in 'disasters' or inequality etc., is it ethically justifiable to move forward with synthetic biology given the possibility that the technology might cause worse disasters or greater inequality?

In practice, given the economic and political drivers for embracing synthetic biology, the bioethical consideration question tends to a pragmatic form asking:

> What is the best way to move forward, such that risk is mitigated and technological progress is introduced in a socially responsible way?

Therefore it is more common practice to refer to these ethical considerations as risk and societal responsibility issues. In the next sections, the risk and societal responsibility issues that have been, or are likely to be, associated with synthetic biology are signposted.

1.4.1 Playing God and Creating Artificial Life

The concept of playing God[75] represents one of the most polar ethical concerns levelled at synthetic biology. On the one hand, for many theists it is heretical to 'play God', that is to take a part in activities that are within his divine remit. On the other hand, since atheists do not believe in the concept of God, playing God is not a meaningful moral consideration. Dabrock[76] tackles the theological argument against the thesis of 'synthetic biology as playing God' in an engaging and detailed essay. Belt[77] points out that although the accusation of playing God is based on a belief statement (i.e. belief in God) the term 'playing God' allows people to articulate their concern that a moral boundary is being crossed. The origin of these moral issues in an ethical context is discussed by Link.[78]

Ethical concerns about scientists creating artificial life are less polar than those of playing God but certainly share some of the same philosophical paradoxes. If one believes it is the exclusive right of God to create life, then it is morally wrong for humans to create life artificially. Equally, in the absence of any exclusive and divine rights on creating life, there is arguably nothing inherently wrong with humans doing so. Here the definition of life underpins the moral and ethical reaction to the concept of creating life as discussed.[79]

Accusations of playing God by creating artificial life have been levelled at scientists many times in the past, hence recent research at the J. Craig Venter Institute on minimal and synthetic genomes that has led to heated ethical debate[80,81] is reminiscent of controversies from previous eras. Mary Shelley's novel *Frankenstein* draws many of its influences from early scientists who were accused of playing God as they sought to understand what life is – usually by reanimating it with electricity. Late 18th century scientist Luigi Galvani's experiments with electricity and amputated frog legs led the way and demonstrated animation of the inanimate by electricity.[82] Inspired by Galvani, others like Giovanni Aldini[83] (Galvani's nephew) and Andrew Ure were moved to try and reanimate corpses.[84] These experiments ultimately led to the invention of the defibrillator.

The influence of the *Frankenstein* topos has diminished little since the 18th century. Twentieth century scientists Heinz Fraenkel-Conrat, Alexis Carrel, Jacques Loeb, Patrick Steptoe and Robert Edwards have each been characterised as 'playing God' and as creators of artificial life or monsters in the media of their day. Jacques Loeb, a German-born American scientist working out of Woods Hole, Massachusetts, became one of the first scientists to be directly associated with the concept of creating artificial life. Loeb was working on artificial parthenogenesis; in his book *Artificial Parthenogenesis and Fertilization* (1913),[85] Loeb reports his progress in the science of the development of embryos from eggs without sperm. Loeb achieved this using a cocktail of chemicals. Loeb's work was a scientific landmark for several reasons. First it was a clear demonstration that life could be reduced to a set of complex chemical reactions, that is, there was no vital spark required. This marked the beginning of the end of vitalism, a task arguably finished by Venter. Second, it demonstrated that humans could 'create' life with chemicals and third, there was now a scientific observation of virgin birth. At a similar time, Alexis Carrel, pioneer of transplant surgery and the first to transplant a kidney,[86,87] showed that cellular tissues could be sustained out the body. This was the birth of tissue culture and further strengthened the idea that life could be reduced to a set of chemical reactions, since a solution of chemicals could sustain life almost indefinitely. Then in 1957, Heinz Fraenkel-Conrat reassembled a functional tobacco mosaic virus[88] from inanimate protein and RNA parts and in the process proved the existence of genomic RNA. Finally, the first test tube baby, Louise Brown, was conceived *in vitro* and transplanted as an embryo into her mother's womb. Born on 25th July 1978, Louise was a perfectly healthy baby.[89]

With the benefit of historical hindsight, Conrat, Carrel, Loeb and Doctors Steptoe and Edwards have not been credited as the scientists who were the first to create synthetic life and it seems ridiculous to accuse them of playing God and creating monsters. What all of these scientists have in common is that they have redefined what we think of life. When it comes to synthetic biology, the accusations of playing God and creating artificial life are articulations of social taboo and in effect ask if there should not be some aspects of the human experience which should be sacred and left alone. History shows us that articulations of social taboo are social conventions that are dependent on culture and belief systems. As such they tend to be dynamic. Today, we do not think of a heart stopped and then restarted by defibrillation as reanimation and playing God but in the late 18th and early 19th centuries reanimation by electrocution was at best a social taboo and at worst a contract with the devil. Thus it remains to be seen if in 20 years' time the labels 'playing God' and 'creating artificial life' will have any cultural relevance to synthetic biology.

Contrastingly, as synthetic biology extends our understanding of what life is, there is a risk that attitudes to the rights of living things (animal or human) are undermined. If we consider what life is in general terms then there is no consensus view. In macro-organisms, the difference between living and dead is usually straightforward. Indeed, the transition from living to dead can be so sudden that it fits well with theories of vitalism, where vitalism is the view that living objects have within them some extra entity not present within inanimate objects[90] and thus death is the loss of this entity. As noted previously, theories of vitalism have disappeared in mainstream science as we have begun to understand the molecular nature of life and life processes. Instead broad features of life are described, especially when searching for evidence of life on other planets,[91] using terms such as homeostasis, organisation, metabolism, growth, reproduction and response to stimuli. Biophysicists discuss life in terms of the sustenance of negative entropy[92] and in the field of complexity science Kauffman has defined life as

> an autonomous agent or a multi-agent system capable of reproducing itself or them-selves, and of completing at least one thermodynamic work cycle.[93]

What we can be sure of is that macroscopic living systems are sufficiently within the consensus opinion of 'what life is' to be protected by bioethical frameworks. As science deconstructs life to a series of molecular processes the protection of bioethical frameworks is less clear. Nowhere is this more emotive than in embryology, where a handful of cells and a set of complex chemical reactions blur the lines between humanity and the molecular. Recent research breakthroughs in the field of cell replacement therapy using human embryonic stem cells have started to bring this debate to the forefront of synthetic biology since gene editing techniques performed on mammalian cells and zebra fish embryos[94] could be extended to embryonic stem cells. Ethically the concern is that human life becomes instrumentalised through the use of human embryos as research tools. Current UK legislation, in the form of the Human Fertilisation and Embryology Act 2008,[95] permits by licence some research on embryos up to 14 days old. Embryonic stem cells are harvested from pre-implantation embryos 4 to 5 days post fertilisation, a process that requires the embryo to be destroyed[96] and attracts more ethical debate. The spectrum of views in this debate are wide; on one side there stands the so-called conceptionist view, which considers the potential of each human embryo to become a person. On the other side, there is the stance taken that these embryos should not be attributed any status of person. All these views are discussed in detail by Wert and Mummery.[96] A critical condition that permits research on embryos, to protect against instrumentalisation, is that no other suitable alternative methods exist.

1.4.2 Environmental Release of Designed Organisms and Their Potential Ecosystemic Impacts

Many applications of synthetic biology, particularly within the iGEM competition, propose environmental release of 'designed' organisms into the environment to exert positive effects, examples being bacterial sensors of disease and toxins as well as bacteria that detect landmines, remove arsenic from water supplies or digest plastic ocean debris. These applications mark the transition from technology that is derived for sole use within laboratories, within defined containment levels, to genetically modified biological organisms for distribution in the environment. Unsurprisingly these release-dependent applications invoke ethical scrutiny, in particular from environmental groups. Broadly speaking these risk issues are collectively referred to as uncontrolled releases. Whilst some commentators point out that deliberate release is anything but uncontrolled, the crucial point is that the release puts the organism into an environment where the risk mitigation by human intervention becomes unquantifiable; hence it is in this sense that the release is uncontrolled. Synthetic biology, of course, has followed GM food technology and as such enters an ethical and public perception debate[97,98] that is still raging.

There are many historical examples of uncontrolled releases ranging from industrial accidents of the chemical industry such as Bhopal and Chernobyl through to the unintended consequences of the jungle defoliant Agent Orange in the Vietnam War. In the case of Agent Orange,[99] as with many other chemical agents, persisting levels of environmental contamination act as a stark reminder that unforeseen risk issues can linger on long after the initial contaminating event.[100] In a microbiological context, there are other examples of uncontrolled releases such as the escape of foot-and-mouth disease[101] in 2007 from Pirbright Farm in the United Kingdom. The bioethical concerns that are raised above are generally health concerns. However, when organisms are released into the wild ecosystem disruption is a likely consequence, which we know from many macrobiological examples.

For example in the United Kingdom, the import of grey squirrels from the Eastern United States to adorn ornamental gardens has had a devastating effect on native red squirrel populations. The larger grey squirrel has all but displaced the red in most of the country by a probable mix of out-competition, interbreeding and the spread of a non-native virus, lethal to reds and harmless to greys. Currently there are concerns that the red squirrels of continental Europe are also under threat from grey squirrels, as a result of colonisation that originated when a US ambassador gifted a grey squirrel pair to Italy in 1948.[102]

There are many other examples that warn of the unpredictable consequences that will face synthetic biologists if they release their synthetic organisms into the wild. In the United Kingdom, in the 1980s American crayfish[103] were imported for farming and use in the restaurant trade. They quickly escaped and displaced the smaller native UK crayfish. American mink have long been prized for their fur by the fashion industry, a controversial practice; in protest many American mink were released from fur farms by animal rights activists.[104] These mink plus other escapees have had a devastating effect on some European native wildlife such as the British water vole.[105] A large number of other examples exist, where the introduction of non-native species has impacted UK biodiversity as reviewed by Manchester and Bullock.[106] The problem is not restricted to the introduction of mammals into new ecosystems; there are many cases of non-native plant species,[107] invertebrate and fungi,[108] oyster[109] disrupting ecosystems.

The widespread potential for ecosystem destabilization caused by non-native species was recognised by the 1992 Convention on Biological Diversity. The convention has since been adopted by an increasing number of countries who recognise the protocol's assertion that the conservation of biodiversity is a 'common concern of humankind'. One of the protocol's purposes that relates specifically to synthetic biology is to establish principles whereby the benefits arising from genetic resources are shared equitably. Further to this the Cartagena Protocol on Biosafety,[110] adopted in 2000 as a supplementary agreement to the Biodiversity Protocol, set out to protect biodiversity from the impact of genetically modified organisms and ultimately synthetic biology.[111]

As we near a time when synthetic organisms will be released into the environment, either deliberately or accidentally, a number of criticisms from environmental groups are raised. This ethical question takes a more economic perspective when the value of ecosystem services is considered.[112] In answer to this concern, synthetic biologists typically aim to engineer their life forms with limited survivability in the wild.[113]

Of course fears of uncontrolled release are not just centred on the disruption of ecosystems. There are also fears that synthetic biology research might cause the release of potent synthetic viruses with devastating effects on global health. The recent swine flu outbreak in 2009[114] and other avian flu outbreaks demonstrate the magnitude of the global concern that another potential flu pandemic, like Spanish flu,[115] might occur soon. In 2005, synthetic biologists recreated the Spanish flu virus from partially degraded samples in the permafrost.[116] The justification of this sort of research is that by recreating something one can truly understand it – a reference to the proverbial Richard Feynman quote – and devise ways to fight it. By the power of genome synthesis, an array of Spanish flu variants might be synthesised and tested to understand what makes some versions of the influenza virus so potent, so that effective vaccines can be made more rapidly. The accidental release of any such variant would have serious global health consequences. The recent case where out of date containment measures were the cause of the escape of the foot and mouth virus from Pirbright Farm, United Kingdom,[117] is a reminder of the risk involved in this research.

A further biocontainment risk is horizontal gene transfer,[118] that is, the transfer of genetic information by means other than traditional reproduction. For example, it is well known that bacteria can exchange plasmids and thus whilst any genetic variant released into to wild might have limited survivability, the transfer of modified genes into wild-type organisms is a genuine concern.[119] To date there are comparatively few confirmed incidences of genetic containment failure in the scientific literature. The highly publicised case suggesting that transgenic DNA spread to traditional strains of maize in Mexico[120] remains contentious.[121]

1.4.3 Biosecurity, Bioterrorism and Biowarfare

December 2008: the Commission on the Prevention of Weapons of Mass Destruction and Terrorism submits its report titled 'World at Risk'. Set in motion by the terrorist attacks that destroyed the World Trade Centre on September 11, 2001 and the subsequent 'anthrax letters' attacks, the US government took a series of introspective looks at potential sources of terrorist activity and weapons. The report[122] identified biological weapons of mass destruction as more likely to be appropriated by terrorists than nuclear weapons. The Commission also recommended that the US Government move rapidly to limit the spread of biological weapons and hence reduce the prospect of a bioterror attack. Whilst the threat of the misuse of biological technology has been recognised for many years,[123,124] the 'World at Risk' recommendation that the 'US government move aggressively' is an indicator of the seriousness with which such issues are dealt in the post-9/11 era.

An obvious paradox presents itself when considering synthetic biology and biosecurity: on the one hand, as discussed, the political and economic drivers for biotechnological revolution are enormous; on the other hand, there are growing concerns that widespread adoption of biotechnology will make it easier for malevolent applications to be realised. The paradox is most apparent in so-called dual use technologies,[124] that is, technologies intended for benign use that can also be co-opted for malign purpose. In February 2012, the Office of Science Policy – National Institute of Health released official dual use guidelines[125] for life science researchers active in the US that also cited a host of other policy documents. For the most part, these documents define their scope of dual use concern to around 20 different toxic organisms, such as Avian influenza virus and *Bacillus Anthracis*. From a dual use perspective, experiments that seek to understand the virulence of these organisms are of most concern since, through modification and testing, more virulent organisms might inadvertently arise and be misused. Full details can be obtained from the National Science Advisory Board for Biosecurity[126] and the implementation of the new policy in the US is discussed in detail by Wolinetz in 2012.[127]

The abuse of dual use technology for terrorist activity gives rise to a further aspect of biosecurity that impacts the traditional publication requirements for scientists. There has been a long tradition in the sciences for publication of discoveries to contain enough experimental details for others to replicate the discovery, usually by the inclusion of a methods section in a manuscript. However if the discovery has a potential dual use then publication of the methods for reproducing the experiments in a traditional scientific journal becomes an issue of biosecurity. Under such circumstances, as recently occurred with the H5N1 avian virus,[128] dual use policy can result in a research moratorium and restricted circulation of experimental details.[129] The H5N1 case represents a relatively rare occurrence, when scientific research, rather than its application, comes into conflict with society. In such cases, scientists are faced with an ethical dilemma either to follow established professional protocols for publication or recognise their social responsibility. This dilemma causes considerable debate in the scientific community.[130–132] In response to the dual use concern many journals now reserve the right to refer manuscripts to a dual use concern review panel, which may choose to restrict access to the publication, or parts thereof.

The relevance of dual use concern research to synthetic biology is growing. Arguably it was the synthesis of the polio virus 2002[133] that first brought the debate to public attention.[134] However, it is the ever-shrinking price of genome sequencing and the ever-increasing capability for rapid genome synthesis, the read/write steps of synthetic biology, that represent the biggest concern, since future projections suggest these technologies will be accessible and affordable to large numbers of people.

Biowarfare applications of synthetic biology are not just restricted to terrorist activities. As has happened previously with atomic technology and the nuclear arms race, there are fears that synthetic biology will lead to a biological arms race. There is some evidence to support this concern with synthetic biology research being funded by the US Department of Defence and UK Ministry of Defence; however, the majority of funding at the moment comes from the US National Institute for Health and National Science Foundation (NSF) followed by the European Framework Protocol.[3,135] The potential for the misuse of synthetic biology research, and the serious and far-reaching consequences that may follow, remain hot topics of discussion beyond the enclaves of science, as any internet search engine can demonstrate. One of the downsides of this debate is that it overshadows some very genuine ethical concerns that can dramatically impact on the lives of the world's poorest people.

1.4.4 Global Justice and Intellectual Property

The potential of synthetic biology to usurp traditional synthetic chemistry[136] in natural product synthesis and drug discovery was demonstrated by Jay Keasling's work on biosynthesis of artemisinic acid. Artemisinic acid is a chemical precursor to artemisinin, one of the few remaining chemical defences against the malaria parasite. Artemisinin, a sesquiterpene, has for the duration of its history in traditional Chinese medicine been extracted from *Artemisia annua*, the sweet wormwood plant. The chemical structure of artemisinin, shown in Figure 1.5, contains an epoxide bridge, which as well as being unusual in naturally occurring compounds, is difficult and therefore expensive to synthesise by chemical means.

Until relatively recently the global supply of artemisinin was dependent on the global harvest of *A. annua*. However, synthetic biology pioneer Jay Keasling and collaborators were able to reproduce the biosynthetic pathway from the *A. annua* plant in both *E. coli*[137] and more effectively in *Saccharomyces cerevisiae*.[55] Keasling's work is clearly a milestone in synthetic biology research; economically the technique produced semisynthetic artemisinin at a fraction of the cost afforded by chemical syntheses.[138] In fact such was the success that many scientists have had their interest piqued by the possibility of synthesising end products, or advanced stage precursors to feed into existing chemical syntheses, through a modified bacterium, examples being taxol[139] and polyketide compounds[140] and very recently surfactant-like compounds.[141]

Historically, ethical concerns regarding development of biotechnological routes to synthesise compounds traditionally produced by agricultural means have been associated with the pharmaceutical industry. From debating these concerns concepts such as 'global fairness' have been adopted. For instance, any practice that created a developing-world dependency on developed-world technology would run counter to many people's concept of global fairness. The term 'biopiracy'[142] has also been coined to describe scenarios where individuals or organisations derive profit from indigenous people's knowledge without

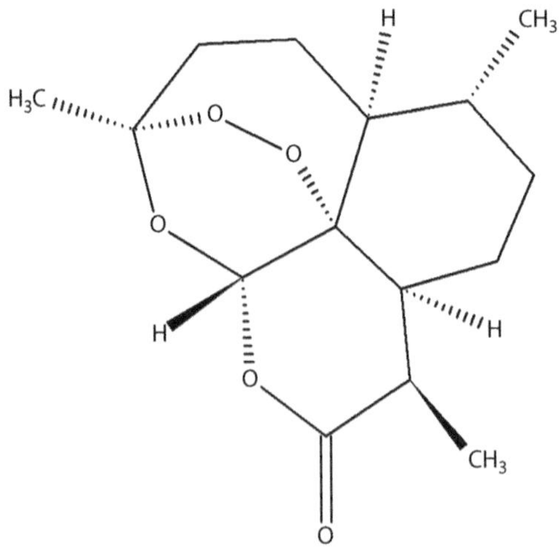

FIGURE 1.5
The molecular structure of artemisinin. The epoxide (O–O) bridge is particularly tricky to introduce by conventional chemical syntheses; however, the introduction of the biosynthetic pathway for artemisinin into a chassis cell means synthesis is comparatively straightforward.

commercial recognition of the role that indigenous people have had in developing that knowledge. In the case of artemisinin, globally *A. aunua* is raised from seedlings in China and Vietnam (70% global production) and African countries (20% global production) and it takes around 8 months to reach harvestable maturity. Artemisinin prices per kilo fluctuate wildly from $120 to $1,200, due to changeable harvest yields and demands.[143] Therefore production of antimalarial treatments containing artemisinin is costly and in principle available only to those who can afford it. The semi-synthesis of artemisinin in yeast or bacterial expression systems has several economic advantages; production costs are potentially lower, and transportation costs can be reduced by strategically placed batch reactors. Finally a stable supply of the material will stop price fluctuations and meet the expanding needs of people with malaria. Whilst at first glance it may seem that these are all positive outcomes for those who need artemisinin-based therapies the most, there are two broad ethical concerns raised. The first is simply expressed as the concern that it is unethical to make a developing nation dependent on a developed nation. The second concerns the fairness of developed nations exploiting and profiting from a technological advantage that potentially destroys the livelihoods of the many developing nation farmers who scrape a living from the earth growing small crops like *A. Annua* to feed their families.

For the moment at least it seems that *A. annua* farmers need not worry. The Bill and Melinda Gates Foundation have had a long involvement with Jay Keasling and Amyris, the spin-out company that manufactures the antimalarial compound artemisinin using yeast. Amyris has made a commitment that the semi-synthetic artemisinin they produce will only be used to meet shortfalls in artemisinin from agricultural sources. Globally, however, the ramifications of a technology that can synthesise natural products usually derived from plant sources in bacterial culture are enormous with the biggest losers likely to be subsistence farmers and some of the poorest people in the world.[144] A 2011 report carried out by 'The International Civil Society Working Group on Synthetic Biology' looks at the potential impacts of synthetic biology on the conservation and sustainable use of biodiversity.[111] The report examines in detail the level of interest in synthetic biology by multinational corporations, who typically are interested in the application of synthetic biology to replace existing chemical feedstocks or biofuels by alternative synthetic routes. Aside from artemisinin, other case studies are presented; one is the potential impact of a microbial biosynthesis of vanillin on the vanilla farmers of Madagascar, where an estimated 80,000 families cultivate vanilla orchids. Evolva Inc. are currently scaling up the microbial production of vanillin for commercialisation, with production located in Denmark and Switzerland. The impact on the estimated 200,000 people involved in vanilla production in developing countries is unknown. The report suggests that the vanilla example is one of many ethically dubious impacts of synthetic biology, others being production of liquorice, palm oil, natural rubber, pyrethrin and many more. As is the case with vanilla, production of these natural extracts in microbial reactors will occur largely in developed countries, but will impact undeveloped countries.

Of course ethical issues are never black and white and it should be pointed out that there are potentially positive environmental impacts of microbial production of natural products. To give just one example, the production of palm oil in countries like Indonesia requires the replacement of tracts of jungle with plantations of palm trees. This practice, growing because of the global interest in biofuels, displaces native species such as the orangutan from their jungle habitat and increasingly conflicts between plantation workers and orangutan occur as shown in Figure 1.6. Should extraction of palm oil from trees be replaced by a microbial method, it is likely that the orangutan will be one beneficiary.

FIGURE 1.6
Stranded orangutan mother rescued from palm oil plantation. Across Indonesia the already endangered orangutans find themselves in conflict with palm oil farmers and plantation owners. Hanna was found with bullet wounds in an area of jungle in Northern Sumatra cleared for a new palm oil plantation. (Reproduced from Hanna Adcock©. With permission.)

After its triumphant sequencing of the human genome in 2000, Celera Genomics, a J Craig Venter Company, filed for preliminary patents on 6,500 partial or whole genes.[145] This is around one quarter of the genes in the human genome. By 2000, American President Bill Clinton had announced that the human genome should not be patented, claiming that it should be made freely available to all researchers.[146] Within two days, Celera stock had plummeted and 50 million dollars were wiped off the biotechnology sector's market capitalisation on the Nasdaq exchange. The estimated cost of sequencing the genome to Celera Genomics was about 100 million dollars, but the collapse of its share price was caused by a clear indication from central government that genes may no longer be subject to patent law. The message to investors in biotechnology companies was simple; the financial revenues previously seen for successfully applied gene patents were likely to be much smaller than hoped.

Patent law is intended to foster technological advance by rewarding the discoverer of new processes with the exclusive exploitation rights to their invention for a period of time. In effect, this is intended to guarantee that the inventor can recoup the cost of their initial investment plus a healthy financial return. Whilst Celera Genomics could be argued to have made excessive patent claims, it was simply the latest in a series of biotechnology companies hoping to get an exclusive licence on gene sequences.[147] Legally the precedent for making gene patents appears to start in the 1970s where a patent on the cDNA sequence for human growth hormone (US Patent 4,363,877) was filed for and granted in 1982.[148] Rimmer[149] provides a fascinating insight into the origins and motivations for gene patenting in the biotechnology industry and the ethics of patenting DNA are considered in a Nuffield Council on Bioethics discussion paper.[150] In 1980, the US Patent Office issued the first of three patents known as the Cohen–Boyer recombinant DNA cloning patents.[151] Before they ran out in 1997[152] it is estimated the patents generated about 200 million dollars for the holders, University of California San Francisco and Stanford University. A full and fascinating account of the impact of these Cohen–Boyer patents on biotechnology is given by Hughes.[153] In short the profitability of gene patents, even after the costs of gene isolation and characterisation, have become sufficient allure to claim them. Critics of gene patenting point out that a patent is to protect a human invention and human genes are not human inventions.[147] In addition the involvement of patent lawyers in biotechnology has allegedly stifled invention – the very process patents are supposed to foster, which occurs largely through the disputation of the range of technologies covered in a patent.[147] For a couple of years biotech companies awaited the decision of the US Supreme Court where the rights of Myriad Genetics to hold patents on the *BRCA1* and *BRCA2* genes was challenged.[154] The patents to the *BRCA* genes effectively guaranteed Myriad the exclusive right to test for hereditary markers of ovarian and breast cancer in patients. The Supreme Court's decision in June 2013 was that the *BRCA 1 and 2* genes were products of nature and, given that Myriad had used the gene unmodified, could not be patented as an invention.[155]

For some synthetic biologists the dispute about gene ownership is clear; genes are to be shared not owned. And it is with this approach in mind that the BioBricks Foundation was formed.[38,156] The BioBricks Foundation is an open source repository for genes and works through the BioBricks User Agreement,[157] which allows users to freely take parts from the registry. In addition, users may submit parts. With the expansion of the iGEM competition, the BioBricks part registry has grown to the extent where some 15,000 genetic sequences that fulfil specific gene coding roles, when incorporated into bacterial 'chassis', are stored. In terms of patent law, this open source approach is at odds[158] with the prevailing practice and the US Supreme Court decision on the *BRCA* genes[155] clarifies the situation to a certain extent, since a naturally occurring gene that is subsequently modified can be classed as an invention and is hence patentable. Several articles detail the intellectual property complications raised by synthetic biology.[158,159] The economics of intellectual property rights (IPR), open source and monopolies with respect to synthetic biology are also discussed.[160]

The open source approach to synthetic biology established by the BioBrick Foundation was created by a group of pioneering synthetic biologists including Tom Knight and Drew Endy at MIT. A recent special issue of the *Biosocieties* journal provides a detailed profile of the open source approach to biology with both political[161] context and historical[162] insights, as summarised in the guest editorial.[163] The diversity in IP approaches, from open source (sharing) through to the aggressive gene patenting (ownership), is contextualised by Calvert.[164] The ideology that synthetic biology should be explored for the benefit of all humanity is an approach that has inspired strong public engagement activities and inspired another pioneer – the biohacker.

1.4.5 Biopunks, Biohackers and Do-It-Yourself Science

Biopunk as a movement has its origins in sci-fi classics such as William Gibson's *Neuromancer*[165] published in 1984. *Neuromancer* tells the story of a washed-up computer hacker who, in a dystopian future, is employed by a mysterious ex-military man to carry out a series of computer-hacking tasks. The novel breaks into new ground when computer hacking and biometric implants become entwined to enhance the breadth of human capability. Biopunk literature as a sub-genre emerged with novels like *Ribofunk*[166] and has become firmly ensconced at the fringes of popular culture. The genre tends to focus on the unintended consequences of recombinant DNA technology and the ensuing biotech revolution. What is surprising is that these futuristic visions have, in part, inspired a breed of DIY bioscientists or biohackers. Use of the word hacker is deliberate to evoke the romance of the Silicon Valley hacker groups that gave rise to a silicon revolution about 50 years ago. In a 2010 interview Bill Gates, one of the original 'hackers', gave the biohacking movement more credence when he announced to *Wired* magazine[167] 'that if he was 13 again he would be hacking biology instead of computers'. An obvious parallel can be drawn between the rise of biohacking, perhaps the beginning of the democratisation of biotechnology, and the computer hacking movement, which marked the beginning of the democratisation of computer technology, a point also raised by Freeman Dyson in *Our Biotech Future.*[25]

So who are the biohackers and biopunks? Typically biohackers and biopunks are portrayed as amateur biologists doing DIY science in garages, kitchens and bedrooms[168,169] – the term amateur is misleading. In his 2010 book *Biopunk,*[170] Marcus Wohlsen describes the rise of the biohacker movement. Many of the early biohackers are professional people and amateur only in the sense that they are not usually trained to advanced levels in the biosciences. What tends to unite biohackers are several concepts: a genuine interest in biology and the beliefs that open source protocols enable more creativity and the tools to carry out biology should be widely available to all. One of the pioneers of biohacking is Meredith Paterson, whose *Biopunk Manifesto*[171] proposes the movement's beliefs.

Whilst the biohacking movement has been met with a certain degree of scepticism from established scientists and academic institutions, it is worth remembering that such scepticisms were also levelled at the computer hackers who went on to create companies like Apple and Microsoft. Currently biohacking is expanding; rather than staying underground in garages and bedrooms, the movement is being financed by entrepreneurs that provide public hack spaces such as BioCurious[172] in California. These are open access laboratories where members of the public can go and learn about biology and carry out their own research. In December 2012, DIYbio Europe announced its intention of creating a network of amateur researchers committed to exploring biology.[173] This European network aims to bring together isolated groups of biohackers, in many of the major European cities, to attempt to overcome many of the problems, such as financing, that they face as separate groups.

One of the reasons that the amateur biologists can begin to explore the hacking of the genetic code is due to the increasing availability of DNA both as sequenced information and as writeable code,[174] the same technological advances that have enabled synthetic biologists to make significant advances. Unsurprisingly, therefore, biohackers have begun to move into the field of synthetic biology. This diffusion of synthetic biology into popular culture brings credible biosafety risks.[175] Amateur biologists might, for example, be responsible for uncontrolled releases of biological material and subsequent ecosystem disruptions. In addition, the biohacker movement raises concerns about biosecurity and bioterrorism in the sense that people with malevolent intentions might join or set up such groups and use the collective knowledge and facilities for harmful purposes. Under the dual use biosecurity recommendations there are specific documents that relate to amateur biology groups.

In *Biopunk,*[170] Wohlsen describes the role of the FBI in building a relationship with biohackers to balance the needs of Homeland Security in the wake of September 11, 2001, whilst fostering a lively entrepreneurial environment. Wohlsen discusses what level of scrutiny US biohacker communities warrant from security agencies and whether this should be guided either by the current practical capabilities of biohackers, their future biotechnological aspirations or both. It is, however, very easy to dismiss the aspirations of amateur biologists on the grounds that the technology gap between the professionals and the amateurs is too great. As an exemplar, it is possible to characterise the projects devised by members of the London BioHackspace into three principle types:

- Type 1: Retrofitting – Reproduction of basic laboratory processes, often in an imaginative way using household chemicals, typically introduction of fluorescent protein genes is a feature.
- Type 2: Reach for the stars – Highly aspirational projects that are currently inactive.
- Type 3: Enabling – Attempts to bridge the technology gap in order to resurrect Type 2 projects.

Type 3 projects have to date been most successful and in achieving this biohackers have begun to shrink the technology gap between them and the professional scientists. For example, DIY gel electrophoresis kits are made using a power pack and small plastic container. Bacterial incubators have been produced using polystyrene containers, a small thermostatically controlled heating mat (available from a pet shop for keeping reptile pets) and computer fans to circulate the air. For sterile cell culture work in academic and industrial settings typically a laminar flow hood is used that drives filtered air over an enclosed work area for manipulating cell growth vessels and equipment. The purpose of these laminar flow devices is to prevent cell cultures from being infected by airborne microbes present in the exterior environment. Biohacker groups have assembled laminar flow devices from large polystyrene boxes, high-efficiency particulate air (HEPA) filters and air blowers. In fact, amateur mushroom cultivators have been using these homemade laminar flow hoods for many years.[176] Small centrifuge devices such as the Dremelfuge[177] are open source 3D printer templates that enable anyone with a drill or rotary tool to achieve separation of DNA, cells and debris from solution. The polymerase chain reaction (PCR) machine, the workhorse of many a life science laboratory, might cost thousands to buy new but a cheap version can be made from a light bulb and a computer fan. Plans for the light bulb PCR,[178] as it is known, can be downloaded from the Internet. Another DIY PCR machine, the OpenPCR,[179] also exists and can be made by the user or purchased for a fraction of the cost of a 'professional' thermocycler.

All these inventions ignore the very obvious way that amateur scientists can equip their labs, which is to buy secondhand equipment from life science labs that are upgrading. Therefore it would seem that biohackers are solving Type 3 problems and perhaps progressing away from the Type 1 projects to the Type 2 projects that really inspire their interest in biotechnology. The scientific successes and biosafety risks that are likely to originate from these biohacker groups are as yet unknown; their existence argues for a governance methodology appropriate to synthetic biology that covers the professional and amateur scientist alike.

One of the interesting observations that emerges from biopunk and biohacker culture is the willingness of some humans to embrace the integration of themselves with technology and presumably biotechnology. Lepht Anonym is a Berlin-based biohacker who has been experimenting with body implantation devices on herself. Her most well-known project involves implanting small niobium magnets into her fingertips so that she can detect electric fields in the immediate environment. These are sensed as tiny induced currents in the magnets that set the nerves in her fingers tingling.[180] The culture of nonmedical magnetic finger implants and magnetic sense appears to stem from the United States where

American body modification artist Steve Haworth has over the last decade implanted thousands of magnets into customers' fingers at his body modification clinic in Phoenix, Arizona. Whilst this is clearly a long way from synthetic biology in a recognisable form it demonstrates a willingness among some people to enhance their experience through implant technology. Synthetic biology might provide a new wave of body modification art and its potential has not been missed by those interested in expanding human capability.

Transhumanism[181] is a movement that embraces the use of technology to transform the human condition. In particular, the transhumanist movement is interested in eliminating aging and improving human intellectual capability through technology. There is much speculation on how synthetic biology might aid the realisation of transhumanist ideals[35] and the subsequent bioethical debates are examined by Jotterand[182] and McNamee.[183]

1.5 Governance of Synthetic Biology

A key governance question that is under scrutiny is whether or not synthetic biology requires any new regulation or even its own branch of bioethics.[184] In essence, this question asks whether or not the existing control measures and regulatory framework such as the Genetically Modified Organisms Act, the Control of Substances Hazardous to Health (COSHH) and Special Animal Pathogen Order (SAPO) are sufficient to cover the risks posed by synthetic biology. Opinions on the answer to this question vary widely.[81] The Presidential Commission on Bioethical Issues found that current US legislation was sufficient to cover synthetic biology but recommended that a series of voluntary guidelines be adopted for now.[185] These are focused on ensuring that new technologies meet standards of public beneficence, responsible stewardship, intellectual freedom and responsibility, democratic deliberation, and justice and fairness.

Synthetic biologist George Church has been one of the leading figures calling for the licence to practice approach to synthetic biology. European safety organisations, Research Councils and academics are currently discussing regulatory practices[186,187] with a view to the best approach for synthetic biology.[188–191] Many organisations such as Friends of the Earth and the ETC are calling for a voluntary moratorium on synthetic biology research, as occurred with recombinant DNA technology in the 1970s, until the risks are better understood.[28] Such calls for a moratorium have not thus far been acted upon by the governments that fund significant scientific research programmes.

The SYNBIOSAFE project was a European Union funded project that researched the safety and ethical aspects of synthetic biology. Several research outputs[192] resulted from the collaboration that involved experts in biosafety, synthetic biology and ethics. The SYNBIOSAFE e-conference was one of the group's initiatives where an online community discussion around ethics, safety, governance, security and IPR was hosted. The e-conference findings were published[193] and provide an excellent overview of the governance debates in synthetic biology. SYNBIOSAFE also published a book that provides an excellent overview of the societal and governance issues surrounding synthetic biology.[192,194]

Several organisations have polled the public as to their knowledge of and concerns over synthetic biology, both in the United States[195] and United Kingdom.[196] These surveys show several features, first that awareness of synthetic biology as a concept in popular culture is growing and as a result awareness of the societal risk issues is also growing. A major concern for the US public is that synthetic biology could be used for the production of bioweapons,[195] whereas in the United Kingdom there is a clear picture that the public wants a greater level of openness about the benefits, applications and responsibility.[196]

The strength of public attitudes to science should not be underestimated; in the United Kingdom the anti-GM movement has largely dominated the public side of the debate and as a result GMed foods and GMed crops are prohibited. This issue is discussed by Tait[2] with particular reference to the public perception challenge likely to face synthetic biology when the political drivers for economic progress and social harmony come into conflict.

1.6 Conclusions

As presented synthetic biology is a branch of science and engineering with the potential to impact society and the environment in very significant ways. Social commentators, ethicists and non-governmental organisations (NGOs) have likened synthetic biology to a Pandora's Box. Similar comparisons were made when atomic technology was developed and the proverbial 'genie out of the bottle' reference was made to warn against the irreversible nature of some technological developments. A selection of quotes serves to illustrate this point:

> The genie is out of the bottle. Our challenge is to ethically master the machines we are creating.[72]
>
> **Julian Savulescu**

> This is a Pandora's Box moment – like the splitting of the atom or the cloning of Dolly the sheep, we will all have to deal with the fall-out from this alarming experiment.[29]
>
> **Pat Mooney**
> *ETC*

> Venter is creaking open the most profound door in humanity's history, potentially peeking into its destiny.[29]
>
> **Julian Savulescu**

> What is really dangerous is these scientists' ambitions for total and unrestrained control over nature, which many people describe as 'playing God'.[197]
>
> **Dr. David King**
> *Human Genetics Alert*

Such quotes often suggest that scientists alone are responsible for the negative effects of their discoveries. Continuing the comparison between synthetic biology and the atomic age, the physicist Otto Hahn expressed profound regret with respect to his discovery of nuclear fission, the technology which ultimately led to the production of atomic bombs and the destruction of Hiroshima.[198,199] Yet in June 1945 the Franck Report,[200] written by a commission comprised of a group of US scientists and headed by Nobel Laureate James Franck, recommended that the US government not use atomic bombs in warfare – a few days later the first atomic bomb was used in warfare. In this instance, history reminds us that politicians, not terrorists, amateurs or misguided scientists, have unleashed the most destructive powers of new technology on the world. How synthetic biology fits into this picture is still an open question. There is however clear evidence that a significant proportion of synthetic biological research is funded by defence organisations in the United Kingdom and the United States; however the biggest funders up to 2012[3,135] are the National Institutes of Health (NIH), the NSF in the United States and the European Framework programme. The research themes being explored promise new recombinant drugs, cheap medicines, food security, pollution control, biofuels, accelerated development in the developing world, longer life and the end of inherited disease. Against this progressive backdrop the ethical and social responsibility issues of synthetic biology beg caution.

Acknowledgements

Much of the material in this chapter originates from lectures written by the author for the ethics component of a synthetic biology module at the University of Southampton, United Kingdom. The author wishes to thank Dr. Ali Tavassoli, Dr. Eugen Stulz and Professor George Attard for their open minded approach to teaching synthetic biology, which allowed him the opportunity to blend science, ethics and popular culture. Thanks are also due to Lisa O'Rourke, for reviewing several drafts of this chapter, and Hanna Adcock for image contribution. Finally many thanks to Alexandra Daisy Ginsberg for her interest, comments and image contribution.

References

1. Gibson, D. G., Glass, J. I., Lartigue, C., et al. 2010. Creation of a bacterial cell controlled by a chemically synthesized genome. *Science*. 329: 52–56.
2. Tait, J. 2009. Upstream engagement and the governance of science. *EMBO Rep*. 10: S18–S22.
3. Oldham, P., Hall, S. and Burton, G. 2012. Synthetic biology: Mapping the scientific landscape. *PLoS ONE*. 7: e34368.
4. Woolfson, D. N. and Bromley, E. H. C. 2011. Synthetic biology. *Biochemist*. 33: 19–25.
5. Tavassoli, A. 2010. Synthetic biology. *Org. Biomol. Chem.* 8: 24–28.
6. Szybalski, W. 1974. *In Vivo* and *in Vitro* Initiation of Transcription. In *Control of Gene Expression*, eds. Kohn, A. and Shatkay A., pp. 404–405. New York, NY: Plenum Press.
7. Szybalski, W. and Skalka, A. 1978. Nobel prizes and restriction enzymes. *Gene*. 4: 181–182.
8. Leduc, S. 1912. *La Biologie Synthetique*. Paris: A. Poinat.
9. Yeh, B. J. and Lim, W. A. 2007. Synthetic biology: Lessons from the history of synthetic organic chemistry. *Nat. Chem. Biol.* 3: 521–525.
10. Wöhler, F. 1828. Ueber künstliche Bildung des Harnstoffs. *Annalen der Physik und Chemie*. 88: 253–256.
11. McKie, D. 1944. Wöhler's 'synthetic' urea and the rejection of vitalism: A chemical legend. *Nature*. 153: 608–610.
12. Kronberger, N., Holtz, P. and Wagner, W. 2012. Consequences of media information uptake and deliberation: Focus groups' symbolic coping with synthetic biology. *Public Underst. Sci.* 21: 174–187.

13. Orwell, G. 1949. *Nineteen Eighty-Four*. London, UK: Secker and Warburg.

14. Huxley, A. 1992. *Brave New World*. New York, NY: HarperCollins.

15. Wells, H. G. 1914. *The World Set Free*. London, UK: Macmillan and Co.

16. Wells, H. G. 1896. *The Island of Dr Moreau*. London, UK: Heinemann.

17. Shelley, M. 1998. *Frankenstein (1831)*. Oxford: Oxford University Press.

18. Natali, V., 2009. *Splice*. Warner Brothers, US, Film.

19. Anderson, P. W. S., 2002. *Resident Evil*. Constantin Film, Film.

20. Berg, P. 2008. Meetings that changed the world: Asilomar 1975: DNA modification secured. *Nature*. 455: 290–291.

21. Berg, P. and Singer, M. F. 1995. The recombinant DNA controversy: Twenty years later. *Proc. Natl. Acad. Sci. U. S. A.* 92: 9011–9013.

22. Stephen, P. S. 1978. The Recombinant DNA Debate. *Philos Public Aff.* 7: 187–205.

23. Crichton, M. 1969. *The Andromeda Strain*. New York, NY: Knopf.

24. Robb, J. W. 1982. Reflections: Ethics and the recombinant DNA debate. *J. Craniofac. Genet. Dev. Biol.* 2: 51–63.

25. Dyson, F. 2007. Our biotech future. *New York Rev Books*. 54. http://www.nybooks.com/issues/2007/jul19

26. Woese, C. R. 2004. A new biology for a new century. *Microbiol. Mol. Biol. Rev.* 68: 173–186.

27. ETC Group. 2007. *Extreme Genetic Engineering: An Introduction to Synthetic Biology*. http://www.etcgroup.org/content/extreme-genetic-engineering-introduction-synthetic-biology (accessed 31 January, 2013).

28. Pennisi, E. 2012. *111 Organizations Call for Synthetic Biology Moratorium*. http://news.sciencemag.org/scienceinsider/2012/03/111-organizations-call-for-synth.html (accessed 31 January, 2013).

29. Macrae, F. 2010. Scientist accused of playing God after creating artificial life by making designer microbe from scratch – but could it wipe out humanity? *Daily Mail Online*. http://www.dailymail.co.uk/sciencetech/article-1279988/Artificial-life-created-Craig-Venter--wipe-humanity.html (accessed 12 December, 2012).

30. Hanlon, M. 2010. Has he created a monster? *Daily Mail Online*. http://www.dailymail.co.uk/sciencetech/article-1279988/Artificial-life-created-Craig-Venter--wipe-humanity.html (accessed 12 December, 2012).

31. Matheson, R. 1954. *I Am Legend*. New York, NY: Gold Medal.

32. Sun, T. 2011. *Scientist 'Creates First Man-Made Cell'*. http://www.thesun.co.uk/sol/homepage/news/2981138/Scientist-creates-first-man-made-cell.html (accessed 15 January, 2013).

33. Turney, J. 2000. *Frankenstein's Footsteps: Science, Genetics and Popular Culture*. New Haven, CT: Yale University Press.

34. Hellsten, I. 2003. Focus on metaphors: The case of "Frankenfood" on the web. *J. Comput. Mediated Commun.* 8. doi: 10.1111/j.1083-6101.2003.tb00218.x

35. Church, G. and Regis, E. 2012. *Regenesis: How Synthetic Biology will Reinvent Nature and Ourselves*. New York, NY: Basic Books.

36. Torgersen, H. 2009. Synthetic biology in society: Learning from past experience? *Syst. Synth. Biol.* 3: 9–17.

37. Jones, C. 2007. *How to Make Life*. http://www.esquire.com/features/best-brightest-2007/synthbiol1207 (accessed 22 December, 2012).

38. Smolke, C. D. 2009. Building outside of the box: iGEM and the BioBricks Foundation. *Nat. Biotechnol.* 27: 1099–1102.

39. Goodman, C. 2008. Engineering ingenuity at iGEM. *Nat. Chem. Biol.* 4: 13.

40. Brown, J. 2007. The iGEM competition: Building with biology. *Synthetic Biology, IET.* 1: 3–6.

41. Campos, L. 2012. The BioBrick Road. *BioSocieties*. 7: 115–139.

42. Guan, Z., Schmidt, M., Pei, L., Wei, W. and Ma, K. 2013. Biosafety considerations of synthetic biology in the International Genetically Engineered Machine (iGEM) competition. *Bioscience*. 63: 25–34.

43. Ginsberg, A. D. 2014. *Synthetic Aesthetics: Investigating Synthetic Biology's Designs on Nature*. Cambridge, MA: MIT Press.

44. Ginsberg, A. D. 2013. *Personal Communication.*

45. The Guardian. 2010. ABCs: National daily newspaper circulation April 2010. *The Guardian.* http://www.guardian.co.uk/media/table/2010/may/14/abcs-national-newspapers (accessed 31 January, 2013).

46. Connor, S. 2010. Dr Craig Venter: So, Doctor, how does it feel to have created artificial life? *The Independent.* http://www.independent.co.uk/news/people/profiles/dr-craig-venter-so-doctor-how-does-it-feel-to-have-created-artificial-life-1978873.html (accessed 31 January, 2013).

47. Pauwels, E. and Ifrim, I. 2008. *Trends in American and European Press Coverage of Synthetic Biology, Tracking the Last Five Years of Coverage.* Woodrow Wilson Centre. http://www.synbioproject. org/library/publications/archive/why_scientists_should_care/ (accessed 31 January, 2013).

48. Pauwels, E., Lovell, A. and Rouge, E. 2012. *Trends in American and European Press Coverage of Synthetic Biology, Tracking the Years 2008–2011.* Woodrow Wilson Centre. http://www. synbioproject.org/library/publications/archive/6636/ (accessed 31 January, 2013).

49. Gschmeidler, B. and Seiringer, A. 2012. "Knight in shining armour" or "Frankenstein's creation"? The coverage of synthetic biology in German-language media. *Public Underst. Sci.* 21: 163–173.

50. Khalil, A. S. and Collins, J.J. 2010. Synthetic biology: Applications come of age. *Nat. Rev. Genet.* 11: 367–379

51. Hodgman, C. E. and Jewett, M. C. 2012. Cell-free synthetic biology: Thinking outside the cell. *Metab. Eng.* 14: 261–269.

52. Alterovitz, G., Muso, T. and Ramoni, M. F. 2010. The challenges of informatics in synthetic biology: from biomolecular networks to artificial organisms. *Brief. Bioinform.* 11: 80–95.

53. Bashor, C. J., Horwitz, A. A., Peisajovich, S. G. and Lim, W. A. 2010. Rewiring cells: Synthetic biology as a tool to interrogate the organizational principles of living systems. *Ann. Rev. Biophys.* 39: 515–537.

54. Levskaya, A., Chevalier, A. A., Tabor, J. J., et al. 2005. Synthetic biology: Engineering *Escherichia coli* to see light. *Nature.* 438: 441–442.

55. Ro, D.-K., Paradise, E. M., Ouellet, M., et al. 2006. Production of the antimalarial drug precursor artemisinic acid in engineered yeast. *Nature.* 440: 940–943.

56. Aubel, D. and Fussenegger, M. 2010. Mammalian synthetic biology—From tools to therapies. *Bioessays.* 32: 332–345.

57. Gibson, D. G., Benders, G. A., Andrews-Pfannkoch, C., et al. 2008. Complete chemical synthesis, assembly, and cloning of a Mycoplasma genitalium genome. *Science.* 319: 1215–1220.

58. Polacek, N. 2011. The ribosome meets synthetic biology. *Chembiochem.* 12: 2122–2124.

59. Grunberg, R. and Serrano, L. 2010. Strategies for protein synthetic biology. *Nucleic Acids Res.* 38: 2663–2675.

60. Wang, L., Brock, A., Herberich, B. and Schultz, P. G. 2001. Expanding the genetic code of *Escherichia coli. Science.* 292: 498–500.

61. Wang, L., Xie, J. and Schultz, P. G. 2006. Expanding the genetic code. *Annu. Rev. Biophys. Biomol. Struct.* 35: 225–249.

62. Corsi, J., Dymond, M. K., Ces, O., et al. 2008. DNA that is dispersed in the liquid crystalline phases of phospholipids is actively transcribed. *Chem. Commun.* 20: 2307–2309.

63. Wilson, R. J., Tyas, S. R., Black, C. F., Dymond, M. K. and Attard, G. S. 2010. Partitioning of ssRNA molecules between preformed monolithic HII liquid crystalline phases of lipids and supernatant isotropic phases. *Biomacromolecules.* 11: 3022–3027.

64. Black, C. F., Wilson, R. J., Nylander, T., Dymond, M. K. and Attard, G. S. 2010. Linear dsDNA partitions spontaneously into the inverse hexagonal lyotropic liquid crystalline phases of phospholipids. *J. Am. Chem. Soc.* 132: 9728–9732.

65. Tsaloglou, M.-N., Attard, G. S. and Dymond, M. K. 2011. The effect of lipids on the enzymatic activity of 6-phosphofructo-1-kinase from B. stearothermophilus. *Chem. Phys. Lipids.* 164: 713–721.

66. Luisi, P. L. 2006. *The Emergence of Life: From Chemical Origins to Synthetic Biology.* Cambridge: Cambridge University Press.

67. Nelkin, D. and Lindee, S. 1995. *The DNA Mystique: The Gene as a Cultural Icon.* New York, NY: Reeman.
68. Bryant, J., Baggott la Velle L. and Searle, J. 2005. *Introduction to Bioethics.* Chichester, UK: John Wiley & Sons.
69. Balmer, A and Martin, P. 2008. *Synthetic Biology: Social and Ethical Challenges.* Nottingham, UK: Institute for Science and Society, University of Nottingham.
70. Boldt, J. and Müller, O. 2008. Newtons of the leaves of grass. *Nat. Biotechnol.* 4: 387–389.
71. de S Cameron, N. M. and Caplan, A. 2009. Our synthetic future. *Nat. Biotechnol.* 27: 1103–1105.
72. Savulescu, J. 2012. Master the new loom before life's tapestry unravels at our hands. *Times Higher Education.* http://www.timeshighereducation.co.uk/story.asp?storycode=419685 (accessed 31 January, 2013).
73. Douglas, T. and Savulescu, J. 2010. Synthetic biology and the ethics of knowledge. *J. Med. Ethics.* 36: 687–693.
74. Nuffield Council on Bioethics. 2012. *Emerging Biotechnologies: Technology, Choice and the Public Good.* London, UK: Nuffield Council on Bioethics.
75. Coady, C. A. J. 2009. Playing God. In *Human Enhancement,* eds. Bostrom, N. and Savulescu, J., pp. 155–180. Oxford: Oxford University Press.
76. Dabrock, P. 2009. Playing God? Synthetic biology as a theological and ethical challenge. *Syst. Synth. Biol.* 3: 47–54.
77. Belt, H. 2009. Playing God in Frankenstein's Footsteps: Synthetic biology and the meaning of life. *NanoEthics.* 3: 257–268.
78. Link, H. J. 2012. Playing God and the intrinsic value of life: Moral problems for synthetic biology? *Sci. Eng. Ethics.* 19, 435–448.
79. Deplazes-Zemp, A. 2012. The conception of life in synthetic biology. *Sci. Eng. Ethics.* 18: 757–774.
80. Silver, L. 2007. Scientists push the Boundaries of Human Life, *Newsweek* (International Editions). http://www.newsweek.com/scientists-push-boundaries-human-life-101723.
81. Cho, M. K. and Relman, D. A. 2010. Synthetic "Life," ethics, national security, and public discourse. *Science.* 329: 38–39.
82. Galvani, L., 2000. De viribus electricitatis in motu musculari commentarius (1791). ed. M. Bresadola. Bologna, Accademia delle Scienze.
83. Parent, A. 2004. Giovanni Aldini: From animal electricity to human brain stimulation. *Can. J. Neurol. Sci.* 331: 576–584.
84. Ure, A. 1819. An account of some experiments made on the body of a criminal immediately after execution, with physiological and practical observations. *Quart. J. Sci.* 6: 283–294.
85. Loeb, J. 1913. *Artificial Parthenogenesis and Fertilization.* Chicago, IL: The University of Chicago Press.
86. Sade, R. M. 2005. Transplantation at 100 years: Alexis Carrel, pioneer surgeon. *Ann. Thorac. Surg.* 80: 2415–2418.
87. Carrel, A. and Guthrie C. C. 1905. Functions of a transplanted kidney. *Science.* 22: 473.
88. Fraenkel-Conrat, H. and Singer. B. 1957. Virus reconstitution. II. Combination of protein and nucleic acid from different strains. *Biochim. Biophys. Acta.* 24: 540–548.
89. Henig, R. M. 2004. *Pandora's Baby.* New York, NY: Cold Spring Harbour Laboratory Press.
90. Greco, M. 2005. On the vitality of vitalism. *Theor Cult Soc.* 22: 15–27.
91. McKay, C. P. 2004. What is life—and how do we search for it in other worlds? *PLoS Biol.* 2: 302.
92. Margulis, L. and Sagan, D. 1995. *What Is Life?* Berkley and Los Angeles: University of California Press.
93. Kauffman, S. 2004. Autonomous Agents. In *Science and Ultimate Reality: Quantum Theory, Cosmology, and Complexity,* eds. Barrow, J. D., Davies, P. C. W. and Harper, C. L. Cambridge: Cambridge University Press.
94. Chang, N., Sun, C., Gao, L., et al. 2013. Genome editing with RNA-guided Cas9 nuclease in Zebrafish embryos. *Cell Res.* 23: 465–472.
95. Department of Health. 2008. *Human Fertilisation and Embryology Act.* London, UK: The Stationary Office.

96. Wert, G. D. and Mummery, C. 2003. Human embryonic stem cells: Research, ethics and policy. *Hum. Reprod.* 18: 672–682.

97. Cook, G., Pieri, E. and Robbins, P. T. 2004. 'The scientists think and the public feels': Expert perceptions of the discourse of GM food. *Discourse Soc.* 15: 433–449.

98. Shaw, A. 2002. "It just goes against the grain." Public understandings of genetically modified (GM) food in the UK. *Public Underst. Sci.* 11: 273–291.

99. Ngo, A. D., Taylor, R., Roberts, C. L. and Nguyen, T. V. 2006. Association between Agent Orange and birth defects: Systematic review and meta-analysis. *Int. J. Epidemiol.* 35: 1220–1230.

100. Schecter, A., Cao Dai, L., Päpke, O., et al. 2001. Recent dioxin contamination from Agent Orange in residents of a southern Vietnam city. *J. Occup. Environ. Med.* 43: 435–443.

101. Cottam, E. M., Wadsworth, J., Shaw, A. E., et al. 2008. Transmission pathways of foot-and-mouth disease virus in the United Kingdom in 2007. *PLoS Pathog.* 4: e1000050.

102. Bertolino, S. and Genovesi, P. 2003. Spread and attempted eradication of the grey squirrel (Sciurus carolinensis) in Italy, and consequences for the red squirrel (Sciurus vulgaris) in Eurasia. *Biol. Conserv.* 109: 351–358.

103. Lodge, D. M., Taylor, C. A., Holdich, D. M. and Skurdal, J. 2000. Nonindigenous crayfishes threaten North American freshwater biodiversity: Lessons from Europe. *Fisheries.* 25: 7–20.

104. 1998. UK Animal rights group claims responsibility for mink release. http://news.bbc.co.uk/1/hi/uk/148120.stm (accessed 26 December, 2012).

105. Woodroffe, G. L., Lawton, J. H. and Davidson, W. L. 1990. The impact of feral mink Mustela vison on water voles Arvicola terrestris in the North Yorkshire Moors National Park. *Biol. Conserv.* 51: 49–62.

106. Manchester, S. J. and Bullock, J. M. 2000. The impacts of non-native species on UK biodiversity and the effectiveness of control. *J. Appl. Ecol.* 37: 845–864.

107. Gordon, D. R. 1998. Effects of invasive, non-indigenous plant species on ecosystem processes. *Ecol. Appl.* 8: 975–989.

108. Langor, D. and Sweeney, J. 2009. Ecological impacts of non-native invertebrates and fungi on terrestrial ecosystems. *Biol. Invasions.* 11: 1–3.

109. Ruesink, J. L., Lenihan, H. S., Trimble, A. C., et al. 2005. Introduction of non-native oysters: Ecosystem effects and restoration implications. *Annu. Rev. Ecol. Evol. Syst.* 36: 643–489.

110. Mackenzie, R. and Ascencio, A. 2003. An explanatory guide to the Cartagena Protocol on Biosafety. *IUCN.* 46.

111. The International Civil Society Working Group on Synthetic Biology. 2011. *A Submission to the Convention on Biological Diversity's Subsidiary Body on Scientific, Technical and Technological Advice (SBSTTA) on the Potential Impacts of Synthetic Biology on the Conservation and Sustainable Use of Biodiversity.* http://www.cbd.int/doc/emerging-issues/Int-Civil-Soc-WG-Synthetic-Biology-2011-013-en.pdf (accessed 31 January, 2013).

112. Harrison, R. M. and Hester, R. E., eds., 2010 *Ecosystem Services.* London: RSC Publishing.

113. Russ, Z. N. 2008. Synthetic biology: Enormous possibility, exaggerated perils. *J. Biol. Eng.* 2: 7.

114. Cohen, J. and Enserink, M. 2009. As swine flu circles globe, scientists grapple with basic questions. *Science.* 324: 572–573.

115. Johnson, N. P. and Mueller, J. 2002. Updating the accounts: Global mortality of the 1918–1920 "Spanish" influenza pandemic. *Bull. Hist. Med.* 76: 105–115.

116. Tumpey, T. M., Basler, C. F., Aguilar, P. V., et al. 2005. Characterization of the reconstructed 1918 Spanish influenza pandemic virus. *Science.* 310: 77–80.

117. Anderson, I. 2008. *Foot and Mouth Disease 2007: A Review and Lessons Learned.* London, UK: The Stationery Office. http://www.official-documents.gov.uk/document/hc0708/hc03/0312/0312.pdf (accessed 26 December, 2012).

118. Syvanen, M. and Kado, C. I., eds., 2002 *Horizontal Gene Transfer.* San Diego, CA: Academic Press.

119. Wright, O., Stan, G.-B. and Ellis, T. 2013. Building-in biosafety for synthetic biology. *Microbiology.* 159: 1221–1235.

120. Quist, D. and Chapela, I. H. 2001. Transgenic DNA introgressed into traditional maize landraces in Oaxaca, Mexico. *Nature.* 414: 541–543.

121. Christou, P. 2002. No credible scientific evidence is presented to support claims that transgenic DNA was introgressed into traditional maize landraces in Oaxaca, Mexico. *Transgenic Res.* 11: 3–5.

122. Graham, B. B., ed., 2010 *Prevention of WMD Proliferation and Terrorism Report Card*. Darby, PA: Diane Publishing Company.

123. National Research Council. 2004. *Biotechnology Research in an Age of Terrorism*. The National Academies Press. http://www.nap.edu/catalog.php?record_id=10827 (accessed 26 December, 2012).

124. National Research Council. 2006. *Globalization, Biosecurity, and the Future of the Life Sciences. Committee on Advances in Technology and the Prevention of Their Application to Next Generation Biowarfare Threats*, Washington, DC: The National Academies Press.

125. National Science Advisory Board on Biosecurity. 2012. *Enhancing Responsible Science Considerations for the Development and Dissemination of Codes of Conduct for Dual Use Research*. National Science Advisory Board on Biosecurity.

126. National Science Advisory Board for Biosecurity. 2012. *About NSABB*. Washington DC: National Science Advisory Board for Biosecurity.

127. Wolinetz, C. D. 2012. Implementing the new U.S. dual-use policy. *Science*. 336: 1525–1527.

128. Fauci, A. S. and Collins, F. S. 2012. Benefits and risks of influenza research: Lessons learned. *Science*. 336: 1522–1523.

129. Frankel, M. S. 2012. Regulating the boundaries of dual-use research. *Science*. 336: 1523–1525.

130. Selgedid, M. J. 2007. A tale of two studies: Ethics, bioterrorism, and the censorship of science. *Hastings Cent. Rep.* 37: 35–43.

131. Selgelid, M. 2009. Dual-use research codes of conduct: Lessons from the life sciences. *NanoEthics*. 3: 175–183.

132. Schubert, C. 2008. Bioterror experts split on recommendations for 'dual use' research. *Nat. Med.* 14: 893–893.

133. Cello, J., Paul, A. V. and Wimmer, E. 2002. Chemical synthesis of poliovirus cDNA: Generation of infectious virus in the absence of natural template. *Science*. 297: 1016–1018.

134. Wimmer, E. 2006. The test-tube synthesis of a chemical called poliovirus. *EMBO Rep.* 7: S3–S9.

135. Synthetic Biology Project. 2010. *Research Brief 1: Trends in Synthetic Biology Research Funding in the United States and Europe*. Woodrow Wilson International Center for Scholars. http://www.synbioproject.org/process/assets/files/6420/final_synbio_funding_web2.pdf (accessed 26 December, 2012).

136. Weber, W. and Fussenegger, M. 2009. The impact of synthetic biology on drug discovery. *Drug Discov. Today*. 14: 956–963.

137. Martin, V. J., Pitera, D. J., Withers, S. T., Newman, J. D. and Keasling, J. D. 2003. Engineering a mevalonate pathway in Escherichia coli for production of terpenoids. *Nat. Biotechnol.* 21: 796–802.

138. Schmid, G. and Hofheinz, W. 1983. Total synthesis of qinghaosu. *J. Am. Chem. Soc.* 105: 624–625.

139. Ajikumar, P. K., Xiao, W.-H., Tyo, K. E. J., et al. 2010. Isoprenoid pathway optimization for taxol precursor overproduction in *Escherichia coli*. *Science*. 330: 70–74.

140. Neumann, H. and Neumann-Staubitz, P. 2010. Synthetic biology approaches in drug discovery and pharmaceutical biotechnology. *Appl. Microbiol. Biotechnol.* 87: 75–86.

141. Akhtar, M. K., Turner, N. J. and Jones, P. R. 2012. Carboxylic acid reductase is a versatile enzyme for the conversion of fatty acids into fuels and chemical commodities. *Proc. Natl. Acad. Sci. U. S. A.*

142. Shiva, V. 1997. *Biopiracy: The Plunder of Nature and Knowledge*. Cambridge, MA: South End Press.

143. Artemisinin Enterprise. 2008. *Meeting the Malaria Treatment Challenge: Effective Introduction of New Technologies for a Sustainable Supply of ACTs*. http://r4d.dfid.gov.uk/output/180080/default.aspx (accessed 26 December, 2012).

144. Thomas, J. 2011. *The Sins of Syn Bio*. http://www.slate.com/articles/technology/future_tense/2011/02/the_sins_of_syn_bio.html (accessed 15 January, 2013).

145. BBC News. 1999. *Human Gene Patents Defended*. http://news.bbc.co.uk/1/hi/sci/tech/487773. stm (accessed 6 January, 2013).

146. Gorner, P. 2000. Clinton, Blair urge sharing of gene data. *Chicago Tribune*. http://articles .chicagotribune.com/2000-03-15/news/0003150079_1_gene-data-genome-data-human-genome-project.

147. Salzberg, S. L. 2012. The perils of gene patents. *Clin. Pharmacol. Ther.* 91: 969–971.

148. Goodman, H. M., Shine, J. and Seeburg, P. H. 1982. Recombinant DNA transfer vectors. United States Patent Office 4363877.

149. Rimmer, M. 2002. Genentech and the stolen gene: Patent law and pioneer inventions. *Bio-Science Law Review*. 5: 198–211.

150. Nuffield Council on Bioethics. 2002. *The Ethics of Patenting DNA*. London, UK: Nuffield Council on Bioethics.

151. Cohen, N. S. and Boyer, H. N. 1980. Process for Producing Biologically Functional Molecular Chimeras. United States Patent Office 4237224.

152. Rauber, C. 1997. $200M patent runs out. *San Francisco Business Times*. http://www.bizjournals. com/sanfrancisco/stories/1997/11/24/story2.html?page=all (accessed 6 January, 2013).

153. Hughes, S. S. 2001. Making dollars out of DNA. *Isis*. 92: 541–575.

154. 2012. *US Supreme Court to Decide on Gene Patents in Myriad Case*. http://blogs.nature.com/ news/2012/11/us-supreme-court-to-decide-on-gene-patents-in-myriad-case.html (accessed 17 January, 2013).

155. US Supreme Court. 2013. *Association for Molecular Pathology et al. v. Myriad Genetics, Inc. et al.*

156. Coghlan, A. 2012. Synthetic biology can build us a better future. *New Sci.* 216: 29.

157. BioBricks Foundation. 2013. *The BioBrick™ User Agreement*. https://biobricks.org/bpa/users/ agreement/ (accessed 8 January, 2013).

158. Rai, A. and Boyle, J. 2007. Synthetic biology: Caught between property rights, the public domain, and the commons. *PLoS Biol.* 5: e58.

159. Kumar, S. and Rai, A 2007. Synthetic biology: The intellectual property puzzle. *Texas Law Review*. 85: 1745–1768.

160. Henkel, J. and Maurer, S. M. 2007. The economics of synthetic biology. *Mol. Syst. Biol.* 3. doi: 10.1038/msb4100161

161. Hilgartner, S. 2012. Novel constitutions? New regimes of openness in synthetic biology. *BioSocieties*. 7: 188–207.

162. Kelly, C. M. 2012. This is not an article: Model organism newsletters and the question of 'open science'. *BioSocieties*. 7, 140–168.

163. Pottage, A. and Marris, C. 2012. The cut that makes a part. *BioSocieties*. 7: 103–114.

164. Calvert, J. 2012. Ownership and sharing in synthetic biology: A 'diverse ecology' of the open and the proprietary? *BioSocieties*. 7: 169–187.

165. Gibson, W. 1984. *Neuromancer*. New York, NY: Ace Books.

166. De Filippo, P. 1996. *Ribofunk*. New York, NY: Four Walls Eight Windows.

167. Levy, S. 2010. Geek power: Steven levy revisits tech titans, Hackers, Idealists. *Wired*. 19.

168. Ledford, H. 2010. Garage biotech: Life hackers. *Nature*. 467: 650–652.

169. Bennett, G., Gilman, N., Stavrianakis, A. and Rabinow, P. 2009. From synthetic biology to bio-hacking: Are we prepared? *Nat. Biotechnol.* 27: 1109–1111.

170. Wohlsen, M. 2010. *Biopunk*. New York, NY: Penguin.

171. Patterson, M. L. 2010. *A Biopunk Manifesto*. http://maradydd.livejournal.com/496085.html (accessed 10 January, 2013).

172. Biocurious. 2012. http://biocurious.org/ (accessed 22 January, 2013).

173. DIYBio Europe. 2012. http://www.diybio.eu (accessed 22 July, 2013).

174. Mueller, S., Coleman, J. R. and Wimmer, E. 2009. Putting synthesis into biology: A viral view of genetic engineering through de novo gene and genome synthesis. *Chem. Biol.* 16: 337–347.

175. Schmidt, M. 2008. Diffusion of synthetic biology: A challenge to biosafety. *Syst. Synth. Biol.* 2: 1–6.

176. Stamets, P. and Chilton, J.S. 1985. *The Mushroom Cultivator: A Practical Guide to Growing Mushrooms at Home*. London, UK: Richmond Publishing Co Ltd.

177. Jonathan Cline. 2010. *DremelFuge DIY-centrifuge Spins the Best!* http://diybio.org/2010/03/21/dremelfuge/ (accessed 22 January, 2013).

178. Durret, R. 2011. *The LightBulb PCR Machine.* http://citizensciencequarterly.com/2011/04/10/the-light-bulb-pcr-machine/ (accessed 22 January, 2013).

179. OpenPCR. 2012. http://openpcr.org/ (accessed 22 January, 2013).

180. Borland, J. 2010. Transcending the Human, DIY Style. *Wired.* http://www.wired.com/threatlevel/2010/12/transcending-the-human-diy-style/).

181. Huxley, J. 1968. Transhumanism. *J Humanist Psychol.* 8: 73–76.

182. Jotterand, F. 2010. Human dignity and transhumanism: Do anthro-technological devices have moral status? *Am J Bioethics.* 10: 45–52.

183. McNamee, M. J. and Edwards, S. D. 2006. Transhumanism, medical technology and slippery slopes. *J. Med. Ethics.* 32: 513–518.

184. Thompson, P. B. 2012. Synthetic biology needs a synthetic bioethics. *Ethics Policy Environ.* 15: 1–20.

185. Presidential Commission for the Study of Bioethical Issues. 2010. *New Directions: The Ethics of Synthetic Biology and Emerging Technologies.* http://www.bioethics.gov/documents/synthetic-biology/PCSBI-Synthetic-Biology-Report-12.16.10.pdf (accessed 26 December, 2012).

186. Joy, Y. Z., Claire, M. and Nikolas, R. 2011. *The Transnational Governance of Synthetic Biology Scientific uncertainty, Cross-Borderness and the 'Art' of Governance.* BIOS: London School of Economics and Political Science.

187. International Risk Governance Council. 2009. *Risk Governance of Synthetic Biology.* http://www.irgc.org/issues/synthetic-biology/ (accessed 10 January, 2013).

188. Richard, K. 2010. *Synthetic Biology From Science to Governance. Directorate-General for Health & Consumers, European Commission.* http://ec.europa.eu/health/dialogue_collaboration/docs/synbio_workshop_report_en.pdf (accessed 31 January, 2013).

189. Ethics of Synthetic Biology. 2009. *Ethics of Synthetic Biology: Opinion No. 25. European Group on Ethics in Science and New Technologies to the European Commission.* http://ec.europa.eu/bepa/european-group-ethics/docs/opinion25_en.pdf (accessed 31 January, 2013).

190. Tait, J. 2012. Adaptive governance of synthetic biology. *EMBO Rep.* 13: 579–579.

191. Tait, J. and Castle, D. 2012. Artificial microbes: Balanced regulation of synthetic biology. *Nature.* 484: 37–37.

192. Schmidt, M., Kelle, A., Ganguli-Mitra, A. and Vriend, H., eds., 2009 *Synthetic Biology: The Technoscience and Its Societal Consequences.* Dordrecht: Springer.

193. Schmidt, M., Torgersen, H., Ganguli-Mitra, A., et al. 2008. SYNBIOSAFE e-conference: online community discussion on the societal aspects of synthetic biology. *Syst. Synth. Biol.* 2: 7–17.

194. Tait, J. 2009. Governing synthetic biology: Processes and outcomes. In *Synthetic Biology: The Technoscience and Its Societal Consequences,* eds. Schmidt, M., Kelle, A., Ganguli-Mitra, A. and Vriend, H., pp. 141–154. Dordrecht: Springer.

195. Peter, D. Hart Research Associates, Inc. 2008. *Awareness of and Attitudes toward Nanotechnology and Synthetic Biology.* The Woodrow Wilson International Center for Scholars. http://www.nanotechproject.org/process/assets/files/7040/final-synbioreport.pdf (accessed 31 January, 2013).

196. Hilary, S. 2011. *Learning Communication Lessons from GM, Asbestos and Other Techno-Disasters.* Matter, Making New Technologies Work for Us All. http://www.matterforall.org/pdf/MATTER-What-does-the-public-want-re-nano-Final.pdf (accessed 10 January, 2013).

197. Alleyne, R. 2010. Scientist Craig Venter creates life for first time in laboratory sparking debate about 'playing god'. *The Telegraph.* http://www.telegraph.co.uk/science/7745868/Scientist-Craig-Venter-creates-life-for-first-time-in-laboratory-sparking-debate-about-playing-god.html (accessed 10 January, 2013).

198. Hahn, O. 1970. *My Life.* Colchester: TBS The Book Service Ltd.

199. Hoffmann, K. 2001. *Otto Hahn: Achievement and Responsibility.* New York, NY: Springer-Verlag.

200. Bert, A. C., ed., 2011 *Franck Report*: Chromo Publishing.

2

Synthetic Biology Standards and Methods of DNA Assembly

Philip D. Weyman and Yo Suzuki

CONTENTS

2.1 Introduction

DNA assembly is the workhorse of molecular biology. Most molecular biology projects begin with joining some combination of DNA pieces together in the service of some larger goal such as testing a biological hypothesis. For 20 years or so, DNA assembly was largely limited to techniques involving *in vitro* restriction digest and ligation.[1] While effective, these techniques often suffered from several limitations. One limitation was the existence of sequence constraints in the final products, as well as construction intermediates, imposed by the recognition sequences of restriction enzymes. A second limitation was the requirement for multiple steps such as incubation with enzymes and purification of DNA between enzyme reactions. A third limitation was the relative inefficiency of the process resulting in low probability of success, especially among inexperienced researchers.

Fortunately, in the past 10 years, there has been an explosion of alternatives to the traditional DNA assembly techniques. Some of the new techniques are improvements of the old techniques, making them more robust and versatile, while others employ entirely new strategies and chemistries. These novel options for DNA assembly have enabled the launch of the discipline of synthetic biology. The 'bottom-up' approach of this discipline is decisively different from molecular genetics, but even the methodology for synthetic biology can be differentiated from genetic engineering by a series of key features. The first feature is the much greater flexibility in the DNA sequence of the assembled product offered by synthetic biology tools. Structural elements of genes can be precisely designed and constructed at a minimal expense of reagents and time. These elements include promoters, protein-coding regions, terminators and other regulators and are sometimes described as 'biological parts' that can be assembled together in different combinations to construct ever more complex 'devices'.[2] The ability to specify the exact sequence between assembled parts allows for greater reproducibility in the function of those parts and devices function as they are applied in ever more complicated biological machines. Such precise assembly has been enabled by new techniques as well as a commercial infrastructure. The availability of overnight delivery of low-cost DNA oligonucleotides, or increasingly rapid turnaround for synthesis of double-stranded DNA (dsDNA) with increasing accuracy and length accelerates synthetic biology research.

The second feature differentiating synthetic biology is the ability to measure the activities of devices using standardised conditions and high-throughput techniques. The ability to reliably predict how a given device will behave enables a third feature of synthetic biology: abstraction. Mathematical models can be translated into biological parts that have a reasonable chance of functioning as designed. Together, these features compose the engineering cycle of 'design, build and test' that enables the construction of ever more complicated biological devices. In this chapter, we focus on the 'build' component of the cycle. We begin by outlining some of the issues and considerations of choosing a DNA assembly method and then describe the methods in greater detail.

2.2 Theory and Considerations in Double-Stranded DNA Assembly

As described in Section 2.3, synthetic biologists now have an incredibly wide variety of assembly tools at their disposal, and this repertoire of tools is constantly growing. Choosing a method for a particular application can be a daunting task. In this section, we attempt to outline some general issues to consider when selecting an assembly method for a project. One of the most basic issues to consider is whether single or multiple inserts will be assembled in a single step. If multiple inserts are to be assembled, methods are available for serial or parallel strategies (Figure 2.1). A serial strategy often permits only a single insert and vector to be combined in a single step, and assembly of multiple inserts would require multiple iterative steps to create a multi-insert construct. Assembly methods that permit only a single insert to be combined with a vector in each step of the process benefit from being methodical in approach but also can be more time consuming.[3]

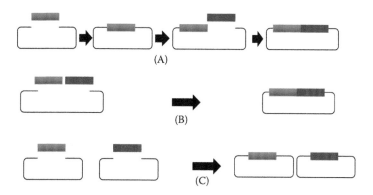

FIGURE 2.1
Comparison of serial, parallel and distributed strategies in DNA assembly. (A) In the serial strategy, two DNA inserts (red and blue rectangles) are cloned in a stepwise manner (i.e. blue and then red). (B) In the parallel approach, the red and blue inserts are cloned together in a single step. (C) In the distributed method, each insert is cloned into a different, compatible vector that can be transformed into the same cell.

In contrast, parallel construction methods allow for multiple fragments to be assembled in single steps, resulting in time and resource savings, but often with the introduction of a greater chance of error.

The choice is not so important if only one plasmid is to be assembled and the multiple inserts can be assembled in a few iterative steps. But if a large series of constructs is required, multiple assembly steps can become prohibitive from a time and organisational standpoint. Combining multiple inserts in a single assembly step becomes increasingly important as the number of constructs increases. In a multiple-insert assembly, the final assembly can be either a single construct (e.g. multiple inserts cloned into a single plasmid as in the construction of a synthetic operon) or the final assembly can be designed using a distributed construct strategy as in the compatible plasmid systems for *Escherichia coli* or the green monster process in yeast (Figure 2.1). For example, to express two proteins simultaneously in *E. coli*, one could assemble both into one vector in a synthetic operon or each protein could be assembled into a separate, compatible cloning vector in a distributed construct strategy. As discussed in Section 2.4, while a single construct can be more portable and sometimes easier to characterise, a distributed design offers many advantages including ease of debugging.

A second important feature to consider is whether a scar site is permitted between two assembled fragments. Such scar sites are often generated by restriction enzyme digest and ligation and are a design feature of many standard assembly strategies. Many of the polymerase chain reaction (PCR)- and recombination-based strategies allow seamless assembly that does not create a scar site, although strategies that require site-specific recombinases create a scar site between two fragments. Seamless assembly permits many applications including creation of chimeric fusion proteins and precise tailoring of the sequence context of the ribosome-binding site (RBS).[4]

The scale of fragment size in the assembly is an important consideration and assembly can occur at the level of regulatory element, gene, pathway or genome. For example, a small promoter and RBS assembly could be accommodated by any number of assembly methods described later, while a genome-scale fragment assembly may require assembly in yeast or *Bacillus subtilis*.[5,6] Mixing scales (i.e. a small promoter and large genome-scale fragments) in a single assembly often presents a new set of challenges.

The availability of templates is another important consideration. If templates for PCR amplification are available, PCR- or recombination-based strategies might be appropriate. If no templates are available, assembly from oligonucleotides or Integrated DNA Technologies (IDT) 0.5 kb gBlocks™ may be required. Standard biological parts can be prepared from plasmid isolation and digestion without the requirement for PCR amplification and its associated risk of error introduction. Some parts are small enough to be introduced as part of a PCR primer (e.g. an RBS). Using oligonucleotides to assemble DNA sequences of interest is becoming increasingly affordable and reliable.[7] Although the distinctions are becoming blurred, we consider such techniques as DNA synthesis, rather than assembly of dsDNA fragments (referred to in this chapter simply as 'DNA assembly').

Finally, the choice of host organism is often constrained by the characteristics of the DNA to be assembled. For example, *E. coli* is not a good host for replicating low G+C DNA.[8] In this case, alternative hosts may be appropriate, such as yeast. Conversely, yeast has difficulty maintaining plasmids greater than 100–150 kb of high G+C DNA without the presence of additional, interspersed replication origins.[9] *E. coli* does not suffer from this problem. In assembling larger fragments, the probability of encountering a toxic gene may increase in pathway- or chromosome-scale assemblies and the extent of 'toxicity' may be host specific.[10–12] Thus, careful consideration of the tolerance of the host is required when choosing an assembly strategy, especially if the strategy requires maintenance of the assembled product in a particular organism. While most assembly methods described later utilise *E. coli* and yeast, another important host for DNA assembly is *B. subtilis*.[13,14] Mitochondria and chloroplast genomes have been successfully assembled in *B. subtilis*, and this host may be advantageous when large pieces of DNA need to be cloned.[6,15]

2.3 Assembly Methods in Detail

2.3.1 Single Construct Assembly Methods

2.3.1.1 Restriction Digest–Ligation Methods

Classic restriction digest and ligation methods were developed in the early 1970s and were some of the first practical approaches to DNA assembly. They remain to this day as one of the most popular methods of DNA assembly. Improvements in the basic restriction digest and ligation method have been made in the form of standardising methods and design strategies. This collection of improved methods – called standard DNA assembly – combines DNA fragments called BioBricks and is reviewed in greater depth in Section 2.7.[16–18] The standard assembly methods pare down the number of restriction enzymes to just three or four and take advantage of highly optimised protocols to achieve simpler designs, convenient reuse of parts and greater cloning success.

One problem commonly encountered with standard assembly methods is the general requirement to assemble DNA fragments iteratively using a serial rather than parallel strategy. Recent innovations in restriction digest and ligation methods have developed parallel assembly strategies that have great utility in synthetic biology applications. The techniques outlined in the following sections represent great advancements in the application of the 'digest and ligate' motif.

2.3.1.1.1 Golden Gate and MoClo

While BioBrick-based strategies, described in Section 2.7, were designed to facilitate serial and hierarchical assembly strategies,[19,20] other restriction enzyme-ligation based strategies such as Golden Gate facilitate multi-fragment parallel assemblies.[21] With Golden Gate, up to 10 pieces can be assembled successfully in a defined order (Figure 2.2). Golden Gate and related methods use a unique class of restriction enzymes (type IIS) that cut outside their recognition sequence, leaving a single-stranded DNA (ssDNA) overhang of typically 4 nt. These overhangs can be native sequence or user-designed sequence depending on the placement of the type IIS recognition site. The design of the overhang sequences on each DNA fragment determines the order of the fragments in the final assembled product.

Digestion and ligation are performed in a one-tube, two-cycle reaction that allows undigested plasmids or PCR products to be added directly to the assembly reaction. The reaction typically cycles between digestion temperature (37°C) and ligation temperature (20°C) during the course of a 10–20 cycle program. While scar sites are usually always present in BioBrick strategies, scar sites can be avoided in Golden Gate methods by careful planning; however, iterative cycles of Golden Gate almost invariably leave a scar site unless new sites are created each cycle via PCR.

One of the most promising applications of Golden Gate is in the assembly of fragments with repeated sequence. Assembly of direct repeats, such as promoters and terminators repeated throughout a multi-gene assembly, can create problems for some of the recombination-based strategies outlined in Section 2.3.1.3.[22] Accordingly, the Golden Gate method has been used in the assembly of programmable DNA-binding domains known as transcription activator-like effectors (TALEs). TALEs require the assembly of up to 20 virtually identical DNA fragments and construction can be efficiently accomplished using two Golden Gate steps.[23,24] Fragments with repeated sequence would not be assembled properly using most recombination-based methods and the restricted specificity of the four-base sequence overlaps used in Golden Gate is ideal for this application.

An extension of Golden Gate known as modular cloning or MoClo has been used to assemble 11 transcriptional units together in a 33-kb plasmid.[25] The MoClo strategy establishes standard overlaps to use for assembly of multi-part protein expression devices that are typically composed of a promoter, RBS, signal peptide, coding sequence and terminator.

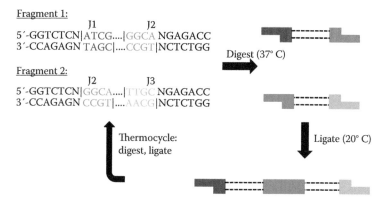

FIGURE 2.2
Golden Gate assembly is based on a series of digestion and ligation cycles. Use of type IIS restriction enzymes allows for the addition of 4-nt overlaps of a unique sequence (Junctions J1, J2, J3).

Thus, the restriction digest-ligation technique has been vastly improved through these methods and adds new tools to the repertoire of assembly methods.

2.3.1.1.2 Methylation-Assisted Tailorable Ends Rationale

Longer DNA fragments have a greater chance that the restriction enzyme recognition sequences will be present within a sequence to be assembled. This problem of off-target restriction digest is common to both traditional and Golden Gate restriction-based cloning methods. A recently developed technique, methylation-assisted tailorable ends rational (MASTER), overcomes this problem by using a type IIS restriction enzyme that digests only at methylated sites ^mCNNR (R=A or G).[26] Thus, primers containing the methylated cytosine can be used in PCR amplification of sequences of interest and designed such that the custom 4-base ssDNA overlaps create a seamless junction between two fragments of interest. Because primers with modified bases can be more expensive, MASTER was designed to use a common set of methylated primers. In this strategy, an initial PCR adds desired overlap regions and a custom sequence tag. This custom sequence tag serves as a recognition sequence for methylated universal primers during the second round of PCR. PCR products are digested with the methyl-recognising restriction enzyme and ligated as in other strategies described in Section 2.3.1.

2.3.1.1.3 Iterative-Capped Assembly

Another interesting twist on the type IIS restriction digest-ligation strategy is iterative-capped assembly (ICA).[27] This strategy joins DNA fragments with 4-base over-hangs created by type IIS restriction digestion as a growing chain attached to a solid sup-port bead. Only three overhang sequences are used (e.g. A, B, C) such that any piece has two of the overlaps, one on each end (e.g. A on the 5'-end and B on the 3'-end). Overhangs are alternated such that piece 1 has A and B, piece 2 has B and C and so on. As a piece (e.g. with A and B overhangs) is added and ligated to the growing chain, a hairpin 'cap' with the C overhang is also added to block incompletely extended chains from previous assembly steps, therefore allowing only the designed sequence to be created. The ICA strategy overcomes the limitations of multiple-fragment ligation by performing the assem-bly in rapid series rather than in a pooled batch. ICA has been used as an alternative to Golden Gate to assemble TALE nucleases (TALENs).[27]

2.3.1.2 PCR-Based Methods

Many assembly methods have been developed that do not require the use of restriction enzymes. Many also do not require a ligation step. All of the methods described in this section, with the exception of chain reaction cloning (CRC), require polymerase extension or amplification and are therefore limited to sequence length constraints imposed by the polymerase. While CRC does not require polymerase activity, it does require annealing of complementary sequences followed by ligation and is thus most similar to PCR-based techniques; it is discussed at the end of this section.

2.3.1.2.1 TA and TOPO-TA Cloning

One of the most basic restriction-free cloning procedures is TA cloning. This basic technique joins a PCR-product insert with a vector and is much more efficient than the combination of T4 polynucleotide kinase and blunt-end ligation. The TA-cloning technique uses the terminal transferase activity of Taq DNA polymerase that adds a 3' deoxyadenosine (A)-overhang to most PCR products.[28] This PCR insert with an

A-overhang is then combined with a linearised vector containing a 3′ T-overhang and the products are joined by ligase. A subsequent modification of this technique involves covalently bound topoisomerase I at the 3′ end of the vector. The 5′-OH on the PCR product insert can then efficiently react with the 3′ end of the vector in a ligase-free reaction.[29] Commercial vectors are sold with the T-tailed vector containing covalently bound topoisomerase.

The A-tailing is unique to Taq polymerase because it lacks 3′–5′ proofreading activity. Since most assembly applications require higher fidelity than can be achieved with Taq polymerase, the A-tail can be added to products in a second reaction before ligation to a 3′ T-tailed vector. TA- and TOPO-TA-assembly strategies both lack the ability to specify directionality of the insert and vector, thus requiring the determination of the orientation before some subsequent applications.

2.3.1.2.2 Sticky-End Polymerase Chain Reaction

A PCR-based approach that mimics restriction digest-ligation but does not actually require restriction digest is called sticky-end PCR.[30] In this technique, a sequence to be ligated is amplified by two different primer pairs (in two separate reactions). The primers are designed to be offset by the sequence of the cohesive end that is compatible with the target vector restriction site. When the two PCR products are mixed together, denatured and annealed, 25% of the reannealed products are a mixture of the two original products and generate the cohesive end. Using 5′-phosphorylated primers saves an extra kinase step but adds some additional expense in oligonucleotide purchase. This strategy is useful if internal restriction sites prevent digestion of the PCR product only at the terminal restriction sites and there is some need to still use restriction digest-ligation procedures. In theory, this method is capable of handling multiple inserts, but the usual limitations of short overlap ligations apply.

2.3.1.2.3 Overlap Extension Polymerase Chain Reaction

A series of techniques have been developed that use PCR to seamlessly join two DNA fragments together that share homologous sequence 'overlaps' at the ends (Figure 2.3). When two DNA fragments that share homology at their 3′ ends are mixed, denatured and annealed, a fraction of the annealed products will have one strand from the first fragment and one DNA strand from the second fragment. This allows the polymerase to extend the products throughout the length of the new template resulting in a seamless joining of the two DNA fragments. This technique can be used to generate chimeric sequences using native conserved sequence as overlaps, or overlap sequences can be added using PCR primers.

This technique was first described as splicing by overlap extension (SOE).[31] The original SOE protocol required joined linear fragments to be assembled into a cloning vector using restriction digestion-ligation techniques. However, a variation on this technique was published that eliminated both restriction digest and ligation steps.[32] The technique, PCR-induced ligase-free subcloning, requires PCR amplification of an insert with terminal homology sequences to the linearised vector. In the published technique, two separate PCR reactions are then performed with primers designed to amplify the two possible overlap extensions: 3′-insert joined to 5′-vector and 5′-insert joined to 3′-vector. The two separate products are mixed, denatured and annealed and the resulting hybrid molecule is transformed into *E. coli* for nick repair. Newer and simpler versions of this technique have since been developed such as integration PCR and circular polymerase extension cloning (CPEC).

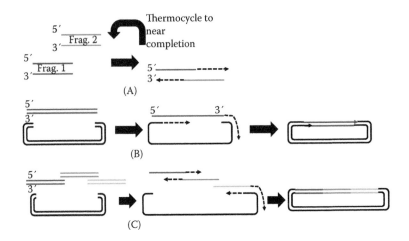

FIGURE 2.3

Comparison of overlap-extension PCR, integration PCR and circular polymerase extension cloning. (A) In overlap extension PCR, two linear fragments to be joined share sequence homology and are cycled through denaturing, annealing and extension steps. (B) In integration PCR, two fragments creating a circular molecule that share homology are cycled together as in A to create a joined molecule. (C) CPEC is a modification of integration PCR to include multiple fragments in a parallel assembly. Dotted lines indicate a newly extended sequence by polymerase after annealing of fragments.

Integration PCR is a variant of overlap extension that allows a PCR-amplified DNA fragment to be inserted at any point in a vector sequence.[33] In this technique, a PCR-amplified insert has homology (*ca.* 20 nt) at each side of the upstream or downstream regions of an insertion site (Figure 2.3). The homology-bearing insert is combined with vector and multiple cycles of denaturing, annealing and extension are performed to seamlessly join the vector and insert sequences. Ligation is not performed and nick repair occurs *in vivo* after transformation of the assembled products in to *E. coli*. To make the procedure more efficient, parental vector template can be removed using DpnI that only digests methylated DNA. Since the parental template vector without the insertion was obtained by purification from routine *E. coli* strains, it contains Dam methylated sequences and will be digested by DpnI. The joined DNA has been completely created by PCR amplification and lacks methylation, thus enriching the assembled product.

A variation on integration PCR has recently been published called **CPEC**.[34] The authors demonstrate that the same basic idea can be used to assemble as many as four PCR products in a single reaction (Figure 2.3). A limitation of this and other extension-based techniques is the size of the pieces that can be assembled. Because each product must be extended across its entire length to be successfully joined to its neighbour, the process will become much less efficient as fragment size increases. The assembly reaction does not involve amplification but merely joins together existing pieces. As a result, mutations are unlikely to accumulate during the assembly reaction itself, but because DNA fragments must be first amplified by PCR to add the homology regions, they may accumulate mutations during this step.

While classic overlap extension-based assembly methods do not require amplification, a variant technique called **inverse fusion PCR cloning (IFPC)** uses amplification to allow assembly with very small amounts of vector (1–10 ng). Purification of vector is also not necessary and the assembly reaction can be performed with insert and *E. coli* extract containing the vector.[35] In this technique, the insert is first amplified with a 5′-homology to the target vector. Then the insert is assembled into the vector by mixing together in a PCR reaction the insert, the reverse primer for the insert and a forward primer for the vector. In the assembly, the reverse strand of the insert anneals to the vector and an extension reaction joins the two sequences together. Then the insert reverse primer and vector forward primer amplify the joined vector-insert. A phosphorylation step is necessary (if primers are not 5′-phosphorylated) and ligase is added to circularise the joined vector-insert.

The **PIPE (polymerase incomplete primer extension)** method takes advantage of incomplete PCR reactions that leave a ssDNA region on the template strand (i.e. a 5′-overlap) (Figure 2.4). The extent of incomplete products in a PCR reaction was tested by treating PCR products with mung bean nuclease to digest ssDNA followed by DNA sequencing.[36] The authors found that 93% of the products were incomplete after their PCR protocol with ssDNA spanning 1–170 nt.[36] This ssDNA region can serve as an overlap if PCR products were designed with 5′ regions of homology (*ca.* 15 bases). The regions of ssDNA overlap allow insert and vector to anneal together. In practice, no ligation reaction is required and gaps and nicks are repaired *in vivo* after transformation.

Another variation on the theme is **successive hybridization assembling (SHA)**.[37] In this technique, each of several fragments to be assembled is first assembled with its neighbour in the final assembly using SOE PCR (see Figure 2.5). The resulting series of products consists of each fragment assembled with its downstream neighbour to allow long sequence overlaps (*ca.* 1 kb) between each piece. The overlapping fragments are mixed, denatured by boiling and slowly cooled to room temperature before being transformed into *E. coli*. The ligation-free reaction requires nick repair from *E. coli* after transformation. A side-by-side comparison with other technologies with shorter overlaps indicated that the longer overlaps yielded a higher frequency of correct multi-fragment assemblies.[37] Thus, the increased effort required to generate the longer overlaps between the pieces may pay off in improved assembly efficiencies.

2.3.1.2.4 Chain Reaction Cloning and Ligase Cycling Reaction

Alternatively, instead of creating PCR products with terminal homologies, assembly of multiple DNA fragments can be guided by adding 'bridge' oligonucleotides in a technique

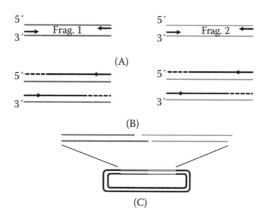

FIGURE 2.4

Polymerase incomplete primer extension. Primer binding (A) and extension during PCR cycling often leads to incompletely extended products (dotted lines in B). These incompletely extended products have 5′ overlaps that can anneal during a subsequent assembly step (C).

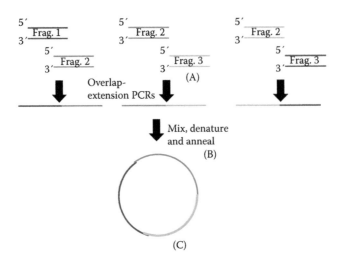

FIGURE 2.5
Successive hybridization assembly. Fragments with overlaps to be assembled (A) are first assembled in pairs
(B) using overlap extension PCR to create long overlaps that facilitate efficient assembly once all fragments are
mixed, denatured and annealed (C).

originally called chain reaction cloning (CRC).[38] This technique has recently been updated
and optimised as ligase cycling reaction (LCR) to allow assembly of up to 20 DNA parts
in a single reaction.[39]

In the basic concept of CRC to assemble two DNA fragments, the bridge oligonucleotides are complementary to fragment 1 (sense-strand) on the 3′-end and to
fragment 2 (sense-strand) on the 5′-end. After the DNA fragments are denatured,
annealing of fragment 1 and 2 to the bridge oligonucleotide allows fragments 1 and 2 to
be joined together by ligase (Figure 2.6). The ligated sense-strand product (1+2) serves
as a template to allow the anti-sense strands of fragments 1 and 2 to be joined by ligase.
Multiple cycles of this process lead to a chain reaction of products that serve as templates
for ligation of additional products.

One attractive feature of this method is that it does not require fragments to have
homologies to adjacent fragments in the final assembly. Thus, fragments can be reused
in different assemblies simply by changing the bridging oligonucleotide. This method
requires efficient annealing of the DNA fragments and this may be problematic with larger
DNA fragments. Thus, this technique is best used for assembly of smaller DNA fragments,
although efficient assemblies of up to 20 kb have been reported.[39] A variant of this method
has been used to assemble ssDNA as well.[40]

2.3.1.3 Recombination-Based Assembly Methods

Recombination-based assembly methods encompass diverse chemistries and strategies
and show some of the best promise for multi-insert, parallel assembly. These methods
utilise enzyme activities including polymerases, ligases, nucleases, DNA-binding proteins,
recombinases and others.[41] This contrasts with the restriction-based and PCR-based methods
which each utilise a relatively narrow spectrum of chemistries (i.e. restriction enzymes and
polymerases, respectively). Recombination-based assembly methods combine many of the
best features of restriction enzyme-based and PCR-based techniques while avoiding many of
their problems. For example, restriction enzyme-based assembly techniques suffer poor DNA
overlap and lack of seamless assembly, and PCR-based techniques are limited by processivity

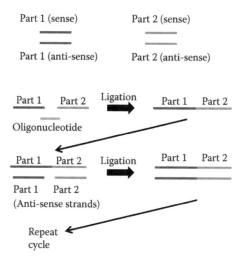

FIGURE 2.6

Chain reaction cloning or ligase cycling reaction. Parts 1 and 2 are denatured and brought together by annealing to a bridging oligonucleotide (orange). After ligation, the joined Parts 1 and 2 can serve as a template for joining additional parts in the same way.

of the polymerase; however, recombination-based techniques offer seamless assembly with longer overlap regions that are effective at assembling fragments up to hundreds of kb.[42]

Recombination-based DNA assembly methods can be divided into two categories. The first class can be characterised as chew-back and anneal (CBA) strategies. These techniques combine DNA fragments that share short sequence overlaps (<50 bp) and feature enzymatic processing of the overlap to produce ssDNA. Complementary ssDNA on separate DNA fragments will anneal leading to ordered, multi-fragment assembly. The second class of recombination-based methods can be characterised as recombinase based. These methods use a site-specific recombinase and require the inclusion of a recombinase recognition sequence. The reactions are performed either *in vivo* or *in vitro* for different methods.

While these methods span a wide range of formats, they all share the need to include homologous sequence at the ends of DNA fragments (either user-specified for CBA methods or specific sequences for recombinase methods). Of the methods outlined so far, these methods have the best potential to be used in parallel strategies and can assemble many fragments together at size scales up to entire prokaryotic genomes. The drawbacks of many of these methods include poor performance when assembling fragments with repeated sequences such as TALE DNA-binding proteins, repeats of clustered regularly interspaced short palindromic repeats (CRISPR) or promoters that are included more than once in an assembly. However, careful choice of junctions between repeats and use of a defined chew-back techniques (e.g. ligation independent cloning of PCR products [LIC-PCR] and USER, defined in the following sections) may alleviate assembly problem associated with repeats (Figure 2.7).

2.3.1.3.1 Ligation Independent Cloning of Polymerase Chain Reaction Products

One of the first CBA methods to be developed was LIC-PCR.[43] This technique uses the 3′–5′ exonuclease activity of T4 polymerase in the presence of only one deoxynucleotide triphosphate (dNTP) (e.g. deoxycytidine triphosphate [dCTP]) to digest the 3′ strand back to the first deoxycytidine monophosphate (dCMP) residue. Vector and/or insert are PCR amplified to have terminal sequence homology to each other. With careful design of the homologies, the exonuclease activity from T4 polymerase can generate ssDNA overhangs

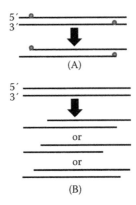

FIGURE 2.7
Defined versus undefined overlap strategies. (A) In defined overlap strategies such as Golden Gate, LIC, SLIC and USER, the double-stranded DNA is chewed back a defined amount to a particular site (either by restriction enzyme digest for Golden Gate, the omitted dNTPs for LIC and SLIC, or the uracil-containing nucleotide for USER). (B) Undefined overlap strategy. While the maximum overlap can be set by sequence design, the extent of single-stranded DNA creation through undefined overlap methods is often not as controllable in methods such as Gibson, PIPE and In-Fusion leading to chew-backs of variable length.

much longer than would be possible with typical type II restriction enzymes. The longer overhangs enable greater fidelity of annealing and also allow the ligation step to be omitted. By using a defined chew-back controlled by the added nucleotide in the T4 polymerase reaction, longer overlaps are created than are available for restriction-based methods such as Golden Gate. In theory, this should allow more efficient assembly of repeated regions by protecting the repeats.

One potential but infrequent problem with this technique emerges when junctions between pieces are sequence constrained. Should one need to clone a sequence consisting of the bases 'GATC' as the last four bases of a DNA fragment, the maximum overlap that could be generated would be three bases. Thus, other techniques such as uracil specific excision reaction (USER, described in the following section) may be more truly sequence independent, but come with the requirement of custom oligonucleotide primers for PCR before assembly.

2.3.1.3.2 Uracil Specific Excision Reaction

Another technique that takes advantage of a well-defined chew-back region is the USER method. Similar to LIC-PCR, this method allows for longer complementary ssDNA regions between two DNA fragments when compared with a restriction digest assembly method. However, the sequence overhangs that can be created by USER are not limited to regions in which one DNA base is underrepresented.

The USER method incorporates deoxyuridine monophosphate (dUMP) residues into primers that are used to PCR amplify insert sequences. The inserts can then be treated with enzymes to excise the uracil residues and generate 3′ overhangs. Careful design of overhang regions allows annealing of the ssDNA and efficient assembly of two DNA fragments. The original description of the technique utilised a uracil DNA glycosylase (UDG) to create abasic sites at the 5′ end of a fragment.[44,45] The resulting abasic sites destabilised base pairing and allowed 3′ overhangs of two similarly treated fragments to anneal. Nick repair and ligation was presumably performed by the *E. coli* cells after transformation. In more contemporary versions of this technique, a second enzyme, an endonuclease, was added to allow cleavage of the phosphodiester backbone at the abasic site to further destabilise the base pairing at the ends of the fragments.[46]

USER DNA assembly reactions can be performed both *in vitro* (ligase-mediated) and *in vivo* (a ligase-free assembly with ligation occurring after transformation into *E. coli*).[47] The USER method has been expanded to allow multiple inserts to be cloned into a vector in a single reaction in a technique called USER Fusion.[48] Three or four inserts and a vector can be efficiently combined in a single, ligase-free reaction.[47,48]

Both the LIC-PCR and USER methods benefit from well-defined excision regions. This can be beneficial when cloning regions with repeated DNA in which too much chew-back could expose alternative sites for assembly and reduce the efficiency of obtaining the desired construct. The next two methods to be described, Gibson and In-Fusion™, both have uncontrolled chew-back steps. However, they both benefit from increased ease of use due to sequence independent junctions and there is no requirement for special, dUMP-containing primers.

2.3.1.3.3 Gibson Assembly

A third CBA-based technique is isothermal assembly or Gibson assembly, after the scientist, Dan Gibson, who developed the technique. The main innovation in this technique is that it allows for one-pot enzymatic activity that includes digestion of 5′ ends to create 3′ overhangs, annealing of homologous regions, high-fidelity polymerase fill-in and ligation. Whereas many of these techniques have been used in other parallel assembly methods, this is the first method to combine them all in a one-step, one-pot reaction. By performing the reaction at 50°C, all of the enzyme activities occur, although their efficiencies can vary. For example, the 5′–3′ digestion by T5 DNA exonuclease must first occur in order for assembly to occur. This enzyme is not optimal at the temperature and will rapidly degrade during the course of the reaction. Thus, overhangs are generated in the initial phase while assembly and ligation continue into the latter phase. Once closed circular molecules are created, the DNA pieces are no longer subject to the T5-exonuclease activity.

After creating ssDNA overhangs, the overhangs can anneal according to their complementarity. It is optimal if these overhangs are longer than 25 nt to allow for efficient annealing at 50°C. After annealing, it is likely that the annealed pieces will need to be filled in if the activity of the T5-exonuclease has digested further than the overlap region. This is typically accomplished by a high-fidelity, proofreading polymerase such as Phusion. While any high-fidelity enzyme could be used in theory, substitution of other enzymes may require buffer reoptimisation. The proofreading 3′–5′ exonuclease activity of the polymerase is able to trim any non-complementary sequence from the 3′ end that may have been generated by restriction digest. The result is a seamless assembly of several DNA fragments in a rapid, one-pot procedure.

By creating a seamless final product, Gibson assembly can be used to clone pieces that may be impractical to efficiently amplify in a single PCR reaction. For example, the entire 14-kb hydrogenase operon including all associated maturation proteins was amplified as four overlapping smaller PCR fragments and efficiently reassembled in an expression vector.[49] The method can also be used to assemble large fragments, (>100 kb) that have sequence homologies.[42] In practice, Gibson assembly efficiently combines several double-stranded fragments in a single reaction; however, assemblies with even greater numbers of starting fragments have been reported. Gibson assembly can also be used to assemble oligonucleotides with sequence overlaps to synthesise dsDNA inserts.[50]

The use of longer sequence overlaps between DNA fragments (30–50 bp) in Gibson assembly makes this technique very efficient at joining multiple fragments. However, the degree of 5′ digestion is controlled by the amount of exonuclease activity and may be more variable than the sequence-determined chew-back regions in LIC-PCR and USER.

Furthermore, if repeat regions are found in the internal regions of DNA fragments, assembly may be confounded by this, although it is not common in most assemblies involving unique pieces. One final note is that the exonuclease activity in the reaction prevents efficient assembly of linear final products. Assembly of DNA fragments into circular molecules is the most efficient configuration to protect the correctly assembled DNA from exonuclease activity.

2.3.1.3.4 In-Fusion

Another method that is similar in concept to Gibson and LIC-PCR is In-Fusion. This method uses poxvirus polymerase to assemble four or more fragments together that overlap in sequence with each other by 15 bp.[51] The ssDNA is generated from the sequence overlap regions by the action of 3′–5′ exonuclease activity from the poxvirus polymerase and no ligation reaction is included.[52] Reagents to perform this method are sold by Clontech as a convenient dried down kit that can be stored at room temperature. In-Fusion has been applied to standard assembly applications (see Section 2.7) with positive results.[53]

2.3.1.3.5 Lambda Red-Based Assembly

Lambda Red has long been used to make insertions, deletions and substitutions in the *E. coli* chromosome and plasmids maintained in *E. coli*.[54] Its use as a strategy for DNA assembly has only recently been explored. Lambda Red occurs through the action of three inducible lambda phage proteins: Gam, Bet and Exo.[55] Gam protects the incoming linear DNA from the RecBCD nuclease, Exo generates ssDNA overlaps by degrading one strand of dsDNA, and Bet binds to the ssDNA to protect and facilitate recombination with the target DNA.

Several groups have used this system for DNA assembly as well. For example, it can be used to assemble a fragment of up to 80 kb from a larger bacterial artificial chromosomes (BAC) clone in a subcloning vector.[56] This technique takes advantage of being able to control the junctions precisely rather than searching for restriction enzymes to yield a fragment that approximates the desired fragment. These techniques were further developed in a technique known as **assisted large fragment insertion by red/ET-recombination (ALFIRE)**.[57] One of the first true assembly-based efforts using Lambda Red was combining large fragments of the *Haemophilus influenzae* chromosome in attempts to reassemble the chromosome in *E. coli*. This method, called iterative clone recombination (ICR),[10] allowed adjacent fragments from BAC libraries to be assembled into larger contiguous fragments. Overlapping sequence ranged from 0.5 to 10 kb and selection required antibiotic swapping in the fosmid vector.

Using Lambda Red as an assembly method would require it to be used in series as the process for assembly has not been applied in parallel. However, recent advances in using Lambda Red systems for genome editing highlight its potential for use in parallel assembly.[58,59] Further advances in this technique may be related to new reports that highlight improved host conditions for this process by removal of endogenous nucleases.[60] Whether these advances facilitate multi-fragment *in vivo* assembly in *E. coli* cells remains to be demonstrated.

An extension of Lambda Red-based assembly can be found in a technique called **mating-assisted genetically integrated cloning (MAGIC)**.[61] This technique was developed to streamline vector swaps so that coding regions of interest could be easily tagged with green fluorescent protein (GFP), 6xHis, or other vector-based modifications. The assembly is performed *in vivo* which conserves time and resources by omitting extensive plasmid preparation steps and allows for automation of the process across an entire set of coding regions for a given organism. In this technique, a coding region of interest is maintained on a 'donor' plasmid containing an oriT for conjugative transfer. This donor plasmid is introduced into a recipient strain that contains the Lambda Red *gam, bet* and *exo* genes as well as the arabinose-inducible I-SceI gene. Upon induction by arabinose, the I-SceI homing endonuclease is expressed and cuts at designed sites in the recipient plasmid and at sites upstream and downstream of the coding region in the donor plasmid. Correct assembly is selected by replacement of a dominant negative selectable gene in the recipient plasmid and by a conditional origin of replication on the donor plasmid that does not replicate in the recipient host.

2.3.1.3.6 RecA-Mediated Assembly In Vitro (Sequence and Ligation Independent Cloning)

A recent innovation has improved the T4 polymerase mediated-CBA assembly strategy by adding the homologous recombination stimulating protein, RecA, to the assembly reaction.[62] The technique, called **sequence and ligation independent cloning (SLIC)**, has been reported to improve efficiency of assembly of a vector and insert by 500-fold by inclusion of the RecA protein in the assembly reaction. Fragments with 20–40 bp sequence overlaps are created by PCR and ssDNA is created by the 3′–5′ exonuclease activity of T4 polymerase (Figure 2.8). Alternatively, ssDNA can be created at the ends of fragments by employing incomplete PCR methods that generate 5′ overhangs.[62] The fragments containing the overhangs are combined and anneal to each other according to sequence complementarity. Gaps may still remain in the annealed pieces and are repaired in the *E. coli* cells after transformation. SLIC can accommodate multiple inserts plus vector, and 5–10 fragments can be efficiently assembled in a single reaction. Nearly 20% of colonies in the 10-piece assemblies were correct demonstrating the power of this improved technique.[62]

2.3.1.3.7 RecA-Independent Assembly In Vitro

While RecA has been shown to stimulate assembly *in vitro*, a RecA-independent mechanism of assembly has also been discovered. The method, **seamless ligation cloning extract (SLiCE),** allows DNA to be assembled *in vitro* by recombination with an *E. coli* extract.[63] Common RecA-lacking *E. coli* strains such as DH10b and JM109 can be used to make the recombinogenic extracts. Efficiency of assembly can be increased even further by using *E. coli* strains that express the Lambda Red recombination genes. By using only a simply prepared cell extract from *E. coli*, significant savings can be achieved in enzymes associated with DNA assembly.

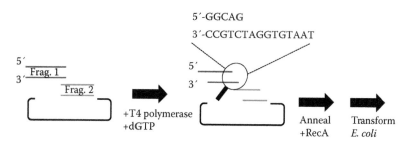

FIGURE 2.8

Sequence and ligation independent cloning. Fragments containing sequence overlap are first treated with T4 polymerase to 'chew-back' the 3′ end to the first encountered instance of the included dNTP. These fragments are then annealed in the presence of RecA and transformed to *E. coli*.

2.3.1.3.8 Cre/Lox-Based Systems

Several recombination-based methods have been developed that use phage recombinases. Much as restriction enzyme-based DNA assembly strategies require a specific DNA sequence at the junction between DNA fragments to be assembled, recombinase-based strategies also require specific DNA sequences at the junctions. The first of these is based on the Cre recombinase system from bacteriophage P1 that recognises the loxP site for activity.[64] Called the **univector plasmid-fusion system (UPS)**, it eliminated the need for restriction digest and ligation and was an early attempt at simplifying the DNA assembly process in subcloning DNA fragments. The first iteration of this technology included a single loxP site upstream of a cloned insert and a single loxP site on a vector. *In vitro* recombination led to a composite vector that could be selected by antibiotic resistance contributed by both insert and vector partners. Directional cloning was later established by the inclusion of a 5′ loxP site for use with the Cre recombinase and a 3′ R recombinase site (RS) for use with the R recombinase in a second step. While innovative, the UPS system has been largely replaced by the more versatile Gateway® system described in the following section.

Cre/Lox-based DNA assembly methods are increasingly being used in plant and animal systems to assemble DNA in chromosomal locations in an iterative fashion.[65–67] While this assembly process can be somewhat slow, especially in relatively long generation systems like plants, the process enables multiple genes of interest to be assembled and expressed in a precise, context-defined region of a chromosome.

2.3.1.3.9 Gateway Cloning and Related Techniques

Another example of a recombinase-based DNA assembly method is Gateway cloning. Gateway cloning was an early expedition into standardisation, borne of the need to expedite the process of systematically testing (by over-expression) thousands of open reading frames (ORFs) of unknown function uncovered by genome sequence projects.[68] Gateway cloning has been used widely in functional genomics projects to investigate the proteome.[69]

The Gateway cloning system uses modified lambda phage proteins to first allow a sequence of interest to be inserted by recombination into an 'entry vector'. From this entry vector, a second recombination reaction can be performed to introduce the cloned sequence of interest into any number of 'destination' vectors, each with its own purpose in functional characterisation of the coding region of interest.[70] For example, one destination vector might allow for expression in yeast while another might add a carboxy-terminal GFP tag. Directional cloning was introduced into the system by mutating the recombination sites so that a unique recombination reaction could be specified at the 5′ and 3′ sides of the coding region.

During the first cloning reaction into the entry vector, the sites of recombination (attB1 and attB2) are included on PCR primers at the 5′ and 3′ ends of a DNA sequence of interest. The attB1 and attB2 sites recombine with the attP1 and attP2 sites, respectively, on the vector. This allows for directional assembly and creates two new sites, attL1 and attL2, on either side of the sequence of interest. The attB + attP reaction is performed *in vitro* with an enzyme mixture called 'BP Clonase' that contains a phage integrase and the bacterial protein integration host factor (IHF). The sequence of interest is now in an entry vector and can be moved to a destination vector through a second recombination reaction. This second reaction uses the attL1 and attL2 sites that were created through the previous recombination of the attB + attP sites. The attL1 and attL2 sites flanking the sequence of interest in the entry vector recombine with the attR1 and attR2 sites, respectively, in a destination vector in a reaction mediated by 'LR Clonase' which contains the integrase, IHF and the phage enzyme excisionase (Xis). The result is the sequence of interest is assembled into the destination vector and is flanked with attB1 and attB2 sites. Destination vectors are typically used in some functional test, such as expression or localisation.

Since the reaction conditions to introduce the coding region of interest remain constant despite disparate insert sequences, the process can be performed in parallel in 96 well plates. Scaling up the process allows for every ORF in a genome (the ORFeome) to be targeted and assembled into expression vectors.[69,71,72] Despite the ease of performing a specific set of reactions in high throughput manner, the original Gateway cloning strategy lacks the iterative capability of other standardised formats such as BioBricks (see Section 2.7). The requirement of specific sequences to mediate the recombinations necessitates a scar site, in contrast to other sequence-independent DNA assembly methods such as Gibson assembly.

Although the original Gateway concept involved a single insert and a single vector, a newer method called **Gateway MultiSite** allows for multiple inserts through the use of additional unique recombination sites (i.e. attL3 + attR3, etc.) Current versions allow up to four inserts to be combined with a destination vector.[73] Multi-fragment assembly permits greater flexibility by allowing additional domains to be added to a translational fusion destination vector without the need to specifically re-engineer a destination vector for a multiple-step Gateway cloning project.[73]

Another innovative use of the Gateway recombinases for parallel assembly was recently published and is called **single-selective-marker recombination assembly system (SRAS).**[74] While the original Gateway system and Gateway MultiSite systems[73] use *in vitro* recombinase reactions, the SRAS and related techniques use *in vivo* recombinase reactions.[74] By targeting the assembly reaction to the chromosome, recombination is only required at one RS to add a fragment compared with two RSs required to add a fragment to a plasmid. Assembly of multiple inserts occurs iteratively and hierarchically. The authors also describe plans for sequence-independent assembly strategy that does not leave a scar site.

2.3.1.3.10 *Yeast-Based DNA Assembly*

The yeast *Saccharomyces cerevisiae* has an efficient mechanism for recombining sequences with homology of as little as 35 bp.[75–77] This feature has been exploited in multiple DNA assembly methods. For example, in a technique known as transformation-associated recombination (TAR) cloning, a linearised vector that has flanking sequences with homology to an insert can be introduced into yeast with an insert fragment or a larger DNA molecule containing the insert fragment to generate a new plasmid with precisely tailored junctions.[78–81] Using a similar strategy, multiple inserts can be assembled into a vector when homologous sequences are present at overlapping junctions. This method was used to combine nine fragments including a vector to clone all genes required for D-xylose utilisation and zeaxanthin biosynthesis in one step.[82] Yeast was also used to assemble complete bacterial genomes of sizes 592 kb and 1 Mb from several genomic pieces.[8,83] For the assembly of the 592-kb genome, as many as 25 fragments were recombined at once to make the complete genome.[84] Yeast is even capable of assembling overlapping ssDNA oligonucleotides. Thirty-eight 60-mer oligonucleotides, each with 30 bases complementary to each of the upstream and downstream oligonucleotides of the opposite strand, were assembled together with a vector sequence to make a 1,170-bp synthetic DNA fragment that was stably cloned in the provided vector.[85] Moreover, the complete coverage of both strands is not necessary. For example, twenty-eight 60-mers with only 20 bases annealing to oligonucleotides of the opposite strand (and therefore with

20 bases of non-paired region) produced a 1,140-bp synthetic DNA fragment that could be stably replicated.[85] These techniques demonstrate the extraordinary power of yeast in assembling DNA containing overlapping or complementary sequence.

2.4 Distributed Construct Assembly Methods

2.4.1 Multi-Construct Strategy for Generating Diversity

Generation of a single DNA molecule containing all required elements may not be desirable when one is uncertain about the final design of the construct. Swapping genes within a 'construct' can be easily achieved when the unit of the construct is a cell and when the genes are introduced into the cell as separate plasmid molecules. Then, replacing one element can be achieved by simply adding a different plasmid in the mix, without the need for any recombinant DNA work. Such 'distributed' assembly methods are worth discussing as alternatives to single-construct assembly of multiple fragments. Examples of distributed assemble for multiple-construct expression include the compatible plasmid systems in *E. coli*, serial cloning and assembly in *B. subtilis*, and the Green Monster process in yeast.

2.4.2 Multi-Plasmid Systems in *E. coli*

To express multiple genes in *E. coli*, this chapter has explored many ways to assemble the desired genes and regulatory elements into a single vector. However, alternative strategies for expression may be equally convenient and are worth mentioning. One useful plasmid system is the Duet series of vectors by Novagen that features multiple compatible expression vector plasmids.[86] These plasmids offer great flexibility to augment an experiment with the expression of additional genes without needing to redesign the existing expression plasmids. This enables simple debugging of individual components of a multi-gene expression experiment. The multiple plasmid systems offer great flexibility in multi-gene heterologous expression projects.[87,88] Such multiple plasmid systems have been exploited in strategies to tune expression levels based on the relative copy numbers of each plasmid.[89] Alternatively, split antibiotic selection systems can be used to select for maintenance of multiple plasmids using a single selectable antibiotic.[90]

2.4.3 *Bacillus Subtilis* Genome Vector Method

Numerous studies dealt with making multiple changes to the genome of an organism, but this method explicitly aims to use a genome as a vehicle for cloning. The genome of *Bacillus subtilis* can accommodate one or more exogenous fragments with the collective size of up to 3.5 Mb.[6] The largest DNA molecule was the entire genome of the cyanobacterium *Synechocystis* sp. PCC 6803 cloned in three pieces introduced into different genomic locations. Once in *B. subtilis*, available genetic tools can be used to further alter or manipulate the cloned DNA.

2.4.4 Green Monster Method for Multi-Gene Engineering

This yeast-based method was developed for addressing the shortage of independently selectable markers available for tracking numerous engineered loci in multi-gene engineering (Figure 2.9). The basic idea is to use one quantitative marker, a GFP reporter

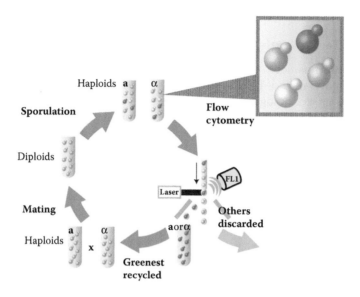

FIGURE 2.9

Sexual cycling scheme of the Green Monster method. Haploid cells carrying a single green fluorescent protein (GFP)-marked alteration are mated with haploid cells of the opposite mating type carrying a different GFP-marked alteration. The resulting diploid cells are allowed to undergo meiosis. The resulting population is a 1:2:1 mixture of zero-GFP cells, one-GFP cells and two-GFP cells. From this mixture, flow cytometry is used to sort the greenest cells that are enriched for the two-GFP cells. The sorted cells are established and confirmed to contain both alterations. Additional changes are introduced into the generated double mutant strain by preparing appropriate strains and repeating the cycle of mating, sporulation, flow cytometric selection and genotype confirmation several times.

gene, to mark all engineered loci.[91] GFP-marked changes are first generated in separate strains and then assembled into a single strain using a repeated cycles of yeast mating, meiosis and flow-cytometric selection. Molecular tools for haploid selection and diploid selection termed the Green Monster Toolkits are integrated into these strains. The result is a strain with a number of changes harboured on different chromosomes.

2.5 Building Combinatorial Complexity during Assembly

An additional consideration of many cloning methods is whether it can be used to build in combinatorial complexity during assembly. For example, multiple alleles of a gene from different species may need to be tested in the context of other genes in a pathway to identify the optimal combination. To include a combinatorial component in an assembly, the method must be efficient enough to generate enough independent assembly events to saturate all combinations of the assembly. For this, Gibson assembly has been successfully used in several cases.[92,93] Golden Gate methods can also be used for combinatorial assembly in a technique known as 'Golden Gate Shuffling'.[21] Several PCR-based assembly methods have been successfully used to generate pools of diversity in a DNA fragment.[94–96] Overall, the choice of DNA assembly method plays a critical role in optimising metabolic processes.

2.6 Emerging Technologies and Conclusions

The last 5–10 years has seen an explosion of innovative DNA assembly technologies. Innovation during the next 5–10 years will likely continue at a rapid pace but may focus on different areas as technologies continue to evolve. The anticipated decrease in cost of DNA synthesis for 1–5 kb fragments may result in more of the smaller assembly projects being shifted to commercial DNA synthesis operations. As a result, assembly reactions performed in research labs will likely prioritise assembly of larger fragments at the gene and pathway scale. Increasing interest in bacterial genome design will continue to catalyse innovation for large fragment assembly.[5,97] DNA assembly methods in the future will certainly incorporate emerging technologies including the genome editing tools of the TALENs and CRISPR-Cas systems. These systems allow for the rapid assembly of customisable nucleases to create dsDNA breaks, and are commonly reported to be used for creation of knock-out mutants in a variety of hosts.[98–103] Their utility in DNA assembly may prove equally beneficial, especially if they facilitate custom junctions in a Golden Gate-style strategy[21] or if they allow efficient recombination of large, bacterial genome-scale clones in a yeast assembly system.[83] It is an exciting time to be involved in DNA assembly and the field is only limited by the creativity of researchers.

2.7 Standardisation of DNA Assembly

Since the creation of the first recombinant plasmid, molecular biologists have been attempting to improve the DNA assembly process. As any new graduate student will testify, learning DNA assembly techniques, historically called 'DNA cloning', is challenging and every new project can present a new batch of time-consuming obstacles to overcome. Given the importance of this technique to most fields of biological inquiry, many improvements to the process have been developed to simplify the workflow and increase the probability of success on the first attempt. One of these improvements, a technique called standard assembly, is the focus of this chapter. Standard assembly specifies a simplified method of DNA assembly for a special case of DNA fragments. These special DNA fragments are referred to as Standard Biological Parts, or BioBricks. In theory, any piece of DNA can be turned into a BioBrick, although in practice this conversion may be easier for some fragments than for others while still preserving the original function of the DNA fragment.

The importance of standardisation on the field of molecular biology and what has emerged as synthetic biology is still only now being fully appreciated. Although improving the methods to join two or more fragments of DNA together is essential, the sequence of the resulting construct is just as important. Many recent examples exist showing that the context of parts is as essential as the part itself.[4,104] For example, consider the prokaryotic RBS. The RBS is an important DNA sequence signalling the ribosome to attach to the mRNA in preparation for the initiation of protein translation. This sequence is often defined as a 6–10 base pair (bp) A+G-rich sequence that is found 6–8 bp before the start codon.[105] Recent work has shown that the context surrounding the RBS sequence itself has tremendous influence on its performance.[4] Therefore, unless a part such as an ORF was used in the same context in which it was tested (i.e. RBS and surrounding sequence is identical), expression of that ORF may not be as expected. Older DNA assembly methods relying upon restriction digest and ligation often prevented exact replication of context. Simply joining together the two pieces of DNA that were required for the experiment was often considered adequate to answer the question at hand. If the junctions were not exactly the same as previous experiments in other labs given the available DNA assembly method, then that was simply accepted as part of doing science. This could be frustrating when sequence features behaved differently when tested in different contexts. Because the same feature might have been cloned in a different reporter vector using different restriction enzymes, results from one lab were sometimes not directly comparable to those in another lab. While sequence features were often robust enough to overcome these context differences, the lack of standardisation made it difficult to distribute the work of testing biological part performance to labs around the world.

To many experienced molecular biologists today, it is still often considered simply part of the job description that construction of plasmids to answer questions in the molecular sciences requires complex, multi-step subcloning schemes. But to a computer scientist and electrical engineer looking in on the field, the *status quo* of DNA assembly was a lost opportunity. What if making the plasmid to answer the research question was not itself a research problem, but instead used a set of simple techniques that were identical every time? This question was answered through work by Thomas Knight and colleagues at the Massachusetts Institute of Technology who developed the concept of standard assembly and the BioBrick Standard Part. Before delving into the innovations of standard assembly of DNA and the Biobrick, some historical perspective on the prior state of the art may help illustrate its novelty.

2.7.1 Life before Standardisation

It was not long after the discovery of DNA ligase[106] and restriction endonucleases[107] that researchers began to see the utility of joining the compatible ends of DNA fragments[108] and recombining restriction fragments from different plasmids to create novel recombinant plasmids.[1] This *in vitro* DNA assembly reaction, first published in 1973, became the standard practice for DNA assembly and is still widely used today. The assembly of DNA fragments using restriction digest and ligation, commonly called 'gene cloning', can be used with DNA fragments at a variety of size scales from tens of bases (e.g. linker ligation) to 100s of kilobases (kb) (e.g. cloning DNA as BACs).

Restriction enzymes, at one time only available through labour intensive, in-house purification, eventually became commercially available. As an enormous diversity of restriction enzyme digestion sites was discovered from different bacteria, the utility of restriction enzymes in DNA assembly was firmly established. The wide variety of restriction enzymes available at reasonably low cost has kept the restriction digest-ligation DNA assembly technique popular. Furthermore, the ability to add restriction enzyme recognition sites to the ends of DNA fragments of interest using PCR has further perpetuated the use of restriction digest-ligation techniques.[109] Since such a great volume of work has been performed using the restriction digest-ligation paradigm, this technique has considerable 'institutional momentum'. That is, researchers are more likely to continue working with techniques that generally work in well-equipped and well-trained hands.

There are many drawbacks to the restriction digest-ligation technique, and several of these concerns have been addressed in revised 'standards of practice' that enable DNA fragments to be assembled with much greater ease. Drawbacks include (1) limitations due to the presence or absence of particular enzyme digestion sites in the vector or insert molecules, (2) high level of technical skill required to achieve success in both project design and execution, (3) the preference of intra-molecular ligation over inter-molecular ligation, (4) the poor ability to control fragment orientation in single-enzyme digestion-ligations, (5) particularly poor performance with multi-piece assemblies, (6) the need for a freezer full of enzymes recognising different sequences for maximum capability and (7) the relatively long time required (2–4 days). Thus, while the technique worked, there was clear motivation to improve upon it.

2.7.2 Standard Assembly and BioBricks

The first true attempt at standardisation occurred with the development of Gateway cloning.[68,70,72] In this system, DNA fragments to be assembled must be flanked by appropriate phage recombination sites. While still in wide use today, this system lacks the iterative assembly features of many restriction enzyme-based DNA assembly techniques and is therefore somewhat limited in the scope of applications that it can perform.

Another attempt at standardisation focused on reducing the numbers of enzymes used in assembly. The hundreds of different restriction enzymes that have been discovered, many of which have unique recognition sites, have been of great utility to molecular biology, but ironically, this has led to one of the great problems with using them in DNA assembly. While many plasmid vectors feature multi-cloning sites with 10 or more restriction sites, it is often difficult to find the appropriate enzymes to subclone a particular DNA fragment into a new plasmid to test a novel function. Lack of compatibility of usable enzymes for DNA assembly is often challenging and prevents the efficient reuse of DNA parts. Standard assembly solves this problem by using the same set of only four restriction enzymes. This greatly simplifies the assembly process by creating a strategy in which ligation of an insert and vector recreates the same restriction sites on either side of the insert. This 'idempotent strategy' allows the same set of operations to be performed to assemble any two pieces of DNA, and these same operations can be performed iteratively to create multi-part DNA assemblies. In the original standard assembly, these enzymes were EcoRI and XbaI on the upstream side and SpeI and PstI on the downstream side.[16] Many newer standard assembly methods have been developed that use different enzymes or assembly chemistries. These assembly standards are disseminated and maintained by the BioBricks foundation[110] and several are described in detail in Section 2.7.3.1.

In order to assemble an insert into a vector using standard assembly, both DNA pieces must take the form of a 'standard biological part' or BioBrick. A BioBrick is a DNA fragment flanked on either side by the standard assembly restriction sites and cloned in a vector. For example, in the original standard, the DNA insert of interest would be flanked by EcoRI and XbaI on the upstream side and SpeI and PstI on the downstream side, and would be contained in a cloning vector. It is essential that the four restriction sites each be unique for a piece of DNA to be considered a standard biological part. That is, if the insert of interest contains any of the four restriction sites, the site must be removed by site-directed mutagenesis. If the insert of interest encodes an ORF, care must be taken to select a suitable alternative codon that removes the offending restriction site but preserves the amino acid sequence of the ORF. If any of the four restriction sites used in assembly is present in a regulatory part (i.e. a promoter or other protein binding site), it may be more troublesome to remove depending on the importance of each of the six bases to the feature of interest. Transcriptional regulatory motifs often have strict sequence specificities for maximal function. The requirement to eliminate certain sequences is one of the limitations of this assembly technique and standard.

Standard assembly proceeds using typical restriction digest and ligation methods but follows a specific sequence of digestions to join an insert in either the upstream or downstream assembly sites (Figure 2.10). To insert a DNA part on the upstream side, the insert would be digested with EcoRI and SpeI while the vector would be digested with EcoRI and XbaI. The EcoRI-digested ends of the insert and vector ligate, and the compatible cohesive ends of SpeI and XbaI (on vector and insert, respectively) ligate to form a scar site between the newly added upstream piece and another piece cloned in the vector downstream of the new piece. Because restriction sites in the original standard are separated by single bases, the scar site between the pieces becomes effectively the 8-bp sequence 5′-TACTAGAG. To add a DNA part in the downstream region, the insert is cut with XbaI and PstI while the vector is cut with SpeI and PstI. After ligation, the PstI ends ligate and the XbaI and SpeI (on the insert and vector, respectively), ligate to form the scar site 5′-TACTAGAG between the previously cloned DNA and the newly added part. Because an upstream insert carries with it the XbaI site remaking the EcoRI-XbaI prefix, the same operation can be used to insert another sequence in the upstream region. Similarly, an insert digested for downstream insertion carries with it the SpeI sequence remaking the SpeI-PstI suffix for subsequent downstream insertions.

The BioBrick assembly is occasionally limited by a design feature encountered by all restriction enzyme-based cloning strategies: the fragments to be assembled must be free of internal restriction enzyme sites that are to be used in the assembly. As a consequence, typically this may require site-directed mutagenesis to convert a DNA fragment into a 'standard biological part'. Thus, the ability to use standard assembly may require considerable work up front to develop the piece for use in the system. But once standard biological parts are created, the planning and work associated with assembly are vastly simplified.

Another potential limitation is that standard assembly lacks efficient 'debugging' strategies. For example, consider building a plasmid containing five BioBrick inserts

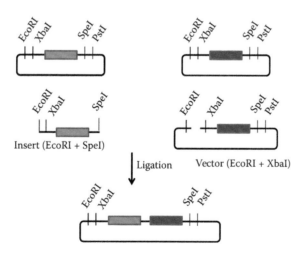

FIGURE 2.10
Standard DNA assembly with BioBricks (BBF RFC 10). Insert plasmid with part to be inserted (blue) is digested with EcoRI and SpeI. Vector plasmid with downstream part (red) is digested with EcoRI and XbaI. After ligation, the new part is inserted upstream of the existing parts. Other configurations to insert parts downstream are similar and described in the text.

using standard assembly methods. The inserts are added, in this hypothetical case, as 3'-'downstream' insertions in consecutive digest-ligation iterations until the final five-insert plasmid has been assembled. When tested, the researcher realises that BioBrick insert number two was incorrect. The only option for the researcher, using standard assembly, is to go back to the plasmid containing one insert and repeat the insertions of BioBricks 2–5 sequentially. Standard assembly does not allow an efficient means to excise an incorrect piece from the middle and efficiently reassemble. Several workarounds are possible, however. BioBricks 3 through 5 could be PCR-amplified and cloned as a new BioBrick. The new composite BioBrick could be assembled behind the correctly cloned BioBrick 2, thus saving iterations. Alternatively, a parallel assembly method (see Sections 2.1 through 2.6) could be used to re-assemble BioBrick 1, the new BioBrick 2 and BioBricks 3–5 in a multi-insert assembly, but the process clearly requires additional skill and knowledge beyond the original standard assembly simplified assembly plan.

BioBricks are further limited by the sequence at the junction between two assembled parts. This is a limitation faced by all type II restriction digest-based assembly methods. As discussed in Section 2.7.3.1, the 'scar site' or DNA sequence resulting between two pieces as a result of the restriction digestion and ligation creates problems for certain applications such as translational fusion proteins or precise RBS context. This has led to the development of alternative standard assemblies such as BglBricks and ePathBrick that are described in Sections 2.7.3.1 and 2.7.3 and offer solutions to this problem for specific applications.

The introduction of alternative assembly standards creates a potential problem for standardisation if the standards themselves are not intercompatible. While it is unlikely that one assembly standard will accommodate all applications, the community is still in flux as to which design standards to maintain and support. As newer, sequence-independent DNA assembly methods are more widely adopted, the role of standards will change. In the past, the two problems of how to make something and what it looked like when completed were necessarily linked. The future of DNA assembly will undoubtedly rely more heavily on sequence-independent assembly and *de novo* DNA synthesis, and the role of assembly standards will need to expand into these techniques. Standardisation is essential for reliable and reproducible measurement of part performance[111] and to permit abstraction, *in silico* modelling of the behaviour of molecular systems, and to achieve the predicted behaviour *in vivo*.[2] Furthermore, at least for the moment, it is usually cheaper for a lab to reassemble a collection of standard parts with the classic restriction digest-ligation methods described below than to synthesise all the sequences of interest. Thus, to allow the widest base of participation in the synthetic biology field from people of many backgrounds and education levels, it is imperative that devices be made in accordance with BioBrick Assembly Standards. This will ensure that the field can capture the ideas and enthusiasm of as many practitioners as possible.

2.7.3 Alternative BioBrick Assembly Standards and Innovations

As molecular techniques continue to develop and improve, many of these new technique have been applied to BioBrick standard assembly methods.[16,53,110] While dozens of newer assembly standards have incorporated more recently developed assembly methods,[110] a common theme of most methods is to preserve the BioBrick compatibility with a standard set of four restriction enzymes for assembly of two pieces. As described in Section 2.7.2, this system was conceived to be 'idempotent' meaning that the final product after restriction digest and ligation of two pieces has the identical restriction sites as the original pieces and therefore, iterative cycles of assembly can be performed in a standardised manner.[16] Maintaining this powerful strategy and extending the BioBrick strategy into new biological systems and with newer technologies has been a goal of the many new assembly standards proposed since the original.[16,110] Several of these innovations are described in the following sections.

2.7.3.1 BglBrick

The original BioBrick assembly standard described earlier was efficient for assembling various genetic elements into a single plasmid.[16] One aspect of that assembly standard that was not supported was the creation of protein fusions, including the addition of peptide tags and fluorescent proteins such as GFP to the beginning or end of a protein. As described earlier, the original standard created an 8-bp scar site between parts that prohibited protein fusions for two reasons: a stop codon was encoded in the scar site and the 8-bp spacing of the added scar site created a translational frame shift. Given the importance of chimeric fusion proteins to molecular biology and biochemistry, adapting a BioBrick standard to address this problem was of immediate importance.

One solution put forward was a new standard commonly called BglBricks.[112,113] While not the first standard to address the problem of creating protein fusions with the original assembly standard, it is one of the most practical. The BglBrick assembly standard uses standard parts that contain a single EcoRI and BglII site at the 5′ end and a single BamHI and XhoI site at the 3′ end of the part (Figure 2.11). Like the original standard, the BglBrick standard assembly process is idempotent and follows the general strategy for assembly outlined earlier. The resulting assembly joins the two standard biological parts

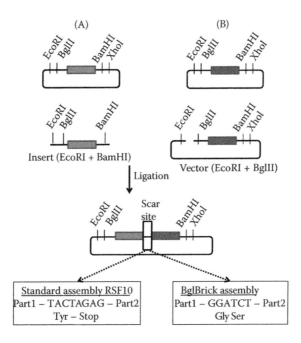

FIGURE 2.11
Alternative assembly strategies allow for new functionality. (A) Original standards (e.g. RFC 10) added a scar site between the two parts that contained a stop codon. (B) Newer standards such as the BglBricks standard (described in text) allow in-frame combination of parts.

and separates them with a 6-bp scar site, 5′-GGATCT, which encodes glycine and serine residues. These two residues are well established linkers and lack much of the chemical functionality of other amino acid residues. Thus, by changing the restriction enzymes used in the assembly, additional functionality is created.

2.7.3.2 BioScaffold

A second solution to the problems associated with scar sites is to create a system in which the sequence between two pieces can be easily modified. This has been accomplished with a technique called BioScaffold.[114] In this method, any two standard biological parts can be assembled with a red fluorescent protein expression cassette separating them. The red fluorescent protein expression cassette has type IIS restriction sites at both sides of the cassette to facilitate the removal of the fluorescent protein scaffold and seamless assembly of a linking DNA fragment of any sequence between the two standard parts. The result is the two standard parts are joined into one modified standard part. This new standard part can still be manipulated as all other standard parts, but the two component parts are now permanently fused. Such a technique allows for optimisation of the RBS sequences upstream of ORFs to precisely tune gene expression and allows much greater flexibility in standard assembly.[114]

2.7.3.3 2ab

Another innovation on the theme of standard assembly is the 2ab assembly method.[19] The 2ab method highlights one of the strong features of standard assembly methods: it is readily amenable to scale-up and automation. The 2ab method was developed using the BglBricks standard, but it could as easily be applied to other standards. To describe the 2ab method, recall that in BglBrick assembly, each standard part consists of single and unique 5′ EcoRI and BglII sites and 3′ BamHI and XhoI sites. The part, which consists usually of a coding region flanked by the restriction enzyme sites, is maintained in a plasmid backbone. In the 2ab method, the XhoI site is simply moved further away from the coding region of the part to include a selectable antibiotic marker after digestion. This innovation in the 2ab assembly method allows for improved selection of properly assembled products. The principle of 2ab is quite simple: because each part has a selectable marker, each assembly of two parts will have two unique selectable markers enabling the correct products to be selected from parent molecules after transformation into *E. coli*.

In practice, the authors of the 2ab method have devised a clever system to enable one pot digestion and ligation. In a 2ab assembly, one of the parts to be assembled, the 'lefty' part, is digested with BamHI and XhoI and the other part to be assembled, the 'righty' part, is digested with BglII and XhoI. The BamHI and BglII sites ligate to create a 6-bp scar site encoding the Gly-Ser linker of the BglBrick assembly standard. Because all parts have the same restriction enzyme sites, correct assembly requires that the lefty part only be cut with BamHI and XhoI and that the righty part only be cut with BglII and XhoI. In order to perform this differential digestion in a one-pot, three-enzyme reaction, the plasmids are prepared from *E. coli* strains that express methylases specific for either BamHI or BglII sites. Thus, by preparing the plasmid that will donate the lefty part from an *E. coli* strain expressing the BglII methylase, and the plasmid that will donate the righty part from a BamHI-methylase expressing *E. coli* strain, the two plasmids can be mixed in the presence of both BamHI and BglII restriction enzymes and will result in a lefty cut only with

BamHI and a righty cut only with BglII. Both lefty and righty parts are also simultaneously digested with XhoI to yield the part linked to a different selectable antibiotic marker. Because the lefty and righty parts do not have the same antibiotics markers, the resulting hybrid assembly product can be selected using the desired combination of antibiotics. This approach greatly decreases the amount of labour required to prepare parts for assembly and is more amenable to automation.

Automation is an increasingly important consideration for DNA assembly. Using the same set of reliable and optimised operations allows for efficient automation of the process. Because the requirement to screen multiple clones is time and resource consuming, especially when performing hundreds of assembly reactions at the same time, the procedures need to be optimised to achieve success in the vast majority of attempts. The 2ab method has proven to be robust under these requirements.[19]

2.7.3.4 ePathBrick

To address many of the shortcomings of standard assembly in the areas of debugging and reuse of sub-assemblies, another standard called ePathBricks has been developed.[89] This method is particularly useful in optimising multi-gene pathways. The general idea of ePathBricks is that four-part protein expression devices are constructed with the standard architecture of promoter-RBS-coding region-terminator.[18] In the ePathBricks standard, each of the parts in the protein expression device is preceded or followed by a unique restriction site (either AvrII, XbaI, SpeI or NheI) that generates a compatible cohesive end with the other three enzymes in the set. For example, the promoter region is preceded by AvrII, the RBS is preceded by XbaI, the terminator is preceded by SpeI, and the whole device is followed by NheI. The four enzymes used in ePathBrick must be unique and standard parts in the ePathBrick standard must lack these enzymes in the internal part regions. Once assembled, the protein-expression devices can be assembled into more complex pathway expression devices by digesting with combinations of the interchangeable four restriction enzymes. This permits two protein expression devices to be joined together on a plasmid in one of three expression strategies: as a polycistron (the second device has only an RBS-coding region), as a pseudo-operon (the second device has a promoter-RBS-coding region), or as a monocistron (the second device has all four parts: promoter-RBS-coding region-terminator). By varying the assembly strategy in the pathway, the level of expression of the pathway can be tuned. Furthermore, in developing the basic architecture of the system, the authors have also developed this for multiple compatible plasmid backbones, allowing for different expression levels based on plasmid copy number and ease of combinatorial swapping by mixing different plasmids together in the same cell. The unique sites separating each part of the protein expression device allow for efficient debugging or modification as necessary. The ePathBrick strategy makes it easier to combine these complex parts in more sophisticated ways than is available with other standard assembly methods.

2.7.4 The Future of Standard Assembly

As the particular specifications of standards continue to evolve, the utility of standard assembly will also grow. The concept of BioBricks has been viewed as so useful that plasmid systems incorporating assembly standards have been created to allow constructs to be used with species other than *E. coli*. For example, shuttle vectors have been constructed that are capable of replication in cyanobacteria[115] and in *Agrobacterium* for use in plant transformation.[116] Newer DNA assembly technologies are also being incorporated into assembly standards. Methods for incorporating the In-Fusion method for PCR-based assembly can allow for multiple parts to be assembled at once.[53] Other parallel DNA assembly methods including USER, Gibson, SLIC and Golden Gate have also been proposed for use with BioBricks and this may help accelerate assembly with standard biological parts.[110] While the methods used to physically assemble pieces of DNA will continue to evolve, the concept of design standardisation will remain an essential consideration in all biotechnology projects.

References

1. Cohen, S., Chang, A., Boyer, H. and Helling, R. Construction of biologically functional bacterial plasmids in vitro. *Proc Natl Acad Sci USA* **70**, 3240–3244 (1973).
2. Endy, D. Foundations for engineering biology. *Nature* **438**, 449–453 (2005).
3. Carr, P. A. and Church, G. M. Genome engineering. *Nat Biotechnol* **27**, 1151–1162 (2009).
4. Salis, H. M., Mirsky, E. A. and Voigt, C. A. Automated design of synthetic ribosome binding sites to control protein expression. *Nat Biotechnol* **27**, 946–950 (2009).
5. Benders, G. A. et al. Cloning whole bacterial genomes in yeast. *Nucleic Acids Res* **38**, 2558–2569 (2010).
6. Watanabe, S., Shiwa, Y., Itaya, M. and Yoshikawa, H. Complete sequence of the first chimera genome constructed by cloning the whole genome of synechocystis strain PCC6803 into the *Bacillus subtilis* 168 genome. *J Bacteriol* **194**, 7007–7007 (2012).
7. Hughes, R. A., Miklos, A. E. and Ellington, A. D. Gene synthesis methods and applications. *Method Enzym* **498**, 277–309 (2011).
8. Gibson, D. G. et al. Complete chemical synthesis, assembly, and cloning of a Mycoplasma genitalium genome. *Science* **319**, 1215–1220 (2008).
9. Noskov, V. N. et al. Assembly of large, high G+C bacterial DNA fragments in yeast. *ACS Synth Biol* **1**, 267–273 (2012).
10. Smailus, D. E., Warren, R. L. and Holt, R. A. Constructing large DNA segments by iterative clone recombination. *Syst Synth Biol* **1**, 139–144 (2007).
11. Karas, B. J., Tagwerker, C., Yonemoto, I., Hutchison III, C. A. and Smith, H. O. Cloning the Acholeplasma laidlawii PG-8A genome in *Saccharomyces cerevisiae* as a yeast centromeric plasmid. *ACS Synth Biol* **1**, 22–28 (2012).
12. Holt, R. A., Warren, R., Flibotte, S., Missirlis, P. I. and Smailus, D. E. Rebuilding microbial genomes. *Bioessays* **29**, 580–590 (2007).
13. Itaya, M., Fujita, K., Kuroki, A. and Tsuge, K. Bottom-up genome assembly using the *Bacillus subtilis* genome vector. *Nat Meth* **5**, 41–43 (2008).
14. Tsuge, K., Matsui, K. and Itaya, M. One step assembly of multiple DNA fragments with a designed order and orientation in *Bacillus subtilis* plasmid. *Nucleic Acids Res* **31**, 133e (2003).
15. Itaya, M., Tsuge, K., Koizumi, M. and Fujita, K. Combining two genomes in one cell: Stable cloning of the Synechocystis PCC6803 genome in the *Bacillus subtilis* 168 genome. *Proc Natl Acad Sci USA* **102**, 15971–15976 (2005).
16. Knight, T. Idempotent vector design for standard assembly of BioBricks. *MIT Artif Intell Lab MIT Synth Biol Work Gr* (2003). http://hdl.handle.net/1721.1/21168
17. Shetty, R. P., Endy, D. and Knight, T. F. Engineering BioBrick vectors from BioBrick parts. *J Biol Eng* **2**, 5 (2008).
18. Canton, B., Labno, A. and Endy, D. Refinement and standardization of synthetic biological parts and devices. *Nat Biotechnol* **26**, 787–793 (2008).

19. Leguia, M., Brophy, J. A., Densmore, D., Asante, A. and Anderson, J. C. 2Ab assembly: A methodology for automatable, high-throughput assembly of standard biological parts. *J Biol Eng* **7**, 2 (2013).

20. Leguia, M., Brophy, J., Densmore, D. and Anderson, J. C. Automated assembly of standard biological parts. *Method Enzym* **498**, 363–397 (2011).

21. Engler, C., Gruetzner, R., Kandzia, R. and Marillonnet, S. Golden gate shuffling: A one-pot DNA shuffling method based on type IIs restriction enzymes. *PLoS One* **4**, e5553 (2009).

22. Temme, K., Zhao, D. and Voigt, C. A. Refactoring the nitrogen fixation gene cluster from *Klebsiella oxytoca*. *Proc Natl Acad Sci USA* **109**, 7085–7090 (2012).

23. Sanjana, N. E. et al. A transcription activator-like effector toolbox for genome engineering. *Nat Protoc* **7**, 171–192 (2012).

24. Zhang, F. et al. Efficient construction of sequence-specific TAL effectors for modulating mammalian transcription. *Nat Biotechnol* **29**, 149–154 (2011).

25. Weber, E., Engler, C., Gruetzner, R., Werner, S. and Marillonnet, S. A modular cloning system for standardized assembly of multigene constructs. *PLoS One* **6**, e16765 (2011).

26. Chen, W.-H., Qin, Z.-J., Wang, J. and Zhao, G.-P. The MASTER (methylation-assisted tailorable ends rational) ligation method for seamless DNA assembly. *Nucleic Acids Res* **41**, e93 (2013).

27. Briggs, A. W. et al. Iterative capped assembly: rapid and scalable synthesis of repeat-module DNA such as TAL effectors from individual monomers. *Nucleic Acids Res* **40**, e117 (2012).

28. Brownstein, M., Carpten, J. and Smith, J. Modulation of non-templated nucleotide addition by Taq DNA polymerase: Primer modifications that facilitate genotyping. *Biotechniques* **20**, 1004–1010 (1996).

29. Shuman, S. Novel approach to molecular cloning and polynucleotide synthesis using vaccinia DNA topoisomerase. *J Biol Chem* **269**, 32678–32684 (1994).

30. Zeng, G. Sticky-end PCR: New method for subcloning. *Biotechniques* **25**, 206–208 (1998).

31. Horton, R., Hunt, H., Ho, S., Pullen, J. and Pease, L. Engineering hybrid genes without the use of restriction enzymes: Gene splicing by overlap extension. *Gene* **77**, 61–68 (1989).

32. Shuldiner, A., Scott, L. and Roth, J. PCR-induced (ligase-free) subcloning: A rapid reliable method to subclone polymerase chain reaction (PCR) products. *Nucleic Acids Res* **18**, 1920 (1990).

33. Geiser, M., Cèbe, R., Drewello, D. and Schmitz, R. Integration of PCR fragments at any specific site within cloning vectors without the use of restriction enzymes and DNA ligase. *Biotechniques* **31**, 88–92 (2001).

34. Quan, J. and Tian, J. Circular polymerase extension cloning of complex gene libraries and pathways. *PLoS One* **4**, e6441 (2009).

35. Spiliotis, M. Inverse fusion PCR cloning. *PLoS One* **7**, e35407 (2012).

36. Klock, H. E., Koesema, E. J., Knuth, M. W. and Lesley, S. A. Combining the polymerase incomplete primer extension method for cloning and mutagenesis with microscreening to accelerate structural genomics efforts. *Proteins* **71**, 982–994 (2008).

37. Jiang, X. et al. In vitro assembly of multiple DNA fragments using successive hybridization. *PLoS One* **7**, e30267 (2012).

38. Pachuk, C. J. et al. Chain reaction cloning: A one-step method for directional ligation of multiple DNA fragments. *Gene* **243**, 19–25 (2000).

39. Kok, S. De et al. Rapid and reliable DNA assembly via ligase cycling reaction. *ACS Synth Biol* **3**, 97–106 (2014).

40. Adams, S. et al. Synthesis of a gene for the HIV transactivator protein TAT by a novel single stranded approach involving in vivo gap repair. *Nucleic Acids Res* **16**, 4287–4298 (1988).

41. Tsvetanova, B. et al. *Primary and Stem Cells: Gene Transfer Technologies and Applications.* DNA Assembly Technologies Based on Homologous Recombination (eds. Lakshmipathy, U. and Thyagarajan, B.) 3–17. Hoboken, NJ: John Wiley & Sons, Inc. (2012).

42. Gibson, D. G. et al. Enzymatic assembly of DNA molecules up to several hundred kilobases. *Nat Meth* **6**, 343–345 (2009).

43. Aslanidis, C. and de Jong, P. J. Ligation-independent cloning of PCR products (LIC-PCR). *Nucleic Acids Res* **18**, 6069–6074 (1990).

44. Nisson, P. E., Rashtchian, A. and Watkins, P. C. Rapid and efficient cloning of Alu-PCR products using uracil DNA glycosylase. *PCR Methods Appl* **1**, 120–123 (1991).
45. Rashtchian, A. Novel methods for cloning and engineering genes using the polymerase chain reaction. *Curr Opin Biotechnol* **6**, 30–36 (1995).
46. Watson, D. and Bennett, G. Cloning and assembly of PCR products using modified primers and DNA repair enzymes. *Biotechniques* **23**, 858–864 (1997).
47. Annaluru, N. et al. Assembling DNA fragments by user fusion. *Methods Mol Biol* **852**, 77–95 (2012).
48. Geu-Flores, F., Nour-Eldin, H. H., Nielsen, M. T. and Halkier, B. A. USER fusion: A rapid and efficient method for simultaneous fusion and cloning of multiple PCR products. *Nucleic Acids Res* **35**, e55 (2007).
49. Weyman, P. D. et al. Heterologous expression of Alteromonas macleodii and Thiocapsa roseopersicina [NiFe] hydrogenases in *Escherichia coli*. *Microbiology* **157**, 1363–1374 (2011).
50. Gibson, D. G. Enzymatic assembly of overlapping DNA fragments. *Method Enzym* **498**, 349–361 (2011).
51. Zhu, B., Cai, G., Hall, E. and Freeman, G. In-Fusion™ assembly: Seamless engineering of multidomain fusion proteins, modular vectors, and mutations. *Biotechniques* **43**, 354–359 (2007).
52. Hamilton, M. D., Nuara, A. A., Gammon, D. B., Buller, R. M. and Evans, D. H. Duplex strand joining reactions catalyzed by vaccinia virus DNA polymerase. *Nucleic Acids Res* **35**, 143–151 (2007).
53. Sleight, S. C., Bartley, B. A., Lieviant, J. A. and Sauro, H. M. In-fusion BioBrick assembly and re-engineering. *Nucleic Acids Res* **38**, 2624–2636 (2010).
54. Yu, D. et al. An efficient recombination system for chromosome engineering in *Escherichia coli*. *Proc Natl Acad Sci USA* **97**, 5978–5983 (2000).
55. Mosberg, J. A., Lajoie, M. J. and Church, G. M. Lambda red recombineering in *Escherichia coli* occurs through a fully single-stranded intermediate. *Genetics* **186**, 791–799 (2010).
56. Lee, E. et al. A highly efficient *Escherichia coli*-based chromosome engineering system adapted for recombinogenic targeting and subcloning of BAC DNA. *Genomics* **73**, 56–65 (2001).
57. Rivero-Müller, A., Lajić, S. and Huhtaniemi, I. Assisted large fragment insertion by Red/ET-recombination (ALFIRE)—an alternative and enhanced method for large fragment recombineering. *Nucleic Acids Res* **35**, e78 (2007).
58. Isaacs, F. J. et al. Precise manipulation of chromosomes in vivo enables genome-wide codon replacement. *Science* **333**, 348–353 (2011).
59. Wang, H. H. et al. Programming cells by multiplex genome engineering and accelerated evolution. *Nature* **460**, 894–898 (2009).
60. Mosberg, J. A., Gregg, C. J., Lajoie, M. J., Wang, H. H. and Church, G. M. Improving lambda red genome engineering in *Escherichia coli* via rational removal of endogenous nucleases. *PLoS One* **7**, e44638 (2012).
61. Li, M. Z. and Elledge, S. J. MAGIC, an in vivo genetic method for the rapid construction of recombinant DNA molecules. *Nat Genet* **37**, 311–319 (2005).
62. Li, M. Z. and Elledge, S. J. Harnessing homologous recombination in vitro to generate recombinant DNA via SLIC. *Nat Meth* **4**, 251–256 (2007).
63. Zhang, Y., Werling, U. and Edelmann, W. SLiCE: A novel bacterial cell extract-based DNA cloning method. *Nucleic Acids Res* **40**, e55 (2012). doi:10.1093/nar/gkr1288
64. Liu, Q., Li, M. Z., Leibham, D., Cortez, D. and Elledge, S. J. The univector plasmid-fusion system, a method for rapid construction of recombinant DNA without restriction enzymes. *Curr Biol* **8**, 1300–1309 (1998).
65. Srivastava, V. and Ow, D. W. Marker-free site-specific gene integration in plants. *Trends Biotechnol* **22**, 627–629 (2004).
66. Livet, J. et al. Transgenic strategies for combinatorial expression of fluorescent proteins in the nervous system. *Nature* **450**, 56–62 (2007).
67. Gaeta, R. T., Masonbrink, R. E., Krishnaswamy, L., Zhao, C. and Birchler, J. A. Synthetic chromosome platforms in plants. *Annu Rev Plant Biol* **63**, 307–330 (2012).

68. Walhout, A. et al. Gateway recombinational cloning: Application to the cloning of large numbers of open reading frames or ORFeomes. *Method Enzym* **328**, 575–592 (2000).

69. Matsuyama, A. and Yoshida, M. Systematic cloning of an ORFeome using the Gateway system. *Method Mol Biol* **577**, 11–24 (2009).

70. Hartley, J. L., Temple, G. and Brasch, M. DNA cloning using in vitro site-specific recombination. *Genome Res* **10**, 1788–1795 (2000).

71. Reboul, J. et al. C. elegans ORFeome version 1.1: Experimental verification of the genome annotation and resource for proteome-scale protein expression. *Nat Genet* **34**, 35–41 (2003).

72. Rual, J. et al. Human ORFeome version 1.1: A platform for reverse proteomics. *Genome Res* **14**, 2128–2135 (2004).

73. Petersen, L. K. and Stowers, R. S. A Gateway MultiSite recombination cloning toolkit. *PLoS One* **6**, e24531 (2011).

74. Shi, Z., Wedd, A. G. and Gras, S. L. Parallel in vivo DNA assembly by recombination: Experimental demonstration and theoretical approaches. *PLoS One* **8**, e56854 (2013).

75. Baudin, A., Ozier-Kalogeropoulos, O., Denouel, A., Lacroute, F. and Cullin, C. A simple and efficient method for direct gene deletion in *Saccharomyces cerevisiae*. *Nucleic Acids Res* **21**, 3329–3330 (1993).

76. Längle-Rouault, F. and Jacobs, E. A method for performing precise alterations in the yeast genome using a recyclable selectable marker. *Nucleic Acids Res* **23**, 3079–3081 (1995).

77. Manivasakam, P., Weber, S. C., McElver, J. and Schiestl, R. H. Micro-homology mediated PCR targeting in *Saccharomyces cerevisiae*. *Nucleic Acids Res* **23**, 2799–2800 (1995).

78. Kouprina, N., Noskov, V. N. and Larionov, V. Selective isolation of large chromosomal regions by transformation-associated recombination cloning for structural and functional analysis of mammalian genomes. *Method Mol Biol* **349**, 85 (2006).

79. Kouprina, N. et al. Functional copies of a human gene can be directly isolated by transformation-associated recombination cloning with a small 3′ end target sequence. *Proc Natl Acad Sci USA* **95**, 4469–4474 (1998).

80. Kouprina, N. and Larionov, V. TAR cloning: Insights into gene function, long-range haplotypes and genome structure and evolution. *Nat Rev Genet* **7**, 805–812 (2006).

81. Larionov, V., Kouprina, N., Solomon, G., Barrett, J. C. and Resnick, M. A. Direct isolation of human BRCA2 gene by transformation-associated recombination in yeast. *Proc Natl Acad Sci USA* **94**, 7384–7387 (1997).

82. Shao, Z., Zhao, H. and Zhao, H. DNA assembler, an in vivo genetic method for rapid construction of biochemical pathways. *Nucleic Acids Res.* **37**, e16 (2009).

83. Gibson, D. G. et al. Creation of a bacterial cell controlled by a chemically synthesized genome. *Science* **329**, 52–56 (2010).

84. Gibson, D. G. et al. One-step assembly in yeast of 25 overlapping DNA fragments to form a complete synthetic *Mycoplasma genitalium* genome. *Proc Natl Acad Sci USA* **105**, 20404–20409 (2008).

85. Gibson, D. G. Synthesis of DNA fragments in yeast by one-step assembly of overlapping oligonucleotides. *Nucleic Acids Res* **37**, 6984–6990 (2009).

86. Tolia, N. H. and Joshua-Tor, L. Strategies for protein coexpression in *Escherichia coli*. *Nat Meth* **3**, 55–64 (2006).

87. Tseng, H.-C., Martin, C. H., Nielsen, D. R. and Prather, K. L. J. Metabolic engineering of *Escherichia coli* for enhanced production of (R)- and (S)-3-hydroxybutyrate. *Appl Env Microbiol* **75**, 3137–3145 (2009).

88. Sun, J., Hopkins, R. C., Jenney Jr., F. E., McTernan, P. M. and Adams, M. W. W. Heterologous expression and maturation of an NADP-dependent [NiFe]-hydrogenase: A key enzyme in biofuel production. *PLoS One* **5**, e10526 (2010).

89. Xu, P., Vansiri, A., Bhan, N. and Koffas, M. A. G. ePathBrick: A synthetic biology platform for engineering metabolic pathways in *E. coli*. *ACS Synth Biol* **1**, 256–266 (2012).

90. Schmidt, C. M., Shis, D. L., Nguyen-huu, T. D. and Bennett, M. R. Stable maintenance of multiple plasmids in *E. coli* using a single selective marker. *ACS Synth Biol* **1**, 445–450 (2012).

91. Suzuki, Y. et al. Knocking out multigene redundancies via cycles of sexual assortment and fluorescence selection. *Nat Meth* **8**, 159–164 (2011).

92. Ramon, A. and Smith, H. O. Single-step linker-based combinatorial assembly of promoter and gene cassettes for pathway engineering. *Biotechnol Lett* **33**, 549–555 (2011).

93. Merryman, C. and Gibson, D. G. Methods and applications for assembling large DNA constructs. *Metab Eng* **14**, 196–204 (2012).

94. Zhao, H., Giver, L., Shao, Z., Affholter, J. and Arnold, F. Molecular evolution by staggered extension process (StEP) in vitro recombination. *Nat Biotechnol* **16**, 258–261 (1998).

95. Stemmer, W. DNA shuffling by random fragmentation and reassembly: In vitro recombination for molecular evolution. *Proc Natl Acad Sci USA* **91**, 10747–10751 (1994).

96. Abécassis, V., Pompon, D. and Truan, G. High efficiency family shuffling based on multi-step PCR and in vivo DNA recombination in yeast: Statistical and functional analysis of a combinatorial library between human cytochrome P450 1A1 and 1A2. *Nucleic Acids Res* **28**, E88 (2000).

97. Karas, B. et al. Direct transfer of whole genomes from bacteria to yeast. *Nat Meth* **10**, 410–412 (2013).

98. Huang, P. et al. Heritable gene targeting in zebrafish using customized TALENs. *Nat Biotechnol* **29**, 699–700 (2011).

99. Li, T., Liu, B., Spalding, M. H., Weeks, D. P. and Yang, B. High-efficiency TALEN-based gene editing produces disease-resistant rice. *Nat Biotechnol* **30**, 390–392 (2012).

100. Wood, A. J. et al. Targeted genome editing across species using ZFNs and TALENs. *Science* **333**, 307 (2011).

101. Hockemeyer, D. et al. Genetic engineering of human pluripotent cells using TALE nucleases. *Nat Biotechnol* **29**, 731–734 (2011).

102. Jiang, W., Bikard, D., Cox, D., Zhang, F. and Marraffini, L. A. RNA-guided editing of bacterial genomes using CRISPR-Cas systems. *Nat Biotechnol* **31**, 233–239 (2013). doi:10.1038/nbt.2508

103. DiCarlo, J. E. et al. Genome engineering in Saccharomyces cerevisiae using CRISPR-Cas systems. *Nucl Acids Res* **41**, 4336–4343 (2013).

104. Ungerer, J. L., Pratte, B. S. and Thiel, T. RNA processing of nitrogenase transcripts in the Cyanobacterium Anabaena variabilis. *J Bacteriol* **192**, 3311–3320 (2010).

105. Shine, J. and Dalgarno, L. Determinant of cistron specificity in bacterial ribosomes. *Nature* **254**, 34–38 (1975).

106. Gellert, M. Formation of covalent circles of lambda DNA by *E. coli* extracts. *Proc Natl Acad Sci USA* **57**, 148–155 (1967).

107. Smith, H. and Wilcox, K. A restriction enzyme from Haemophilus influenzae. I. Purification and general properties. *J Mol Biol* **51**, 379–391 (1970).

108. Lobban, P. and Kaiser, A. Enzymatic end-to-end joining of DNA molecules. *J Mol Biol* **78**, 453–460 (1973).

109. Jung, V., Pestka, S. B. and Pestka, S. Efficient cloning of PCR generated DNA containing terminal restriction endonuclease recognition sites. *Nucleic Acids Res* **18**, 6156 (1990).

110. http://openwetware.org/wiki/The_BioBricks_Foundation:RFC.

111. Mutalik, V. K. et al. Quantitative estimation of activity and quality for collections of functional genetic elements. *Nat Meth* **10**, 347–353 (2013).

112. Lee, T. S. et al. BglBrick vectors and datasheets: A synthetic biology platform for gene expression. *J Biol Eng* **5**, 12 (2011).

113. Anderson, J. C. et al. BglBricks: A flexible standard for biological part assembly. *J Biol Eng* **4**, 1 (2010).

114. Norville, J. E. et al. Introduction of customized inserts for streamlined assembly and optimization of BioBrick synthetic genetic circuits. *J Biol Eng* **4**, 17 (2010).

115. Huang, H.-H., Camsund, D., Lindblad, P. and Heidorn, T. Design and characterization of molecular tools for a synthetic biology approach towards developing cyanobacterial biotechnology. *Nucleic Acids Res* **38**, 2577–2593 (2010).

116. Boyle, P. M. et al. A BioBrick compatible strategy for genetic modification of plants. *J Biol Eng* **6**, 8 (2012).

3

Standardised Genetic Output Measurement

Marcus K. Dymond

CONTENTS

3.1 Introduction

As the ambition of synthetic biologists grows, systems and synthetic approaches to biology are beginning to be used concomitantly. Nowhere is this more apparent than in metabolic engineering applications where synthetic biologists have staked much on the promise of synthetic biology to build cellular factories producing cheap abundant biofuels and diverse natural products. To achieve these challenging goals, it is necessary to standardise metabolic network engineering so that product yields can be maximised against the natural tolerances of host organisms for non-native genes and biochemical pathways. Such standardisation requires gene measurement to be tuned and characterised and in this chapter current progress towards these goals is reviewed. The emphasis is on simple single-gene measurement techniques since these are the most accessible to those new to synthetic biology; however, complex multiple-gene expression profiling methods are mentioned. Reviews on the many different areas of gene measurement are signposted and progress towards their standardisation is discussed. Where appropriate, applications from the synthetic biology literature are presented and synthetic biology's quest for a set of biological engineering standards and democratic tools for measuring gene expression is discussed.

3.2 Standardisation in Conventional Biology

Standardisation in biology is not a new concept; crystallographic databases have acted as a driver for formalising standard methods of reporting and submitting protein structures, resulting in a set of guidelines that have been widely adopted by publishers.[1] Similarly, the Standards for Reporting Enzyme Activity Data (STRENDA)[2] are another example of a widely adopted standard. Standardisation is, in many ways, the endpoint of the development of any analytical technique, which will typically evolve through a qualitative stage, then a relative quantitative stage before internal or external standards allow absolute quantification in the International System of Units (SI). After quantification but prior to the adoption of a standard methodology, sets of minimal guidelines are often reported. The purpose of such sets of minimal guidelines is to lay down explicitly the basic details needed to achieve satisfactory scientific standards in the reporting of complex and diverse experimental procedures.

There are a number of repositories of minimal information guidelines in existence. For example, a series of minimal guidelines have been published for proteomic research by the Human Proteome Organisation.[3] Another effort is the 'Minimum Information about a Biomedical or Biological Investigation' repository, where a large number of minimal information guidelines are reported for processes such as cellular assays, DNA sequencing and much more.[3] Finally, the Bioshare consortium is driving standardisation of biological data repositories (Biobanks).[4] Many of these online resources are directly relevant to the gene measurement techniques discussed here.

One of the motivations for these minimal information efforts is the concern that the interdisciplinary nature of modern science has led to drop in quality of reporting biological processes in the literature, a process which can lead to apparently conflicting results, false positives and unnecessary repetition of experiments across different laboratories. The process through which standardisation evolves from minimal guidelines is broadly similar. Initially, a standard set of 'rules' is established that achieve appropriate reproducibility of a quantity (or quality). Standard rules are decided upon by experts in the field (manufacturers, end users, etc.) and agreed within a document, that is, a standard that has an agreed uniform set of technical specifications, methods and/or processes for a procedure. Critical qualifiers of quality are that results should be comparable between samples, at both an intra-laboratory and inter-laboratory level, and consistency in language terminology and data storage should be agreed.[5] There are several issuers of standards, such as the International Organisation for Standardisation (ISO), the European Committee for Standardisation (CEN) and the American National Standards Institute (ANSI). Although only indirectly relevant to standardised measurement of gene output, Lyon and Horobin[6] discuss in great detail the methodology of standard adoption, maintenance and regulation in the context of standardisation of stains and dyes for histology.

In the development of standardised language for the naming and data storage of biological systems several initiatives are in progress. The motivation here is that with a consistent set of conventions in place the comparison of data between experiments is quality assured and the lifetime of data from biological experiments is extended long after their initial collection and to other users. The Open Biological and Biomedical Ontologies[7] foundry is dedicated to formalising data standards for biological methods. One aspect of the foundry especially relevant to synthetic biology is the 'gene ontology' which covers the domains of cellular components, molecular functions and biological processes. In systems biology, the systems biology markup language (SBML),[8] a free open-source web-based (XML) format, sets a standard for the democratic interchange of information between biological models, other standardised language approaches are reviewed.[9]

3.2.1 The Central Dogma and Information Transfer

The central dogma of molecular biology describes the flow of genetic information between biopolymers, such that information is directionally transferred from DNA to RNA to protein. There are nine theoretical ways that information might be transferred between the biopolymers DNA, RNA and protein: three are general in the sense that they are believed to occur in all cells, three are special transfers, special in the sense that they only occur under specific conditions, and three are not known to occur.

The three general transfers are DNA to DNA (replication), DNA to RNA (transcription) and RNA to protein (translation). The three special transfers are RNA to DNA (reverse transcription), RNA to RNA (RNA replication) and DNA to protein (direct translation). The three less-characterised transfers are protein to protein (amyloids), protein to RNA and protein to DNA. In gene measurement, the idea of information transfer, as demonstrated by the central dogma, underlies many of the techniques that have emerged for studying gene expression.

An important aspect of the central dogma is the concept of information transfer via hybridisation and molecular recognition by base pairs. Base pair relationships are the fundamental molecular interaction through which information transfers from gene to RNA transcripts and ultimately through to proteins. The canonical base pairings, the so-called Watson–Crick motifs, are adenosine–thymidine (A–T) (or adenosine–uridine (A–U) in RNA) and guanosine–cytidine (G–C) and it is these molecular interactions (or variations thereof) that form the basis of many of the techniques for measuring the sequence-specific properties of products of gene expression. The intricacies of molecular recognition through base pairing has been reviewed recently[10] and the prospects for new base pair genetic codes are discussed.[11]

An important terminological convention for DNA is that of the assigning the label 'sense' or 'antisense' to a given DNA strand (see Figure 3.1). By convention, double-stranded DNA sequence information is depicted with the 5' ribose carbon of the upper strand at the left, the 3' at the right. For the lower strand, the 3' ribose carbon is on the left, the 5' on the right, as is required due to the strands running antiparallel to each other. The sense strand is the non-coding strand, running in the 3' to 5' direction, whilst the strand transcribed into mRNA, running 5' to 3', is the antisense strand. Therefore transcript mRNA is a copy of the sense strand (since it is the transcript of antisense DNA) and any base pairing nucleic acid sequences which are complementary to the mRNA are thus antisense. This is the origin of antisense technologies,[12] like antisense RNA and antisense oligonucleotides, which are frequently used within the field of nucleic acid research. Gene expression is most commonly measured as the quantity of transcribed (mRNA) or translated product (protein); however,

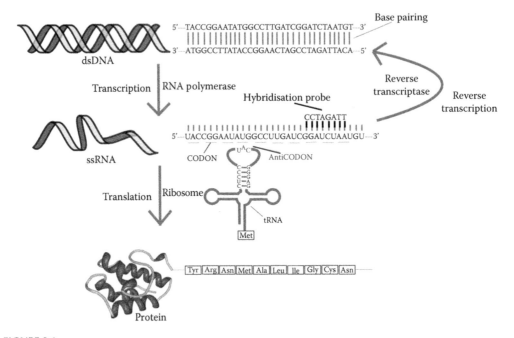

FIGURE 3.1

Measuring genetic information flow. Summary of the processes and terminology of gene measurement: transcription, reverse transcription, translation, base-pairing and hybridisation. The transfer of information from DNA to RNA and protein enables gene activity measurement by multiple methods.

the quality of the copy, that is, its fidelity, is also an important measure. Transcript sequencing is still the most effective method of assessing fidelity and methodologies have been reviewed[13–16] extensively. The central dogma is discussed in great detail in molecular biology textbooks as are the canonical base pairings.[17] Figure 3.1 summarises the central dogma, base pairing, hybridisation and the sense–antisense terminology.

3.3 Standardisation in Synthetic Biology

One of the grand challenges in synthetic biology, which stems from the notion of engineering biological systems,[18] is the creation of a set of biological engineering standards.[5] A first step in creating these standards is the BioBricks parts registry, where a part is defined as 'a genetically encoded object that performs a biological function and that has been engineered to meet specified design or performance requirements'. Thus the BioBricks protocol standardises gene assembly[19] and gene regulation.[20] Another standardised idempotent approach to gene assembly, BglBricks,[21] has also been developed. From a gene measurement perspective, standardisation of parts is desirable because it enables the prediction of part behaviour. Thus in principle any BioBricks part, specifically its gene sequence, can be imported into a new application and that application can be tested with the confidence that the registry part will be reliable. Müller and Arndt recently discussed the BioBricks standard for DNA vector assembly.[5] In other engineering disciplines, components are provided with datasheets that characterise their behaviour; these datasheets allow designers to interface components with predictable outcomes. One suggestion by synthetic biologists is the introduction of datasheets for biological parts as discussed[22] and demonstrated in principle by Canton et al.[23] for a BioBrick part.

There is a general trend towards the standardisation of gene measurement, which is driven by the developers and principle users of these technologies. As a broad generalisation, those techniques that require purification of the products of gene expression have progressed most towards absolute quantification and standardisation, whilst those techniques that rely on characterisation *in vivo* have progressed least. This observation is especially pertinent to synthetic biology since the vast majority of applications are *in vivo* and thus the most challenging to standardise. The reasons for this are the inherent phenotypic variability within any species and the heterogeneity of gene transcripts and translated protein within different tissues, which coupled with the potential for unwanted intermolecular interactions and secondary metabolism makes standardisation difficult. The knock-on effect to synthetic biology is that the characterisation and standardisation of gene measurement within host cells is highly ambitious. Equally as each of the gene measurement techniques becomes standardised by its own community of users there is a tendency for diverse sets of gene measurement standards to emerge. From the perspective of synthetic biology, which takes genetic sequences and protocols from a variety of sources, another challenge arises and that is to find approaches that bring together all the different standardised methodologies under one umbrella.[5]

The first significant step in this process of global standardised gene measurement was facilitated by the BioBricks Foundation and its work with the international Genetically Engineered Machine (iGEM) competition and the registry of Standard Biological Parts. Critically one of the tenets of the BioBricks Foundation is that engineering standards in biology should be democratic. Thus in contrast to ISO standards and published literature detailing minimal guidelines which are purchased by end users and thus only available to those with the funds, democratic standards are open access. Müller and Arndt present a detailed overview of the standardisation and development of the BioBricks assembly protocols.[5]

In addition to standardised approaches to DNA assembly there have been a number of initiatives to develop language standards for data in synthetic biology. The Synthetic Biology Open Language (SBOL) is one proposed language standard, BioBricks are being made available in SBOL.[24] And synthetic biological extensions to the Digital Imaging and Communications in Medicine (DICOM) standard are in development.[25]

Despite this progress, one of the biggest challenges to the goal of engineering biological systems is predicting the behaviour of the final functioning of the biological system. Currently, at the design phase, the logical construction of modular parts within a genetic circuit to yield a functional synthetic cell appears straightforward; however, within the complexity of a biological host the potential for unexpected intermolecular interactions and secondary metabolism[26] can be significant enough to prevent the envisioned design from working efficiently. This lack of predictability has been described as one of the five hard truths for synthetic biology[27] and it is easy to recognise that standardised gene measurement techniques useable *in vivo* would be of benefit here.

One approach being developed by synthetic biologists to increase design predictability is the BioCAD approach, as reviewed.[28–30] The aim here is to use computer aided design to develop models that contain enough part-specific information to predict the behaviour of combined parts within a biological host, allowing the incompatible parts to be replaced or redesigned. Here, the idea of standard datasheets[23] for biological parts fits well with the BioCAD design strategy. Within the recent literature, a number of BioCAD models have been reported; central to all of these is the modular treatment of biological molecules or sets of molecules (parts), usually BioBricks, each with a defined set of characteristics. A few of the commonly used models are CellDesigner,[31] Bio-Tapestry[32] and TinkerCell.[33]

The successful application of prediction algorithms for the design of synthetic biological devices requires a set of standard measures of gene output. Whilst currently there are only suggestions of what these need to be to make the design process deterministic, there is a broad agreement on which biological characteristics are necessary. Logically, the first is the DNA sequence that codes for the part, since this can be used to troubleshoot complications due to hybridisation between mRNA gene products and the coding DNA sequence. The translated protein sequence can also in principle provide information on inhibitory protein–protein and protein–nucleic acid interactions. Another critical measure is promoter strength,[34] which can to a certain extent be considered to be the affinity constant between the polymerase and the DNA sequence that initiates transcription. Similarly, the strengths of ribosome binding are also useful measures. The variance of both polymerase and ribosome binding under different environmental conditions is also a useful measure to include in any design algorithm, as is a measure of the competition (or cooperativity) between promoters and transcription factors (proteins that up-regulate or down-regulate transcription) and small interfering RNAs.

Varying the subtle interplay between transcription factors, promoters and ribosomes allows gene output to be tuned. Ang et al.[35] review current progress in the area of tuning gene output, which relies on finding ways to moderate the response curves of translated gene output systems in response to transcription factor inputs. Whilst these response curves are a useful way to characterise the behaviour of parts and build datasheets, disentangling the complex interactions that underpin tuning is challenging, since, for example, the same output signals (concentration of translated protein) can in theory be the result of different inputs within different environments. Whilst characterising parts in all possible environments is a possibility, it is complex and time consuming. Thus it is desirable to have direct measures of the promoter and ribosome binding and transcription factor interactions. The polymerases per second (PoPS) and ribosome initiations per second (RiPS) units introduced in Section 3.3.3 are measures that begin to take many of the variability factors into account.

3.3.1 Gene Expression: Power Is Nothing without Control

Synthetic biology in its most recognisable form is concerned with making genetic modifications to existing life forms; sometimes only a single, foreign, gene is added, however, usually multiple genes are introduced. The range of projects at the iGEM competition is a good example of the scope of these modifications.[20] Projects range from biosensors of toxic heavy metals through to oil spill cleanup and biological visual display units. Usually a microorganism, such as *E. coli* or *B. subtilis*, is chosen as the chassis for modification and genes are transfected into the microorganism's intracellular environment through plasmid vectors in a set of expression cassettes.[36] In these expression cassettes DNA segments, genes, promoters, codons and so on can be taken from multiple organisms and assembled[37] into one or more plasmids to transfect host cells. Alternatively, non-plasmid methods of genome modification also exist, such as the total genome syntheses pioneered by Craig Venter[38] and the Church Lab's multiplex automated genome engineering (MAGE) approach.[39] Inherent in the process of genetic modification is the ability to control the introduced genes and their interaction with endogenous gene networks. When novel genetic circuits are introduced, gene control systems are usually taken from existing natural transcriptional control systems and generally fall into two categories, positive and negative feedback control.[40] Target strategies for gene control fall into three basic areas: transcription control, that is control at the level of the production of the mRNA transcript; post-transcriptional control, that is at the level of targeting specific gene transcripts and finally post-translational regulation where the fully formed protein is targeted by a regulatory method. There are three ways that measuring gene expression proves to be invaluable, listed as follows:

First, measuring gene expression demonstrates that the introduced genes can be transcribed and/or translated in the host, a critical step confirming successful transfection.

Second, quantifying the level of gene expression allows the efficiency of the transcription or translation process to be optimised. For example, unexpected protein–protein or protein–nucleic acid interactions can occur when genes from multiple organisms are combined in a new organism; quantifying the interaction between genes, promoters, transcripts and translated products is vital when troubleshooting these issues.

Third, following the subcellular distribution of transcription and translation products can provide vital clues to the origin of up- or downstream problems.

In this chapter, we discuss spectrophotometric and electrophoretic methods to quantify total DNA, RNA and total protein loads within samples and tissues. Such techniques cannot be carried out *in vivo* and therefore isolation and purification of RNA or protein is required prior to using these techniques. Discussion of the large number of techniques that exist for the isolation and purification of gene products is outside the scope of this book, but they are the subject of a number of excellent reviews for the interested reader.[41–45]

3.3.2 Turning the Dials of Gene Expression

One of the guiding principles of synthetic biology is the reconstruction of biological processes with a view to understanding a process by recreating it. Researchers, inspired by this philosophy and the idea that *in vivo* the condensation of chromatin[46,47] is a contributory factor to the regulation of transcription, have tried to make synthetic nucleosomes. Their approach was to take linear dsDNA, coding for the luciferase protein, under control of the T7 promoter, complexed to polyamidoamine dendrimers. Transcriptional accessibility of the T7 RNA polymerase was correlated to the degree of compaction in the aggregate DNA: dendrimer complex.[48,49] Condensation/compaction of the coding dsDNA was controlled by increasing the number of positive charges on the dendrimer surface, which has the effect of binding the negatively charged DNA more tightly. Transcription and translation was monitored using a coupled assay where the luminescence of the luciferase was quantified by comparison with a standard curve and the yield of translated protein correlated with DNA compaction. In similar work, another spectrophotometric method was used to quantify gene expression. Using dsDNA-containing inverse hexagonal lyotropic liquid crystal phases[50] their potential as cell nuclei mimics[51] was investigated. The transcriptional accessibility of dsDNA, and partitioning of transcript mRNA onto the lyotropic liquid crystal phase,[52] containing the gene for luciferase under the control of the T7 RNA polymerase promoter was quantified by absorbance of the final mRNA transcript yield on a fibre optic UV/Vis spectrophotometer. Additionally, gene transcripts, for the single gene under control of the promoter, were identified by RNA gel electrophoresis.

Conversely, when multiple genes need to be detected, in particular against the background of genomic transcripts, hybridisation probes provide a more specific method of identification. This is demonstrated by Dymond et al.,[53] who assess the suitability of yeast as an alternative cell chassis to *E. coli* for synthetic biology applications, using synthetic chromosome arms to design phenotypic yeast variants. The approach utilises the SCRaMbLE (synthetic chromosome rearrangement and modification by loxP-mediated evolution) system, which enables parts of the yeast genome to be deleted and new synthetic gene sequences to be inserted in their place. Full experimental details of this methodology are provided.[53] Southern blots were used to verify both the deletion of genes and the insertion of new genes.

Rather than opt for replacing genes, as in the previous example, another strategy explores genetic modification of organisms using a non-native riboswitch to turn native genes off and on again when required.[54,55] One interesting application of this approach is the temperature sensitive riboswitch, an RNA sequence that undergoes conformational change with respect to temperature change. In the examples that occur naturally a complementary base pair sequence of RNA binds to genes at the ribosome binding site blocking the access of transcription factors. As the temperature rises, this base-pair region melts and accessibility to the transcription factor is restored. Waldminghaus et al.[56] have explored the synthesis of temperature sensitive riboswitches by probing the *bgaB* sequence that codes for a heat sensitive β-galactosidase. Northern blots were performed targeting *bgaB* transcripts using hybridisation probes coupled to digoxigenin and analysed by chemiluminescence to assess the functionality of the riboswitch.

An alternative approach is to control gene expression using gene knockdown or RNA interference. One application uses siRNA to build tunable switches for regulating gene expression in mammalian systems. Deans et al.[57] used the enhanced green fluorescent protein (EGFP) as a reporter protein under the control of the *lacI* and *tetR* repressor system in Chinese hamster ovary cells. To overcome the leaky nature of the *lacI TetR* system, which makes transcription impossible to turn off completely, the group engineered into the pathway an RNAi element. In the off state, the genetic circuit produces siRNA, which targets and destroys transcripts from the leaky *lacI TetR* system. The end result is a genetic switch that has a much greater level of fine control. Several other examples of siRNAs engineered into eukaryotic systems exist as reviewed.[58]

In another method of tuning gene expression, the ability of chassis cells under genetic modification to withstand the stresses of modification was optimised. In metabolic engineering, it is often necessary to use multiple plasmids to achieve production of the desired metabolite in excess. However, traditional transfection methods were developed using single high copy plasmids to maximise recombinant protein yields. Where multiple high copy plasmids are incorporated for synthetic biology applications the viability of the host cell can be compromised and plasmids can be rejected. To develop a solution to this problem, RNA slot blots were employed by Smolke and Keasling,[59] when they studied the effect of mRNA stability and DNA copy number on protein expression. A dual gene operon with reporter genes *gfp* and *lacZ* controlled by the *araBAD* promoter was incorporated into low copy, F plasmid-based vectors (1 copy per cell) and high copy pMBI-based (100 copies per cell) vectors. The work explores the best compromise between final product yield and low copy plasmids that are indefinitely stable gene expression systems,[60] capable of replicating large pieces of DNA with a low metabolic burden on host cells.[61] Slot blots were used to probe the steady-state levels of mRNA targeted at the transcripts of *gfp* and *lacZ*, which were contrasted to the protein yields of the reporter protein.

Another topical area of synthetic biology is biofuel production, where research is rapidly increasing the yields of bacterial derived biofuels using both metabolic engineering and systems biology approaches to drive development. Goh et al.[62] report engineering a bacterial synthesis of methyl ketones for biofuel production, where the aliphatic diesel fraction (C_{11}–C_{15}) of the methyl ketones was enriched by a factor of around 700 times by inducing overproduction of β-ketoacyl coenzyme A thioesters. This was achieved by modification of the β-oxidation pathway through over-expression of acyl-CoA oxidase and native FadB followed by chromosomal deletion of FadA and over-expression of a native thioesterase (FadM). Whilst the group focused on assessing the yield and composition of the methyl ketones by analysing total cellular extracts using mass spectroscopy, the gene expression profiles of different modified strains of *E. coli* were also investigated. Quantitative RT-PCR and a cDNA hybridisation arrays were used to achieve this and RNA quality was confirmed spectroscopically. Rather than build their own microarrays the group opted for commercially available microarrays that target the genome of the *E. coli* K12 derivative. Fluorescent cDNAs were prepared from mRNA using a fluorescent cDNA labelling kit and after purification the fluorescent cDNA molecules were incubated with the microarray chip. Arrays were scanned and analysed using commercial software.

3.3.3 Common Signal Carriers for Measuring Gene Output

It is the hope of some synthetic biologists that more engineering standards[22,59,63] for biological parts will be developed. Such standards would form the basis of datasheets and go a long way to resolving the so called 'five hard truths'[27] for synthetic biology. The five hard truths have been described as follows: (1) that many of the standard biological parts are undefined or poorly characterised, (2) the combination of parts into genetic circuits is unpredictable, (3) the complexity of synthetic biological systems makes construction unwieldy and time consuming, (4) many parts are incompatible with other parts or the host and (5) variability (in cell growth, environmental conditions, genetic mutation) crashes the system. Kitney and Freemont[64] discuss recent progress in synthetic biology as a riposte to the five hard truths. In development of new synthetic biological engineering standards at least two new measures have been proposed, the PoPS measurement and the RiPS.[18] The PoPS and RiPS measurements in effect look at the rates of formation gene products on a molecular scale, in SI units.

The benefit of PoPS and RiPS measurements stems from the following observation. If in bulk solution the mRNA products of transcription are a given amount (X), then if the amount of polymerase in solution is (Y) the rate of production of mRNA per molecule is $(X/Y)/t$ where t is the duration of the reaction. The limitation of bulk measurements is that they omit the molecular specificity of interactions, such as the polymerase–DNA binding constant and the variance in polymerase efficiency under different experimental/environmental conditions. Attempts to quantify these factors have been performed by Kelly et al.,[34] who provide relative quantification of promoter activity for a standard BioBricks part. This is achieved under a range of different experimental conditions where variance in promoter activity is measured and normalised to the activity of a 'standard' promoter in the same environment. However, the advantage of a PoPS measurement is that it would specifically detail the number of bound polymerases passing a specific point on a DNA molecule. Thus if the PoPS rate was directly measureable for a biological part, it could be quoted in a datasheet, as could its variance under differing experimental conditions. This would enable end users to assess the compatibility of biological parts to their specific end purpose and prior to assembly, improving the predictability of biological engineering. In effect, the PoPS measurement is a standardised unit (molar per second) of polymerase activity or, in other words, it is a common signal carrier for polymerase activity. The RiPS measurement is the equivalent for protein translation. Another advantage of the PoPS and RiPS rates is that they can in principle be dynamic measures.

Thus, in contrast to the total RNA of a system which accumulates over time, the PoPS and RiPS rates vary over time.

PoPS and RiPS rates have been incorporated into a kinetic, ordinary differential equation approach to designing synthetic gene circuits, where Marchisio and Stelling[65] propose further common signal carriers to enable detailed predictive models of genetic circuits to be constructed. These are the factors per second (FaPS) and signals per second (SiPS). FaPS are defined as the quantity of transcription factors (activators or repressors) produced per second inside their corresponding coding regions and SiPS are the environmental chemical signals, inducers or repressors that enter the cell per second. Additionally a measure of small RNAs per second (RNAPS), which can moderate transcription, has been proposed.[66] Figure 3.2 shows a schematic representation of these common signal carriers which when integrated into an ordinary differential equation model have the potential to make the final amounts of produced transcripts deterministic. However, to achieve this, the PoPS, RiPS, FaPS, SiPS and RNAPS data need to be available, and measurable, for each part. To date, there has been little progress towards measuring the common signal carriers experimentally. However it is clear that finding an experimentally measurable common signal carrier for biological processes opens up new avenues in the absolute measurement, standardisation, datasheet representation, prediction and tuning of gene output. Canton et al.[23] report a method to integrate the PoPS signal within a cell to cell communication device. Another interesting aspect of this work is that it opens up opportunities for parts to communicate through common signal carriers rather than through intermediate gene products.

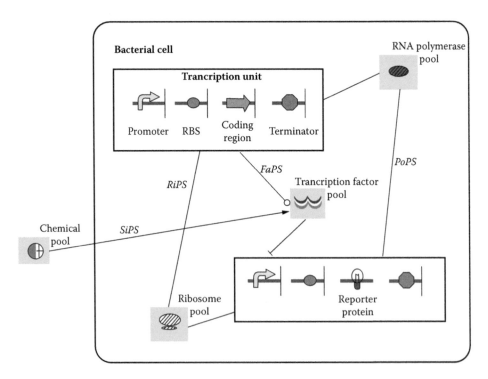

FIGURE 3.2
Common signal carriers comprising a one-step signalling cascade in bacteria. (Reproduced from Marchisio et al., *BMC Syst. Biol.*, 7, 42, 2013. With permission under Creative Commons Attribution License 2.0.)

3.3.4 Towards Standardised Reporter Proteins

A reporter protein is a relatively simple tool used to determine whether, or not, transfected genes have been successfully transcribed and translated and to quantify the degree of transcription and translation activity. The technique works by incorporating a known gene that codes for stably translated protein into the same operon as the gene of interest. This introduced protein is usually detectable in a simple way, typically by fluorescence, luminescence or by a colour change, although in principle any protein can be used if its presence can be confirmed unambiguously. Common examples are fluorescent proteins like the green fluorescent protein (GFP) from the jellyfish *Aequorea Victoria* and EGFP,[67] a modified variant (see Figure 3.3).

Alternatively luminescent systems like luciferase from the firefly *Photinus pyralis* or colour changing reporters like β-galactosidase (β-Gal), which causes an intense blue colour change when incubated with 5-bromo-4-chloro-indolyl-β-D-galactopyranoside (X-gal), can be used. A large number of other reporter gene systems also exist as reviewed.[68] The precise choice of reporter system depends ultimately on application.

The power of reporter genes comes when they are attached to regulatory sequences for other genes. It is in this respect that reporter genes are most used by the synthetic biologist. Often the end product of bacterial biosensor systems is a reporter gene for a fluorescent protein, in effect making the reporter protein the light at the end of the genetic switch. One of the simplest ways to achieve this is to make a fusion protein. In this system, a plasmid comprised of a promoter, then the gene of interest, followed by the reporter gene is constructed and transfected into the host cells. During plasmid construction the stop codon is omitted between the gene of interest and the reporter gene; the translated product of the plasmid is a fusion protein. If the reporter gene is a GFP, for example, then gene expression of the target gene can be inferred *in vivo* when the host cells fluoresce.

An alternative method transfects a gene that codes for a fusion protein plus a peptide tag, which can be used to separate the fusion protein from the total cellular extract. Common

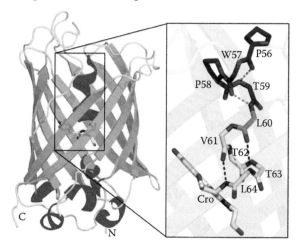

FIGURE 3.3

Structure of EGFP. Ribbon representation of EGFP highlighting beta-strands in green, helices in red and loops in pale green. **Inset**, stick representation of the central helix forming residues in both the 3_{10} conformation (3 residues per turn, 10 atoms in the ring formed by the hydrogen bond), shown in dark red, and in the α conformation (yellow). Residue Leu60 (orange) is involved in hydrogen bonding in both the 3_{10} portion of the helix (grey dashed lines) and the α portion of the helix (black dashed lines). (Adapted from Arpino et al. *PLoS ONE, 7*, e47132, 2012. With permission under a Creative Commons Licence.)

tags are 6-histidine (His) and glutathione-S-transferase (GST) although many others have been developed as reviewed,[69] which are usually fused to the C or N terminus of the protein. Strategies that utilise the peptide tag concept use it as a method to ease the purification of the protein, using a nickel affinity column in the case of the his-tag, a process that is now commonplace. Alternatively, the tag is one that has a specific antibody, which is used to bind it to an antibody affinity resin (immunoaffinity chromatography) for purification as reviewed.[70,71] Quantification of reporter proteins can be performed using spectrophotometric methods; variability between experimental systems can be mitigated for by normalising to an internal control, typically another reporter protein under vector control.

Reporter proteins feature heavily in many synthetic biology applications and are the focus of standardisation discussions, largely because in principle they offer a strategy to detect quantitative gene measurement *in vivo*. However, there are a number of challenges associated with using reporter proteins in synthetic biology. The first stems from the fact that the original BioBricks standard methodology (BBF RFC 10) for joining DNA sequences cannot be used to create translatable fusion proteins. This is due to the fact that in the BioBricks methodology parts are assembled by sequential restriction and ligation steps using Xba I and Spe I restriction sites at the 5' and 3' end of the part. After ligation a scar sequence between the two joined parts is created, which does not contain the complete Xba I and Spe I restriction sites (see Chapter 2 for further discussion of BioBricks).

The newly formed part is flanked by Xba I and Spe I and can be further incorporated into new sequences by successive restriction and ligation steps; with each new step a new scar is introduced. It is this scar region that prevents fusion proteins from being produced because the scar sequence (TACTAGAG) of nucleotides codes for a tyrosine followed by a stop codon.[21] The BglBricks[21] standard assembly protocol addresses this problem with Bgl II and BamH I restriction sites flanking parts. Upon ligation a scar sequence is no longer a barrier to fusion protein expression. The BioBricks fix (BBF RFC 25) is another solution; however, neither fix is perfect since a number of compatibility issues arise with other aspects of the standard assembly protocol. Other problems to overcome when using reporter proteins within synthetic biological systems are the maturation time, that is, the time taken for fluorescent protein to be translated and become fluorescent (in some cases through post-translational modifications), and the comparatively long half-life of reporter proteins; for example, the half-life of wild-type GFP is around 24 hours.[72] This means that dynamic changes in gene expression events are problematic to study. A solution to this problem has been proposed by Anderson et al.[73] who engineered novel GFP variants with C-terminal sequences that rendered the GFP molecule susceptible to cellular proteases, thus reducing the half-life of GFP to between 40 minutes and a few hours.

In development of a standardised set of reporter proteins for synthetic biology that advance upon the relative strengths and weaknesses of existing reporter proteins, the following four suggestions have been made.[74] First, reporter proteins should function within the commonly employed cellular chassis used by synthetic biologists (*E. coli*, *B. subtilis* etc.). Second, reporter proteins should be quantifiable whether weakly or strongly expressed. Third, the dynamic range should be characterised over a range of expression levels. Finally, reporter proteins should be flexible enough to be used in single-cell assays or high-throughput screening methods. Martin et al.[74] demonstrate one way to achieve these standardisation goals for reporter protein by making a Gemini reporter protein that is an active fusion fragment of GFP and β-Gal. Pasotti et al.[75] present some of the obstacles that need to be overcome before standard reporter proteins can be modularised, with the conclusion that promoter activities can vary by up to 22% using different reporters and up to 44% if within a functionally interconnected genetic circuit.

3.4 Adapting Current Techniques as Synthetic Biology Standards

In synthetic biology, the process of measuring gene expression typically begins with choosing the genes that are to be transfected and subsequently measured. The precise methodology for detection will ultimately depend on the intended application and how the gene's expression is regulated. One of the advantages of synthetic biology in formats such as BioBricks is that genetic modification gives the user a choice of gene, promoter, regulatory mechanism and host organism. Robust and reliable performance metrics for these genetic elements could ultimately enable design and materialisation of synthetic cells and systems with wholly predictable and engineerable phenotypes.

Life scientists currently have access to a wealth of methods for gene expression analysis,[76] typically with no urgent burden of expectation to develop or adhere to any quantitative, absolute or agreed standards. Transcription of genes into mRNA is commonly investigated using Northern blot gel electrophoresis, *in situ* hybridisation, an RNase protection assay (RPA) and reverse transcription PCR (RT-PCR). Translated proteins are frequently measured using reporter gene assays, Western blots, *in situ* analysis and enzyme-linked immunosorbent assays (ELISA). DNA–protein interactions can be measured by electrophoretic mobility shift assay (EMSA), DNase I footprinting and chromatin immunoprecipitation (ChIP) assay. Protein–protein interactions are typically investigated using pull-down assays, co-immunoprecipitation and fluorescence resonance energy transfer (FRET). Structural changes to proteins are often measured by X-ray crystallography, NMR, cryoelectron microscopy and atomic force microscopy (AFM). Whilst all of these technologies have potential uses in synthetic biology, development of standards is often some way off. We now discuss a selection of methods for which international standards have been established or are soon to be developed.

3.4.1 Standard Spectrophotometric Measures of Transcription and Translation

Spectrophotometric methods[77] using absorbance or fluorescence are frequently used for the quantification of total amounts of DNA or RNA. Whilst these approaches lack specificity they can provide rapid and accurate quantification of gene expression. The principle of measuring nucleic acid concentration by absorbance stems from the ability of nucleic acids to absorb ultraviolet light. At a wavelength of 260 nm, the average extinction coefficients of dsDNA and ssRNA are 0.020 $(\mu g/ml)^{-1}$ and 0.025 $(\mu g/ml)^{-1}$ respectively. From the Beer–Lambert Law,[78,41] it is straightforward to calculate the concentration of nucleic acid from measurements of optical density. Several adaptations allow the purity of the nucleic acid solution to be estimated. For example, the ratio of absorbance at 260:280 nm for pure DNA is 2 and for pure RNA is 1.8. $A_{260/280}$ ratios that diverge from these indicate sample impurity and a probable contamination by protein.[77,79]

Fluorescence assays rely on the binding of nucleic acids to a fluorescent dye. The first reports utilised ethidium bromide as the fluorophore[80] for quantifying dsDNA. Over time methodology has been optimised to allow ssDNA and ssRNA to be quantified.[81] The basic principle of the method utilises a dye that is fluorescence quenched in aqueous solution and unquenched when bound to the hydrophobic regions of the polynucleic acid. Fluorescence intensity scales with the concentration of nucleic acid in solution. For determination of RNA concentration, quantification is achieved by comparison with a standard curve of fluorescence versus RNA concentration; best results are achieved using the target RNA sequence for the standard curve. Over absorbance, fluorescence offers the advantage that smaller amounts of DNA or RNA can be detected. Due to safety concerns, the popularity of ethidium bromide has waned and a range of commercial kits utilising safer dyes have come into widespread use. SYBR® dyes are one popular alternative[82] which offer greater sensitivity towards RNA; alternative commercial products also exist[83] and have been compared in standard experiments.[84] Modifications to the standard protocol have led to fluorescence assays that can quantify DNA or RNA in the presence of each other[85] and in complex mixtures.[86]

Analysis of protein expression products by spectrophotometric methods is also relatively straightforward. A range of absorbance methods such as the Bradford method[87] or Lowry method[88] have been in use for many years. The principle of both assays is similar although reagents differ. In the Bradford assay, the dye Coomassie Brilliant Blue binds noncovalently to protein molecules, stabilising the blue form of the dye over a red form. This colour difference is measured by a change in absorbance at around 595 nm. Quantification of an unknown concentration of protein is achieved by comparing the absorbance of the unknown with the absorbance of known protein concentrations. In the Lowry method, aromatic protein residues are reduced by Folin–Ciocalteu reagent (a phosphotungstic acid and phosphomolybdic acid mixture) and the resulting change in absorbance at 760 nm is monitored. Complications can arise due to side reactions between non-protein components and assay reagents, the result being artefacts in protein quantification. The Lowry method is particularly prone to these side reactions; however, a number of modified protocols[89] exist which reduce this problem. Fluorimetric methods[90] also exist for protein quantification; typically the fluorescence of a dye like NanoOrange® is enhanced on protein binding and quantification is achieved by reference to a standard concentration curve. A recent review covers the advantages and disadvantages of the many different protein quantification assays.[91] Whilst these fluorometric and absorbance methods are relatively straightforward to perform, their sensitivity is limited by the millilitre volume required by standard spectrometers. A recent advance in the field is the use of fibre optic technology to enable microvolume (1–2 µl) samples to be analysed, often non-destructively.[92,93]

One of the advantages of spectrophotometric methods is that they can be standardised by the calibration of spectrometers to external standards and the use of standard nucleic acid sequences as reference solutions. These considerations are covered within ISO 17025:2005(en) General requirements for the competence of testing and calibration laboratories.[94] Important details for absolute quantification are that the nucleic acid or protein solution is pure and in a spectrophotometrically 'clean' buffer is used. Standardisation methods for spectrophotometrically quantifying viral nucleic acid loads in SI units have been developed[95] and interlaboratory trials of spectrophotometric methods for nucleic acid quantification have been performed. However, the small amounts and mixed nature of biologically derived nucleic acids and proteins means that spectroscopic techniques are not well suited to absolute quantification of these sample types. Another disadvantage of spectrophotometric methods is that they require extraction and purification of expression products and hence provide no detail regarding the subcellular localisation, diversity, quality or quantity of mixtures of gene products from cell extracts. Despite these drawbacks, the low cost and rapid, absolute quantification of nucleic acids or proteins by spectrophotometric methods makes for a popular tool especially when combined with the electrophoresis, hybridisation probe and antibody methods discussed in the next sections.

3.4.2 Standard Electrophoretic Measures of Transcription and Translation

In the simplest of expression systems, perhaps where a single gene is being transcribed, mRNA purity and length can be assessed by agarose gel electrophoresis. The technique commonly utilises a denaturant such as formaldehyde or glyoxal to break down secondary RNA structures; however, non-denaturing methods also exist.[96] A sample of known RNA lengths, the ladder, is run alongside experimental samples to allow estimates of RNA length to be made. If the length of the gene is known then the identity of the transcript RNA can be inferred from its length as estimated against the length of RNA fragments in the ladder. A number of different buffer systems are popular as compared[97] and the basic protocol is discussed.[98] Visualisation of the RNA on the gel is commonly accomplished by the use of the DNA intercalating dyes such as ethidium bromide, although silver staining protocols were once popular.[99] A number of alternatives to ethidium bromide now exist[100] offering greater selectivity to single-stranded polynucleic acids and improved safety features. A variety of protocols exist where the dye (ethidium bromide) is either cast into the gel or added to the RNA samples or completed gels are soaked in a solution of the dye. Typically the gel is imaged in an ultraviolet transilluminator and images are processed based on fluorescence intensity.

In comparison with DNA gel electrophoresis, RNA gel electrophoresis is more problematic; this is due to the almost ubiquitous presence of RNases that digest RNA molecules. The effect of RNase contamination in an RNA gel is a smear of shorter RNA molecules. Sources of RNase contamination can be difficult to identify since total cell extracts are likely to contain significant traces of RNases. Plus RNases naturally present on human skin[101] and hence in airborne dust can be a significant contributor. In extreme cases, this can cause the different RNA bands to smear together, making gene identification difficult and quantification inaccurate. Furthermore, uncertainty over the origin of RNase contamination can cause confusion as to whether the RNA is genuinely degraded at source or degrading by RNAse digestion in the sample preparation/electrophoresis steps. To combat this problem at source, RNase inhibitors are added to RNA containing extracts and all buffers and washing steps are performed with nuclease-free water. Single denaturation/RNase deactivation protocols have also been reported such as the method of Aranda et al.[102] using household bleach.

Quantification of DNA/RNA in electrophoresis gels can be achieved by a fluorescence densitometry trace,[103] calibrated against a series of known concentrations of DNA/RNA. Standard guides for single cell gel electrophoresis (comet assay) of DNA, single-strand damage in eukaryotic cells[104] and detecting irradiated foodstuffs[105] have been developed. One of the main advantages of gel electrophoresis is that it is comparatively cheap to carry out; therefore it can be used, in the interests of cost efficiency, to confirm the quality of RNA samples before they are used in more expensive applications. It is in this respect, as RNA quality determinants, that standard methods for RNA electrophoresis are being developed. Imbeaud et al.[106] report steps towards a standard method using capillary gel electrophoresis for assessing RNA quality, comparing it with spectrophotometric and conventional RNA gel electrophoresis methods.

3.4.3 Standardised Quantitative PCR

The basic premise of quantitative PCR (qPCR) is to use fluorescent molecules within the PCR reaction. After each successive cycle the fluorescence of the sample is measured, which enables the amplification curve to be generated for each individual sample, thus the non-linear calibration effects single end-point determinations can be avoided. There are several different ways of obtaining fluorescence from the PCR mixtures. The simplest is to use an intercalating fluorescence dye, such as ethidium bromide or one of the many similar commercially available products. These intercalating dyes have their fluorescence quenched when solvated by water molecules. The presence of a dsDNA, with its hydrophobic interior, allows the dye to intercalate into the helix where the water that was quenching the fluorescence is now driven off. The result is that fluorescence intensity can be calibrated to dsDNA concentration. The second approach uses a fluorescently labelled hybridisation probe as primer, which is designed to form a hairpin bend and self-quench until it becomes hybridised to ssDNA where the DNA polymerase catalyses the formation of the complementary strand and makes fluorescent dsDNA. The strength of qRT-PCR is that after each successive thermal cycle in the PCR protocol, the fluorescence of each sample is measured. Background fluorescence can be measured prior to thermal cycling; throughout cycling the level of fluorescence rises in the so-called log-phase, until eventually a plateau is reached where each sample has a constant fluorescence. Quantification of initial copy numbers is typically performed in the exponential phase when the fluorescence intensity is around 10 times greater than the background,[107] as shown in Figure 3.4.

The many variations of quantitative RT-PCR (RT-qPCR) have been reviewed.[107] Parallel methods have been developed that enable quantitative RT-PCR to be used to profile multiple-gene expressions.[108] Impressive examples exist where over 2500 primer pairs have been used to simultaneously profile the expression of plant genes.[109] Other high-throughput versions are reviewed by Bookout.[110]

3.4.3.1 RNA Measurement with qPCR

Reverse-transcriptase PCR (RT-PCR) takes an RNA extract and synthesises the reverse transcript or cDNA sequence. This is accomplished using a reverse transcriptase enzyme, many of which are now commercially available. This step is non-specific and as such the

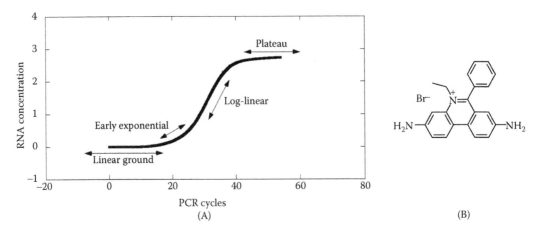

(A)   (B)

FIGURE 3.4

The RT-PCR experiment. (A) The rise in RNA concentration measured from a theoretical RT-PCR experiment. Four phases of the curve are identified; the linear ground phase, where background fluorescence is determined; the early exponential phase, where fluorescence intensity rises above the background; the log-linear phase; and the plateau phase, where the degradation of reagents becomes the limiting factor in amplification. (B) The structure of ethidium bromide, an intercalating dye that fluoresces when bound to the hydrophobic interior of a DNA helix, intercalating dyes or fluorescent hybridisation probes, are most commonly used in quantitative PCR.

cDNA for all of the RNA in the sample is generated. Subsequently, a PCR reaction is carried out on the cDNA that has been generated; usually primers for the gene of interest are included in the sample and, using this methodology, the successful transcription of the gene is inferred. Typically the process is performed in one or two steps. In the single-step process, the reverse transcriptase, PCR primers, RNA, DNA polymerase and buffer reagents are combined. In the two-step process, the RNA, non-specific primers, reverse transcriptase, DNA polymerase and buffer reagents are added. Then in the second step a specific primer to amplify the gene of interest is added and PCR is performed. Full experimental details can be found in reviews.[111] Both the one- and two-step methods have advantages and disadvantages, which in combination with the fragile nature of RNA make the technique powerful but subject to artefacts in the hands of the inexperienced. Several quantitative methods for RT-PCR have been developed such as relative RT-PCR,[112] competitive RT-PCR[113] and comparative RT-PCR.[114,115] The general approach adopted by these methods is to place an internal standard RNA into the RT-PCR mix although there are important differences between methodologies. The response of this internal standard RNA must be well characterised with respect to its amplification in the RT-PCR process. Then, using a calibration curve, the quantity of target transcript can be determined relative to the absolutely quantified standard RNA. To a large extent, these quantitative protocol have been superseded by the availability of fluorescent quantitative RT-PCR (qRT-PCR) protocols.

There are many reviews in the literature that focus on troubleshooting the pitfalls of RT-PCR, which can be magnified when a quantitative step is included,[111] causing artefacts in quantification and data that cannot be reproduced. With the aim of alerting potential new users of RT-PCR to these issues, a few common pitfalls are discussed here. The first step in any RT-PCR protocol is likely to be purification of the RNA. Typically, but not exclusively, this is transcript mRNA. In its native state, mRNA is likely to carry with it, or be found in the vicinity of, its cDNA. Since the RT-PCR procedure initially generates the cDNA for a total RNA extract and then subsequently amplifies this cDNA by PCR, cDNA ineffectively purified out of the initial sample can lead to false identification of gene transcripts or exaggerated expression profiles when quantitative steps are introduced.

Another pitfall arises when attempts are made to correlate the amount of final cDNA between different RT-PCR runs. RT-PCR is a cyclical amplification process, where the products of one cycle are the substrate of the next; thus a PCR reaction will plateau when the formation of products is limited by the availability of substrate. In reality, this plateau is reached more swiftly than predicted (by substrate exhaustion), due to the instability of the PCR substrates (nucleotide bases) and DNA polymerase at the elevated temperature of PCR (95°C). Therefore small changes in the initial conditions of PCR amplification can give rise to very big changes in the amounts of final products observed. Finally, another important consideration is that the rate of amplification is dependent on the DNA sequence, as discussed by Gause and Adamovicz.[112] One of the effects of this differential amplification is that individual primer DNA combinations amplify at different rates, which can give misleading results when comparing between different experiments.

3.4.3.2 qPCR Standards

The standardisation of qPCR and RT-PCR methodologies is one of the areas in gene measurement where steps towards standardisation appear to be advancing most rapidly. The main challenge here is to standardise data from independent biological repeats.[116] Multicentre trials of the analytical accuracy of PCR for the salmonella DNA have been performed[117] and a series of ISO standards for PCR detection in foodstuff have been produced. Examples include ISO 21570:2005 which uses PCR to detect genetically modified foodstuffs[118] and ISO 22118:2011 PCR for the detection of food-borne pathogens.[119] A standardised RT-PCR method for the quantification of hepatitis A and norovirus in food (ISO 15216-1) has also been reported.[120] Many other standardised PCR-related techniques have been proposed in the recent scientific literature and a set of minimal guidelines for the publication of qPCR and RT-qPCR data have been produced.[121]

3.4.4 High-Throughput Systems for Gene Transcript Measurement

The 'Omics' revolution and the emergence of so-called 'Big Data' techniques perhaps hold the most promise for the synthetic biology vision of a fully quantitative biology. Interest in gene expression profiling in the systems biology context plus the success of sequencing projects like the Human Genome Project have led to many high-throughput techniques for gene expression measurement being developed. A snapshot of techniques are briefly highlighted and discussed in Sections 3.4.4.1 and 3.4.4.2 and the interested reader should be aware of a number of excellent reviews of this field.[122–125]

3.4.4.1 Serial Analysis of Gene Expression

Serial analysis of gene expression (SAGE) protocols[126] is used to assign the identity of multiple mRNA transcripts in a single sample. The technique is founded on two principles. First, that eukaryotic RNA is polyadenylated and can be hybridised to a polythimidine hybrid. Second, a sequence of 9–10 nucleotides is enough to uniquely assign the presence of mRNA to the activation of a specific gene. In practice, SAGE starts with an mRNA extract and converts this to cDNA using a reverse transcriptase enzyme (as performed in RT-PCR) before cutting it down to 10 short base sections? In the original method,[127] short cDNAs were synthesised from the mRNA using a biotinylated primer. This cDNA was treated with an endonuclease that leaves a 4 bp sticky end and cuts the cDNA into multiple fragments. The fragment closest to the 3′ end, the biotin labelled end, is bound to streptavidin coated beads and removed by centrifugation. These small cDNA sequences, still bound to the beads, are divided into two sets and ligated with one of two complementary sticky-end linkers containing a PCR primer and the binding site for a tagging enzyme. Crucially the tagging enzyme is type II endonuclease, which typically cuts 20 bp downstream of the binding site. The result is that when excised from the beads, the short cDNA sequences contain an identical number of base pairs from the native mRNA sequence next to an anchoring enzyme site for ligation. These are then ligated together to give a ditag that contains identifiable gene sequences for two genes with a different PCR primer at either end of the sequence. Amplification by PCR is performed and these new ditag sequences are cleaved to remove the primers. Concatenation of the ditags gives a long-DNA sequence that can be cloned and sequenced. Figure 3.5 shows the different stages of the SAGE protocol in more detail.

From the concatemer sequence information, in combination with a database of gene sequences, it is possible to assign the identity of individual gene products that were transcribed. In addition, the number of occurrences of each gene tag in a sequence provides a quantitative measure of transcript abundance. Different combinations of tagging, anchoring, ligating and restriction enzymes have been used depending on the organism being targeted.[126,128,129]

The efficacy of SAGE depends on how efficiently and uniquely gene transcripts can be identified from the 9 to 10 base-pair sequences in the final concatemer. Modifications to

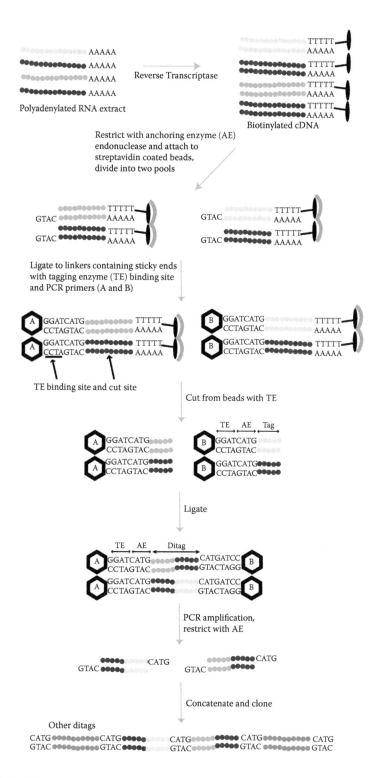

FIGURE 3.5

The stages of the SAGE process. After the final concatemer is made, it is sent for DNA sequencing. Tag sequences are used to assign gene identity and the frequency of tags gives a measure of transcript abundance.

the SAGE technique such as long-SAGE[130] and super-SAGE[131] increase the sample length of gene transcripts and thus improve the accuracy of gene identification. SAGE is most useful when genome wide expression profiles are being targeted and has found a wide number of uses in the biosciences as reviewed.[132]

3.4.4.2 Gene Expression Arrays

One of the most powerful aspects of oligonucleotide probe-based technologies is the opportunity for parallel approaches to gene expression measurement. This is the basis of a large number of array methods such as the cDNA arrays, oligonucleotide arrays and antibody arrays. cDNA libraries are constructed from mRNA sequences. First, the cDNAs are constructed using a reverse transcriptase. An RNase is added to remove the now unwanted mRNA and the cDNA sequences are then cloned into bacterial plasmids. Converting the sscDNA sequences into dsDNA plasmids is accomplished using DNA polymerase. As first developed cDNA synthesis for libraries relied on a free 3'-OH dsDNA to bind effectively and initiate synthesis. cDNAs that readily form hairpin loops to bind the DNA polymerase are hence most easily turned into dsDNA sequences, after restriction at the hairpin by the S1-nuclease. Over time improved methods have been developed[133] for cDNA synthesis that are PCR based[134] or replace the traditional endonuclease approach.[135] cDNA libraries, or arrays, are constructed by depositing the cDNA for different genes onto different regions of a glass slide. This can be performed mechanically with a number of robotic array printers available commercially.[136] Exposure of a total RNA extract to the array allows the complementary RNA to bind to the cDNA and identifies the transcript. Visualisation can be carried out by a number of hybridisation assays. Usually the speed and convenience of fluorescent methodologies makes them the first choice. cDNA arrays can be visualised by densitometry and the fluorescence intensity of individual spots correlated to transcript abundance. Control over array construction and a choice between detection methods is one of the features that make cDNA libraries a powerful and flexible method for gene expression, enabling custom analyses targeted at specific experimental (or organism) models.[136]

One of the challenges in creating cDNA libraries occurs due to disparate levels of abundance of mRNA; in effect if a particular mRNA is expressed infrequently then the corresponding amount of cDNA generated by reverse transcription will be small. Accounting for these differences can make quantification of gene expression problematic unless the level of cDNA is normalised.[137] In synthetic biology, cDNA libraries are particularly useful when tuning the regulation of metabolic networks. The use of antisense oligonucleotides to construct arrays to profile gene expression[138] has also found extensive applications. The standardisation of protocols in cDNA microarray analysis is discussed by Benes and Muckenthaler.[139] One of the most challenging steps in standardising array data is normalisation to a reference gene set; for whole cell expression studies this is accomplished by determining which genes in the sample set are unaffected by the experimental conditions. One of the biggest challenges to standardisation of microarray data is the identification of these genes; a number of strategies exist where a few genes are used as the reference. Alternatively, a global approach makes the assumption that the majority of gene expression profiles do not change and using complex algorithms, normalisation is done to these.[139] A recent study illustrates the difficulty of finding reference genes that are fit for this purpose.[140] Rogers and Cambrosio[141] discuss the development standard microarray methods and the history of the Microarray Gene Expression Data Society (MGED).

3.5 Towards Synthetic Biology Standards for Protein Measurement

Usually in synthetic biology the end focus of the research is application-driven. In most cases, these applications are accomplished by expressed proteins rather than their mRNA transcripts. For example, when Levskaya et al.[142] engineered *E. coli* to see light they made a photoresponsive bacteria using phytochrome normally found in cyanobacteria. This was accomplished by introducing two genes into *E. coli* that code for phycocyanobilin biosynthesis (*ho1* and *pcyA*) in cyanobacteria. When Jay Keasling's group engineered yeast and bacterial[143,144] chassis to produce artemisinic acid, a natural product that is not found in the wild-type organism, they incorporated many different genes into the chassis to make the enzymes and proteins that could carry out the chemical reactions in the modified cell. In both these cases and numerous other synthetic biology examples, success depends on transcription of the introduced genes and their effective translation into functional enzymes. Absolute quantification and standardisation of these protocols presents another level of difficulty due, in part, to the potential for protein–protein, protein–DNA and protein–RNA interactions occurring.

A number of easy ways exist to assess whether translation has occurred, or not; one of the simplest relies on reporter genes and their translated products, reporter proteins. Other methods make use of antibodies to target specific proteins or other intermolecular interactions between proteins and, for example, DNA. The Human Proteome Organisation has produced a set of minimal guidelines for the general reporting of intermolecular interactions and these are broadly relevant to the topic of measuring protein translation.[145]

3.5.1 Western Blotting and *In Situ* Protein Quantification

Western blots are an electrophoresis technique like the Southern and Northern blots; however, Western blots are specific to protein analyses. The name Western maintains the naming convention started by Southern[146] and differences in the physical properties of amino acid derived biopolymers as compared with nucleic acid derived biopolymers necessitate several modifications to the standard gel electrophoresis procedure. In nucleic acid electrophoresis, the net negative charge of the phosphate backbone of polynucleic acids cause molecules to migrate towards the cathode of the electrophoresis cell. In protein gel electrophoresis, however, since the net charge of a protein is dependent on its amino acid sequence and pH, negative charges need to be introduced onto the molecule. This is commonly accomplished using the anionic surfactant sodium dodecyl sulphate (SDS), which denatures protein subunits and binds to hydrophobic patches on the protein molecules. Then using a polyacrylamide gel the protein mixture can be separated electrophoretically by the process known as SDS–PAGE. SDS–PAGE can be both one- and two-dimensional[147] and proteins can be extracted from gels for further analyses as reviewed.[148]

Developing a standardised method for gel electrophoresis requires that molecule mobility is identical under set experimental conditions and therefore reproducible across multiple laboratories. Zakharov et al.[149] assess the reproducibility of macromolecule mobility within both agarose and polyacrylamide gels over a range of concentrations. The motivation for this work is that macromolecule mobility has been linked to a number of physicochemical characteristics such as net surface charge, conformation, and molecular identity. Their results show that reproducibility of mobility, where mobility is defined as migration rate of a band (millimetres travelled per second)/electric field strength (V/cm) is concentration dependent. Typical standard deviations in mobility for agarose gels were

0.2%–3% for globular proteins and 1.4%–5.3% for SDS gels. In polyacrylamide gels, the standard deviations ranged from 11% to 19%. ISO standards for PAGE electrophoresis have been published such as ISO 8981:1993 Wheat – identification of varieties by electrophoresis.[150] The Human Proteome Organisation has published a set of guidelines for reporting the use of gel electrophoresis in proteome research.[151]

Western blots use antibodies to probe blots (nylon membranes) of SDS–PAGE gels and it is the specificity of the antibody–protein interaction that allows proteins to be identified unambiguously. For a review of antibody production, see Schirrmann et al.[152] Antibody mediated detection of proteins can be accomplished by one of two basic methods. These either exclusively use primary antibodies or use both primary and secondary antibodies. In the primary antibody method, reporter proteins such as GFP or β-gal are attached to the primary antibody and contacted with the blot. The occurrence of the target protein and hence gene of interest is then flagged on the gel. In the two antibody system, the primary antibody binds to the target protein and the second antibody binds to the first antibody; in this instance the secondary antibody is tagged with the reporter protein. The technique is reviewed in depth in the literature[153–155] and other protein blotting methods also exist as reviewed.[148] A set of minimal guidelines for reporting antibody experiments have been published.[156] As with the RNA hybridisation probes, protein antibodies for Western blots can also be used *in vivo*. *In situ* protein analysis is the protein–antibody equivalent of RNA *in situ* hybridisation and has the advantage that the spatial distribution of translated protein can be visualised. In essence, the technique, an immunoassay, is similar to the Western blot in that both primary or primary and secondary antibodies can be used. When only primary antibodies are used the technique is called direct immunofluorescence assay and if both primary and secondary antibodies are used the technique is called an indirect immunofluorescence assay as reviewed.[157] Whilst these *in situ* antibody techniques are widespread, they are at best semi-quantitative[158] due to the variability of antibodies diffusing into different regions of samples. An alternative method of *in situ* protein analysis uses fusion proteins, an example being the histone H2B fused to GFP, which is effective for visualising the transcribed chromatin complex.[159] However, the size of the fusion protein can sometimes raise questions about artefacts in the spatial distribution of proteins, especially if the protein tag is much larger than the protein of interest. Minimal guidelines for the publication of *in situ* hybridisation and immunohistochemistry experiments have been published.[160]

Recently, standardised Western blot protocols have been reported; usually these are developed within clinical environments as diagnostic tools and tested across a wide number of patients.[161] Taylor et al.[162] report a defined methodology for reliable quantification of Western blots using chemiluminescent protocols and discuss in depth the challenges to standardisation of the technique. One aspect of the challenge is image manipulation, where it is estimated that 25% of Western blot images in the literature (up to 2006) were inconsistently manipulated.[163]

3.5.2 Enzyme-Linked Immunoabsorbent Assay

These immunoassays quantify the antibody–antigen reaction by combing the sensitivity of spectrophotometric assays with the specificity of the antibody approach. Full experimental details can be found in the literature.[164,165] The important details are summarised here as shown schematically in Figure 3.6.

In ELISA protocols, it is common practice to use a gene fusion protocol such that the protein of interest is fused to a reporter protein like chloramphenicol acetyltransferase (CAT). This has the advantage of standardising many of the stages in the ELISA method.

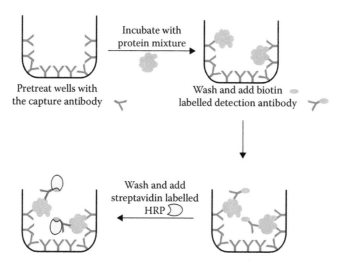

FIGURE 3.6

The simplified ELISA protocol. Once the horseradish peroxidase (HRP) is bound, its substrate is added and HRP activity is measured, typically by absorbance or fluorescence, and the amount of target protein is quantified.

Once CAT tagged proteins have been prepared, they are introduced either as a total cell lysate or a crude protein solution into the wells of a 96-well plate, the internal surface of which is pretreated with antibodies to CAT. This exclusively binds the CAT labelled proteins to the internal surface of the 96-well plate. After rinsing, a set of secondary antibodies that bind to CAT are added; these usually contain a common protein linker such as biotin or digoxigenin, followed by a set of antibodies to digoxigenin but conjugated to a peroxidase enzyme. Finally, a peroxidase substrate is added, which when cleaved by the enzyme results in a coloured product. Using an absorbance detection method and microplate reader, the level of absorbance in the samples can be used to quantify the initial amount of CAT in the sample and hence the amount of target protein is quantifiable. As described, the aforementioned ELISA protocol is a little cumbersome, in particular it is not always convenient or possible to make CAT fused target proteins. One alternative is to use antibodies for the specific target protein to bind it to the 96-well plate; hence there are many adaptations to the ELISA as reviewed.[166–171] Due to their use in clinical applications and disease diagnosis a large number of standardised ELISA protocols have been reported. These stem from observations in the early 1990s that many independent labs had developed in-house ELISA methods, which led to a set of recommended procedures.[172] Two examples are a standardised ELISA test for rabies[173] and a standardised test for allergens as reviewed.[174] ISO standards utilising ELISA protocols have been published, as ISO 11843-5:2008: Methodology in the linear and non-linear calibration cases[175] and ISO 23893-3:2013: Biochemical and physiological measurements on fish – Part 3.[176]

3.6 Conclusions

From the few applications that have been presented it is clear that synthetic biology applications for gene measurement are diverse and many. Work in the field ranges from total genome synthesis[38] and *in vivo* editing through to *in vitro* applications such as cell-free enzyme catalysed synthetic pathways[177] and aggregates of mRNA, DNA and proteins and lipids[50,52,178] as model organelles. Across the range of these examples, many of the techniques of gene measurement usually developed with an emphasis on understanding *in vivo* complexity have been reinvented or reapplied in a new context. The yeast two-hybrid assay[179] is an excellent example of this process of reinvention. For a significant number of years, the yeast two-hybrid assay existed as a diagnostic tool for DNA–protein interactions; however, the engineering perspective of synthetic biologists has re-formalised the assay as a biological AND gate.[180] A significant number of other biological logic gates have now also been designed.[181] Thus it would seem that on paper, at least, molecular biology can be reduced to simple Boolean operators. Cells, however, are a complex network of non-linear molecular interactions. As the complexity of genetic circuits introduced into cells grows, the probability of unwanted protein–protein and DNA–protein interactions also grows and thereby the probability of system failure increases.

The aspects of gene expression measurement that identify protein–protein and DNA–protein interactions are thus likely to become increasingly useful to the synthetic biologist, since their successful application does not only identify, eliminate and tune the novel genetic circuit but also preserves the paradigm of biology digitised. The achievement of quantitative standardised *in vivo* gene measurement is for now a distant goal. It is however easy to see that the design strategies of synthetic biologists utilising datasheets and standardised reporter proteins, PoPS and RiPS and other common signal carriers have the potential to revolutionise standardised gene output. The likely way that this will occur is through the integration of new biological parts (BioBricks or BglBricks) into the current gene measurement techniques. For example, a PoPS counter might be designed to bind to a DNA strand and count the passing polymerases – similar parts might be designed for the other common signal carriers. The Registry of Standard Biological Parts contains a number of BioBrick devices designed to do just this. The potential for PoPS counters to provide standard quantitative information on gene measurement techniques that rely on polymerases such as the yeast two-hybrid assay and PCR is enormous; however, such concepts are also ambitious, and certainly the next decade will be decisive if synthetic biology is to achieve the goal of standardised genetic output.

Acknowledgement

The author wishes to thank Lisa O'Rourke for reviewing several drafts of the chapter and her helpful comments.

References

1. Berman, H.M. 2008. The protein data bank: A historical perspective. *Acta Crystallogr. A.* 64: 88–95.
2. Apweiler, R., Cornish-Bowden, A., Hofmeyr, J.H., et al. 2005. The importance of uniformity in reporting protein-function data. *Trends Biochem. Sci.* 30: 11–2.

3. Gardossi, L., Poulsen, P.B., Ballesteros, A., et al. 2010. Guidelines for reporting of biocatalytic reactions. *Trends Biotechnol.* 28: 171–80.

4. *Bioshare.* [cited 22 September, 2014]. Available from: www.bioshare.eu.

5. Müller, K.M. and Arndt, K.M. 2011. Standardization in synthetic biology. *Methods Mol. Biol.* 813: 23–43.

6. Lyon, H.O. and Horobin, R.W. 2010. Standardization and standards for dyes and stains in used in biology and medicine, in *Special Stains and H & E*, Kumar, G.L. and Kiernan, J.A., Editors, Dako: Carpinteria.

7. Smith, B., Ashburner, M., Rosse, C., et al. 2007. The OBO Foundry: Coordinated evolution of ontologies to support biomedical data integration. *Nat Biotech.* 25: 1251–5.

8. Hucka, M., Finney, A., Sauro, H.M., et al. 2003. The systems biology markup language (SBML): A medium for representation and exchange of biochemical network models. *Bioinformatics.* 19: 524–31.

9. Stromback, L., Hall, D., and Lambrix, P. 2007. A review of standards for data exchange within systems biology. *Proteomics.* 7: 857–67.

10. Sessler, J.L., Lawrence, C.M., and Jayawickramarajah, J. 2007. Molecular recognition via base-pairing. *Chem. Soc. Rev.* 36: 314–25.

11. Wojciechowski, F. and Leumann, C.J. 2011. Alternative DNA base-pairs: From efforts to expand the genetic code to potential material applications. *Chem. Soc. Rev.* 40: 5669–79.

12. Kurreck, J. 2003. Antisense technologies. *Eur. J. Biochem.* 270: 1628–44.

13. Franca, L.T., Carrilho, E., and Kist, T.B. 2002. A review of DNA sequencing techniques. *Q. Rev. Biophys.* 35: 169–200.

14. Pettersson, E., Lundeberg, J., and Ahmadian, A. 2009. Generations of sequencing technologies. *Genomics.* 93: 105–11.

15. Rogers, Y.H. and Venter, J.C. 2005. Genomics: Massively parallel sequencing. *Nature.* 437: 326–7.

16. Xu, M., Fujita, D., and Hanagata, N. 2009. Perspectives and challenges of emerging single-molecule DNA sequencing technologies. *Small.* 5: 2638–49.

17. Allison, L.A. 2007. *Fundamental Molecular Biology.* Oxford: Blackwell Publishing.

18. Endy, D. 2005. Foundations for engineering biology. *Nature.* 438: 449–53.

19. Ho-Shing, O., Lau, K.H., Vernon, W., Eckdahl, T.T., and Campbell, A.M. 2012. Assembly of standardized DNA parts using BioBrick ends in *E. coli*. *Methods Mol. Biol.* 852: 61–76.

20. Smolke, C.D. 2009. Building outside of the box: iGEM and the BioBricks Foundation. *Nat. Biotechnol.* 27: 1099–102.

21. Anderson, J.C., Dueber, J.E., Leguia, M., et al. 2011. BglBricks: A flexible standard for biological part assembly. *J. Biol. Eng.* 4: 1.

22. Arkin, A. 2008. Setting the standard in synthetic biology. *Nat. Biotechnol.* 26: 771–4.

23. Canton, B., Labno, A., and Endy, D. 2008. Refinement and standardization of synthetic biological parts and devices. *Nat. Biotechnol.* 26: 787–93.

24. Galdzicki, M., Rodriguez, C., Chandran, D., Sauro, H.M., and Gennari, J.H. 2011. Standard biological parts knowledgebase. *PLoS ONE.* 6: e17005.

25. DICOM. *Digital Imaging and Communications in Nature.* [cited 23 June, 2014]. Available from: http://dicom.nema.org.

26. Breitling, R., Achcar, F., and Takano, E. 2014. Modeling challenges in the synthetic biology of secondary metabolism. *ACS Synthetic Biology.* 2: 373–8.

27. Kwok, R. 2010. Five Hard Truths for synthetic biology. *Nature.* 463: 288–90.

28. Alterovitz, G., Muso, T., and Ramoni, M.F. 2010. The challenges of informatics in synthetic biology: From biomolecular networks to artificial organisms. *Brief. Bioinform.* 11: 80–95.

29. Clancy, K. and Voigt, C.A. 2010. Programming cells: Towards an automated 'Genetic Compiler'. *Curr. Opin. Biotechnol.* 21: 572–81.

30. MacDonald, J.T., Barnes, C., Kitney, R.I., Freemont, P.S., and Stan, G.B. 2011. Computational design approaches and tools for synthetic biology. *Integr. Biol.* 3: 97–108.

31. Matsuoka, Y., Funahashi, A., Ghosh, S., and Kitano, H. 2014. Modeling and simulation using CellDesigner. *Methods Mol. Biol.* 1164: 121–45.

32. Longabaugh, W.J. 2011. BioTapestry: A tool to visualize the dynamic properties of gene regulatory networks. *Methods Mol. Biol.* 786: 359–94.
33. Chandran, D., Bergmann, F.T., and Sauro, H.M. 2009. TinkerCell: Modular CAD tool for synthetic biology. *J. Biol. Eng.* 3: 19.
34. Kelly, J.R., Rubin, A.J., Davis, J.H., et al. 2009. Measuring the activity of BioBrick promoters using an *in vivo* reference standard. *J. Biol. Eng.* 3: 4.
35. Ang, J., Harris, E., Hussey, B.J., Kil, R., and McMillen, D.R. 2013. Tuning response curves for synthetic biology. *ACS Synthetic Biology.* 2: 547–67.
36. Wang, T., Ma, X., Zhu, H., et al. 2012. Available methods for assembling expression cassettes for synthetic biology. *Appl. Microbiol. Biotechnol.* 93: 1853–63.
37. Ellis, T., Adie, T., and Baldwin, G.S. 2011. DNA assembly for synthetic biology: From parts to pathways and beyond. *Integr. Biol.* 3: 109–18.
38. Gibson, D.G., Benders, G.A., Andrews-Pfannkoch, C., et al. 2008. Complete chemical synthesis, assembly, and cloning of a mycoplasma genitalium genome. *Science.* 319: 1215–20.
39. Wang, H.H., Isaacs, F.J., Carr, P.A., et al. 2009. Programming cells by multiplex genome engineering and accelerated evolution. *Nature.* 460: 894–8.
40. Afroz, T. and Beisel, C.L. 2013. Understanding and exploiting feedback in synthetic biology. *Chem. Eng. Sci.* 103: 79–90.
41. Bastard, J.P., Chambert, S., Ceppa, F., et al. 2002. RNA isolation and purification methods. *Ann. Biol. Clin. (Paris).* 60: 513–23.
42. Charcosset, C. 1998. Review: Purification of proteins by membrane chromatography. *J. Chem. Technol. Biotechnol.* 71: 95–110.
43. Dubin, P.L., Gao, J., and Mattison, K. 1994. Protein purification by selective phase separation with polyelectrolytes. *Separ Purif Rev.* 23: 1–16.
44. Porath, J. 1992. Immobilized metal ion affinity chromatography. *Protein Expr. Purif.* 3: 263–81.
45. Young, C.L., Britton, Z.T., and Robinson, A.S. 2012. Recombinant protein expression and purification: A comprehensive review of affinity tags and microbial applications. *Biotechnol. J.* 7: 620–34.
46. Kulić, I.M. and Schiessel, H. 2008. *Opening and closing DNA: Theories on the Nucleosome, in DNA Interactions with Polymers and Surfactants*, (eds R. Dias and B. Lindman). Hoboken, NJ: John Wiley & Sons, Inc. doi: 10.1002/9780470286364.ch7.
47. Paranjape, S.M., Kamakaka, R.T., and Kadonaga, J.T. 1994. Role of chromatin structure in the regulation of transcription by RNA polymerase II. *Annu. Rev. Biochem.* 63: 265–97.
48. Ainalem, M.-L., Bartles, A., Muck, J., et al. 2014. DNA compaction induced by a cationic polymer or surfactant impact gene expression and DNA degradation. *PLoS ONE.* 9: e92692.
49. Fant, K., Esbjorner, E.K., Lincoln, P., and Norden, B. 2008. DNA condensation by PAMAM dendrimers: Self-assembly characteristics and effect on transcription. *Biochemistry.* 47: 1732–40.
50. Black, C.F., Wilson, R.J., Nylander, T., Dymond, M.K., and Attard, G.S. 2010. Linear dsDNA partitions spontaneously into the inverse hexagonal lyotropic liquid crystalline phases of phospholipids. *J. Am. Chem. Soc.* 132: 9728–32.
51. Corsi, J., Dymond, M.K., Ces, O., et al. 2008. DNA that is dispersed in the liquid crystalline phases of phospholipids is actively transcribed. *Chem. Commun.* 20: 2307–9.
52. Wilson, R.J., Tyas, S.R., Black, C.F., Dymond, M.K., and Attard, G.S. 2010. Partitioning of ssRNA molecules between preformed monolithic HII liquid crystalline phases of lipids and supernatant isotropic phases. *Biomacromolecules.* 11: 3022–7.
53. Dymond, J.S., Richardson, S.M., Coombes, C.E., et al. 2011. Synthetic chromosome arms function in yeast and generate phenotypic diversity by design. *Nature.* 477: 471–6.
54. Bauer, G. and Suess, B. 2006. Engineered riboswitches as novel tools in molecular biology. *J. Biotechnol.* 124: 4–11.
55. Wittmann, A. and Suess, B. 2012. Engineered riboswitches: Expanding researchers' toolbox with synthetic RNA regulators. *FEBS Lett.* 586: 2076–83.
56. Waldminghaus, T., Kortmann, J., Gesing, S., and Narberhaus, F. 2008. Generation of synthetic RNA-based thermosensors. *Biol. Chem.* 389: 1319–26.

57. Deans, T.L., Cantor, C.R., and Collins, J.J. 2007. A tunable genetic switch based on RNAi and repressor proteins for regulating gene expression in mammalian cells. *Cell.* 130: 363–72.

58. Haynes, K.A. and Silver, P.A. 2009. Eukaryotic systems broaden the scope of synthetic biology. *J. Cell Biol.* 187: 589–96.

59. Smolke, C.D. and Keasling, J.D. 2002. Effect of copy number and mRNA processing and stabilization on transcript and protein levels from an engineered dual-gene operon. *Biotechnol. Bioeng.* 78: 412–24.

60. Carrier, T., Jones, K.L., and Keasling, J.D. 1998. mRNA stability and plasmid copy number effects on gene expression from an inducible promoter system. *Biotechnol. Bioeng.* 59: 666–72.

61. Jones, K.L. and Keasling, J.D. 1998. Construction and characterization of F plasmid-based expression vectors. *Biotechnol. Bioeng.* 59: 659–65.

62. Goh, E.-B., Baidoo, E.E.K., Keasling, J.D., and Beller, H.R. 2012. Engineering of bacterial methyl ketone synthesis for biofuels. *Appl. Environ. Microbiol.* 78: 70–80.

63. Linshiz, G., Goldberg, A., Konry, T., and Hillson, N.J. 2012. The fusion of biology, computer science, and engineering: Towards efficient and successful synthetic biology. *Perspect. Biol. Med.* 55: 503–20.

64. Kitney, R. and Freemont, P. 2012. Synthetic biology–the state of play. *FEBS Lett.* 586: 2029–36.

65. Marchisio, M.A. and Stelling, J. 2009. Computational design tools for synthetic biology. *Curr. Opin. Biotechnol.* 20: 479–85.

66. Marchisio, M.A., Colaiacovo, M., Whitehead, E., and Stelling, J. 2013. Modular, rule-based modeling for the design of eukaryotic synthetic gene circuits. *BMC Syst. Biol.* 7: 42.

67. Arpino, J.A.J., Rizkallah, P.J., and Jones, D.D. 2012. Crystal structure of enhanced green fluorescent protein to 1.35 Å resolution reveals alternative conformations for Glu222. *PLoS ONE.* 7: e47132.

68. Ghim, C. M., Lee, S. K., Takayama, S. and Mitchell, R. J. 2010. The art of reporter proteins in science: past, present and future applications. *BMB Rep.* 43: 451–460.

69. Terpe, K. 2003. Overview of tag protein fusions: From molecular and biochemical fundamentals to commercial systems. *Appl. Microbiol. Biotechnol.* 60: 523–33.

70. Fitzgerald, J., Leonard, P., Darcy, E., and O'Kennedy, R. 2011. Immunoaffinity chromatography. *Methods Mol. Biol.* 681: 35–59.

71. Nisnevitch, M. and Firer, M.A. 2001. The solid phase in affinity chromatography: Strategies for antibody attachment. *J. Biochem. Biophys. Methods.* 49: 467–80.

72. Ghim, C.M., Lee, S.K., Takayama, S., and Mitchell, R.J. 2010. The art of reporter proteins in science: Past, present and future applications. *BMB Rep.* 43: 451–60.

73. Andersen, J.B., Sternberg, C., Poulsen, L.K., et al. 1998. New unstable variants of green fluorescent protein for studies of transient gene expression in bacteria. *Appl. Environ. Microbiol.* 64: 2240–6.

74. Martin, L., Che, A., and Endy, D. 2009. Gemini, a bifunctional enzymatic and fluorescent reporter of gene expression. *PLoS ONE.* 4: e7569.

75. Pasotti, L., Politi, N., Zucca, S., Cusella De Angelis, M.G., and Magni, P. 2012. Bottom-up engineering of biological systems through standard bricks: A modularity study on basic parts and devices. *PLoS ONE.* 7: e39407.

76. Roth, C. 2002. Quantifying gene expression. *Curr. Issues Mol. Biol.* 4: 93–100.

77. Gallagher, S.R. and Desjardins, P.R. 2007. Quantitation of DNA and RNA with absorption and fluorescence spectroscopy. *Current Protocols in Human Genetics.* 53:3D:A.3D.1 -A.3D.21.

78. Calloway, D. 1997. Beer-Lambert law. *J. Chem. Educ.* 74: 744.

79. Sambrook, J.J., Russell, D. W. *Molecular Cloning: A Laboratory Manual.* 2001, New York, NY: Cold Spring Harbor Laboratory Press.

80. Bonasera, V., Alberti, S., Sacchetti, A. 2007. Protocol for high-sensitivity/long linear-range spectrofluorimetric DNA quantification using ethidium bromide. *Biotechniques.* 43.

81. Barbas, C.F., Burton, D.R., Scott, J.K., and Silverman, G.J. 2007. Quantitation of DNA and RNA. *Cold Spring Harb. Protoc.* 2007: pdb.ip47.

82. Leggate, J., Allain, R., Isaac, L., and Blais, B. 2006. Microplate fluorescence assay for the quantification of double stranded DNA using SYBR Green I dye. *Biotechnol. Lett.* 28: 1587–94.

83. Jones, L.J., Yue, S.T., Cheung, C.-Y., and Singer, V.L. 1998. RNA quantitation by fluorescence-based solution assay: RiboGreen reagent characterization. *Anal. Biochem.* 265: 368–74.

84. De Mey, M., Lequeux, G., Maertens, J., et al. 2006. Comparison of DNA and RNA quantification methods suitable for parameter estimation in metabolic modeling of microorganisms. *Anal Biochem.* 353: 198–203.

85. Morozkin, E.S., Laktionov, P.P., Rykova, E.Y., and Vlassov, V.V. 2003. Fluorometric quantification of RNA and DNA in solutions containing both nucleic acids. *Anal. Biochem.* 322: 48–50.

86. Dell'Anno, A., Fabiano, M., Duineveld, G.C.A., Kok, A., and Danovaro, R. 1998. Nucleic acid (DNA, RNA) quantification and RNA/DNA ratio determination in marine sediments: Comparison of spectrophotometric, fluorometric, and highperformance liquid chromatography methods and estimation of detrital DNA. *Appl. Environ. Microbiol.* 64: 3238–45.

87. Kruger, N. 2002. *The Bradford method for protein quantitation*, in *The Protein Protocols Handbook*, ed. John Walker. New Jersey: Humana Press. 15–21.

88. Waterborg, J., 1996. *The Lowry Method for Protein Quantitation*, in *The Protein Protocols Handbook*, ed. John walker Humana Press New Jersey. 7–9.

89. Brown, R.E., Jarvis, K.L., and Hyland, K.J. 1989. Protein measurement using bicinchoninic acid: elimination of interfering substances. *Anal. Biochem.* 180: 136–9.

90. Jones, L.J., Haugland, R.P., and Singer, V.L. 2003. Development and characterization of the NanoOrange protein quantitation assay: A fluorescence-based assay of proteins in solution. *Biotechniques.* 34: 850–4.

91. Noble, J.E. and Bailey, M.J. 2009. Quantitation of protein. *Methods Enzymol.* 463: 73–95.

92. Desjardins, R.P. and Conklin, D.S. 2011. Microvolume quantitation of nucleic acids. *Current Protocols in Molecular Biology.* 93:3J:A.3J.1 -A.3J.16.

93. O'Neill, M., McPartlin, J., Arthure, K., Riedel, S., and McMillan, N. 2011. Comparison of the TLDA with the nanodrop and the reference qubit system. *J Phys: Conference Series.* 307: 012047.

94. *ISO/IEC 17025:2005 General requirements for the competence of testing and calibration laboratories.* 2005, Geneva, Switzerland: International Organization for Standardization.

95. Pawlotsky, J.-M., Bouvier-Alias, M., Hezode, C., et al. 2000. Standardization of hepatitis C virus RNA quantification. *Hepatology.* 32: 654–9.

96. Rio, D.C., Ares, M., Jr., Hannon, G.J., and Nilsen, T.W. 2010. Nondenaturing agarose gel electrophoresis of RNA. *Cold Spring Harb. Protoc.* 2010: pdb prot5445.

97. Masek, T., Vopalensky, V., Suchomelova, P., and Pospisek, M. 2005. Denaturing RNA electrophoresis in TAE agarose gels. *Anal. Biochem.* 336: 46–50.

98. Boffey, S.A. 1986. Restriction endonuclease digestion and agarose gel electrophoresis of RNA, in *Experiments in Molecular Biology*, Slater, R.J., Editor, New Jersey, NJ: Humana Press. 121–9.

99. Blum, H., Beier, H., and Gross, H.J. 1987. Improved silver staining of plant proteins, RNA and DNA in polyacrylamide gels. *Electrophoresis.* 8: 93–9.

100. Bourzac, K.M., LaVine, L.J., and Rice, M.S. 2003. Analysis of DAPI and SYBR Green I as alternatives to ethidium bromide for nucleic acid staining in agarose gel electrophoresis. *J. Chem. Edu.* 80: 1292.

101. Probst, J., Brechtel, S., Scheel, B., et al. 2006. Characterization of the ribonuclease activity on the skin surface. *Genet. Vacc. Ther.* 4: 4.

102. Aranda, P.S., LaJoie, D.M., and Jorcyk, C.L. 2012. Bleach gel: A simple agarose gel for analyzing RNA quality. *Electrophoresis.* 33: 366–69.

103. Projan, S.J., Carleton, S., and Novick, R.P. 1983. Determination of plasmid copy number by fluorescence densitometry. *Plasmid.* 9: 182–90.

104. *ASTM E2186-02A(2010) Standard Guide for Determining DNA Single-Strand Damage in Eukaryotic Cells Using the Comet Assay.* 2010, Philadelphia, PA: ASTM International.

105. *BS EN 13784:2002 Foodstuffs. DNA comet assay for the detection of irradiated foodstuffs. Screening Method.* 2002, London: British Standards Institution.

106. Imbeaud, S., Graudens, E., Boulanger, V., et al. 2005. Towards standardization of RNA quality assessment using user-independent classifiers of microcapillary electrophoresis traces. *Nucleic Acids Res.* 33: e56.

107. Wong, M.L. and Medrano, J.F. 2005. Real-time PCR for mRNA quantitation. *Biotechniques.* 39: 75–85.

108. Cassan-Wang, H., Soler, M., Yu, H., et al. 2012. Reference genes for high-throughput quantitative reverse transcription-PCR analysis of gene expression in organs and tissues of eucalyptus grown in various environmental conditions. *Plant Cell Physiol.* 53: 2101–16.

109. Caldana, C., Scheible, W.-R., Mueller-Roeber, B., and Ruzicic, S. 2007. A quantitative RT-PCR platform for high-throughput expression profiling of 2500 rice transcription factors. *Plant Methods.* 3: 7.

110. Bookout, A.L., Cummins, C.L., Mangelsdorf, D.J., Pesola, J.M., and Kramer, M.F. 2001. High-throughput real-time quantitative reverse transcription PCR. *Current Protocols in Molecular Biology.* 73:15.8:15.8.1–15.8.28.

111. Bustin, S.A., Benes, V., Nolan, T., and Pfaffl, M.W. 2005. Quantitative real-time RT-PCR—a perspective. *J. Mol. Endocrinol.* 34: 597–601.

112. Gause, W.C. and Adamovicz, J. 1994. The use of the PCR to quantitate gene expression. *Genome Res.* 3: S123–S35.

113. Tsai, S.J. and Wiltbank, M.C. 1996. Quantification of mRNA using competitive RT-PCR with standard-curve methodology. *Biotechniques.* 21: 862–6.

114. Halford, W.P., Falco, V.C., Gebhardt, B.M., and Carr, D.J. 1999. The inherent quantitative capacity of the reverse transcription-polymerase chain reaction. *Anal. Biochem.* 266: 181–91.

115. Ramakers, C., Ruijter, J.M., Deprez, R.H., and Moorman, A.F. 2003. Assumption-free analysis of quantitative real-time polymerase chain reaction (PCR) data. *Neurosci. Lett.* 339: 62–6.

116. Willems, E., Leyns, L., and Vandesompele, J. 2008. Standardization of real-time PCR gene expression data from independent biological replicates. *Anal. Biochem.* 379: 127–9.

117. Malorny, B., Hoofar, J., Bunge, C., and Helmuth, R. 2003. Multicenter validation of the analytical accuracy of salmonella PCR: Towards an international standard. *Appl. Environ. Microbiol.* 69: 290–6.

118. *ISO 21570: 2005 Method of analysis for the detection of genetically modified organisms and derived products.* 2005, Geneva: ISO standards.

119. *22118:2011 Polymerase Chain Reaction (PCR) for the detection and quantification of food-borne pathogens.* 2011, Geneva: ISO Standards.

120. *ISO 15216-1:2013 Horizontal method for the quantification of hepatitis A virus and norovirus in food using real-time RT-PCR.* 2013, Geneva: ISO Standards.

121. Bustin, S.A., Benes, V., Garson, J.A., et al. 2009. The MIQE Guidelines: Minimum information for publication of quantitative real-time PCR experiments. *Clin. Chem.* 55: 611–22.

122. Bassett, D.E., Jr., Eisen, M.B., and Boguski, M.S. 1999. Gene expression informatics—it's all in your mine. *Nat. Genet.* 21: 51–5.

123. Edgar, R., Domrachev, M., and Lash, A.E. 2002. Gene expression omnibus: NCBI gene expression and hybridization array data repository. *Nucleic Acids Res.* 30: 207–10.

124. Lorkowski, S., Cullen, P.M., ed. *Analysing Gene Expression: Possibilities and Pitfalls – A Handbook of Methods.* Vol. 2. 2003, Weinheim: Wiley-VCH.

125. Moore, J.H. 2007. Bioinformatics. *J. Cell. Physiol.* 213: 365–9.

126. Hu, M. and Polyak, K. 2006. Serial analysis of gene expression. *Nat. Protoc.* 1: 1743–60.

127. Velculescu, V.E., Zhang, L., Vogelstein, B., and Kinzler, K.W. 1995. Serial analysis of gene expression. *Science.* 270: 484–7.

128. Gnatenko, D.V., Dunn, J.J., Schwedes, J., and Bahou, W.F. 2009. Transcript profiling of human platelets using microarray and serial analysis of gene expression (SAGE). *Methods Mol. Biol.* 496: 245–72.

129. Tarasov, K.V., Brugh, S.A., Tarasova, Y.S., and Boheler, K.R. 2007. Serial analysis of gene expression (SAGE): A useful tool to analyze the cardiac transcriptome. *Methods Mol. Biol.* 366: 41–59.

130. Saha, S., Sparks, A.B., Rago, C., et al. 2002. Using the transcriptome to annotate the genome. *Nat. Biotechnol.* 20: 508–12.

131. Matsumura, H., Ito, A., Saitoh, H., et al. 2005. SuperSAGE. *Cell. Microbiol.* 7: 11–8.

132. Yamamoto, M., Wakatsuki, T., Hada, A., and Ryo, A. 2001. Use of serial analysis of gene expression (SAGE) technology. *J. Immunol. Methods.* 250: 45–66.

133. Kimmel, A.R., Berger, S.L., and Shelby L. Berger, A.R.K. 1987. *Preparation of cDNA and the generation of cDNA libraries: Overview*, in *Methods Enzymol.*, Academic Press. 307–16.

134. Belyavsky, A., Vinogradova, T., and Rajewsky, K. 1989. PCR-based cDNA library construction: General cDNA libraries at the level of a few cells. *Nucleic Acids Res.* 17: 2919–32.

135. Gubler, U. and Hoffman, B.J. 1983. A simple and very efficient method for generating cDNA libraries. *Gene.* 25: 263–9.

136. Xiang, C.C. and Chen, Y. 2000. cDNA microarray technology and its applications. *Biotechnol. Adv.* 18: 35–46.

137. Bogdanova, E.A., Shagina, I., Barsova, E.V., et al. 2010. Normalizing cDNA Libraries. *Current Protocols in Molecular Biology.* 90:V:5.12:5.12.1 -5.12.27.

138. Mitchell, L.W. and Hoffman, E.P. 2004. Using oligonucleotide arrays, in *Encyclopedia of Genetics, Genomics, Proteomics and Bioinformatics*, Jorde, L., Little, P., Dunn, M., and Subramaniam, S., Editors, New York, NY: John Wiley & Sons, Ltd.

139. Benes, V. and Muckenthaler, M. 2003. Standardization of protocols in cDNA microarray analysis. *Trends Biochem. Sci.* 28: 244–9.

140. Loven, J., Orlando, D.A., Sigova, A.A., et al. 2012. Revisiting global gene expression analysis. *Cell.* 151: 476–82.

141. Rogers, S. and Cambrosio, A. 2007. Making a new technology work: The standardisation and regulation of microarrays. *Yale J. Biol. Med.* 80: 165–78.

142. Levskaya, A., Chevalier, A.A., Tabor, J.J., et al. 2005. Synthetic biology: Engineering *Escherichia coli* to see light. *Nature.* 438: 441–2.

143. Martin, V.J., Pitera, D.J., Withers, S.T., Newman, J.D., and Keasling, J.D. 2003. Engineering a mevalonate pathway in *Escherichia coli* for production of terpenoids. *Nat. Biotechnol.* 21: 796–802.

144. Ro, D.-K., Paradise, E.M., Ouellet, M., et al. 2006. Production of the antimalarial drug precursor artemisinic acid in engineered yeast. *Nature.* 440: 940–3.

145. Orchard, S., Salwinski, L., Kerrien, S., et al. 2007. The minimum information required for reporting a molecular interaction experiment (MIMIx). *Nat Biotech.* 25: 894–8.

146. Southern, E.M. 1975. Detection of specific sequences among DNA fragments separated by gel electrophoresis. *J. Mol. Biol.* 98: 503–17.

147. Celis, J.E. and Gromov, P. 1999. 2D protein electrophoresis: Can it be perfected? *Curr. Opin. Biotechnol.* 10: 16–21.

148. Kurien, B.T. and Scofield, R.H. 2009. A brief review of other notable protein blotting methods. *Methods Mol. Biol.* 536: 367–84.

149. Zakharov, S.F., Chang, H.-T., and Chrambach, A. 1996. Reproducibility of mobility in gel electrophoresis. *Electrophoresis.* 17: 84–90.

150. Pavlickova, P., Schneider, E.M., and Hug, H. 2004. Advances in recombinant antibody microarrays. *Clin. Chim. Acta.* 343: 17–35.

151. Gibson, F., Anderson, L., Babnigg, G., et al. 2008. Guidelines for reporting the use of gel electrophoresis in proteomics. *Nat Biotech.* 26: 863–4.

152. Schirrmann, T., Al-Halabi, L., Dubel, S., and Hust, M. 2008. Production systems for recombinant antibodies. *Front. Biosci.* 13: 4576–94.

153. Alegria-Schaffer, A., Lodge, A., and Vattem, K. 2009. Performing and optimizing Western blots with an emphasis on chemiluminescent detection. *Methods Enzymol.* 463: 573–99.

154. Crisp, S.J. and Dunn, M.J. 1994. Detection of proteins on protein blots using chemiluminescent systems. *Methods Mol. Biol.* 32: 233–7.

155. Egger, D. and Bienz, K. 1994. Protein (western) blotting. *Mol. Biotechnol.* 1: 289–305.

156. Bourbeillon, J., Orchard, S., Benhar, I., et al. 2010. Minimum information about a protein affinity reagent (MIAPAR). *Nat Biotech.* 28: 650–3.

157. Aoki, V., Sousa, J.X., Jr., Fukumori, L.M., et al. 2010. Direct and indirect immunofluorescence. *An. Bras. Dermatol.* 85: 490–500.
158. McCabe, A., Dolled-Filhart, M., Camp, R.L., and Rimm, D.L. 2005. Automated quantitative analysis (AQUA) of *in situ* protein expression, antibody concentration, and prognosis. *J. Natl. Cancer Inst.* 97: 1808–15.
159. Zink, D., Sadoni, N., and Stelzer, E. 2003. Visualizing chromatin and chromosomes in living cells. *Methods.* 29: 42–50.
160. Deutsch, E.W., Ball, C.A., Berman, J.J., et al. 2008. Minimum information specification for *in situ* hybridization and immunohistochemistry experiments (MISFISHIE). *Nat Biotech.* 26: 305–12.
161. Ribeiro, L.C., Goncalves, C.C., Slater, C.M., Carvalho, S.M., and Puccioni-Sohler, M. 2013. Standardisation of Western blotting to detect HTLV-1 antibodies synthesised in the central nervous system of HAM/TSP patients. *Mem. Inst. Oswaldo Cruz.* 108: 730–4.
162. Taylor, S.C., Berkelman, T., Yadav, G., Hammond, M 2013. A defined methodology for reliable quantification of Western blot data. *Mol. Biotechnol.* 55: 217–26.
163. 2006. A picture worth a thousand words (of explanation). *Nat. Methods.* 3: 237. http://www.nature.com/nmeth/journal/v3/n4/full/nmeth0406-237.html
164. Lai, Y., Feldman, K.L., and Clark, R.S. 2005. Enzyme-linked immunosorbent assays (ELISAs). *Crit. Care Med.* 33: S433–4.
165. Porstmann, T. and Kiessig, S.T. 1992. Enzyme immunoassay techniques. An overview. *J. Immunol. Methods.* 150: 5–21.
166. Butler, J.E. 2000. Enzyme-linked immunosorbent assay. *J. Immunoassay.* 21: 165–209.
167. Hornbeck, P. 2001. Enzyme-linked immunosorbent assays. *Curr. Protoc. Immunol.* Chapter 2: Unit 2 1.
168. Mendoza, L.G., McQuary, P., Mongan, A., et al. 1999. High-throughput microarray-based enzyme-linked immunosorbent assay (ELISA). *Biotechniques.* 27: 778–80, 82–6, 88.
169. Nielsen, U.B. and Geierstanger, B.H. 2004. Multiplexed sandwich assays in microarray format. *J. Immunol. Methods.* 290: 107–20.
170. Paulie, S., Perlmann, H., and Perlmann, P. 2001. *Enzyme-linked immunosorbent assay*, in *eLS*, New york, NY: John Wiley & Sons, Ltd.
171. Wingren, C. and Borrebaeck, C.A. 2006. Antibody microarrays: current status and key technological advances. *OMICS.* 10: 411–27.
172. Wright, P.F., Nilsson, E., Van Rooij, E.M., Lelenta, M., and Jeggo, M.H. 1993. Standardisation and validation of enzyme-linked immunosorbent assay techniques for the detection of antibody in infectious disease diagnosis. *Rev. Sci. Tech.* 12: 435–50.
173. Cliquet, F., Thomas Müller, T., Mutinelli, F., et al. 2003. Standardisation and establishment of a rabies ELISA test in European laboratories for assessing the efficacy of oral fox vaccination campaigns. *Vaccine.* 21: 2986–93.
174. Chapman, M. and Briza, P. 2012. Molecular approaches to allergen standardization. *Curr. Allergy Asthma Rep.* 12: 478–84.
175. *ISO 11843-5: 2008 Capability of Detection—Part 5: Methodology in the linear and non-linear calibration cases.* 2008, Geneva: ISO Standards.
176. *ISO 23893-3: 2013 Water Quality—Biochemical and physiological measurements on fish – Part 3.* 2013, Geneva: ISO Standards.
177. Zhang, Y.H.P., Myung, S., You, C., Zhu, Z., and Rollin, J.A. 2011. Toward low-cost biomanufacturing through *in vitro* synthetic biology: Bottom-up design. *J. Mater. Chem.* 21: 18877–86.
178. Tsaloglou, M.N., Attard, G.S., and Dymond, M.K. 2011. The effect of lipids on the enzymatic activity of 6-phosphofructo-1-kinase from B. stearothermophilus. *Chem. Phys. Lipids.* 164: 713–21.
179. Young, K.H. 1998. Yeast two-hybrid: so many interactions, (in) so little time. *Biol. Reprod.* 58: 302–11.
180. Grünberg, R. and Serrano, L. 2010. Strategies for protein synthetic biology. *Nucleic Acids Res.* 38: 2663–75.
181. Siuti, P., Yazbek, J., and Lu, T.K. 2013. Synthetic circuits integrating logic and memory in living cells. *Nat. Biotechnol.* 31: 448–52.

Section II

Engineering Biology with Legacy Tools

4

Bacterial Cells as Engineered Chassis

Yanika Borg, Alexander Templar, Desmond Schofield and Darren N. Nesbeth

CONTENTS

4.1 Introduction

Since the 1970s *E. coli* has been the standard biological vehicle for propagation of recombinant DNA, transgenic production of proteins and, via expression of recombinant enzymes, biological synthesis of small molecules. The natural utility of *E. coli* has benefitted fundamental research and wider society, via the biotechnology industries, on a scale that is difficult to quantify. Despite this, *E. coli* did not evolve to assist human activity *per se*, beyond its commensal role in the mammalian gut, and most of the exploited properties of *E. coli*, such as rapid growth under laboratory conditions, have been derived from fortuitous evolutionary coincidence.

A truly designed vehicle for recombinant DNA, protein and enzyme production might look considerably different from *E. coli* and would certainly, as a first priority, dispense with all non-essential components that are immunogenic in humans. Because of its pre-eminent position as the system in which most molecular biologists operate, *E. coli* has also been the tool of choice for the majority of synthetic biologists to date (Markson and Elowitz 2014). Well-established cloning techniques for recombinant DNA technology (Voigt 2011), extensive troubleshooting know-how (Cameron et al. 2014), rapid growth (Markson and Elowitz 2014) and cellular robustness (Liu et al. 2013) are just some of the characteristics of *E. coli* that have been exploited by synthetic biologists. Two key synthetic biology aims for *E. coli* are (i) to build new capabilities into the organism as a host chassis, using engineering principles and approaches, and (ii) to exploit *E. coli* as a biomolecular foundry for building new genomes, chassis, genetic codes and genetic circuits for use in bacterial chassis and beyond. In this chapter, we will take a tour through major elements of a typical bacterial cell, starting from the genetic material and working outward, discussing areas of active synthetic biology research or intersection for future research possibilities. Finally this chapter concludes with an examination of attempts to engineer the outermost surface of the bacterial cell.

4.2 Designing and Refactoring Bacterial Genes

In basic research settings a highly effective, but relatively small, list of promoters, selectable markers and replication origins has emerged over time to meet the requirements of researchers carrying out standard molecular biology techniques. While discovery science at the laboratory bench typically requires maximal plasmid copy number and transgene expression, large-scale bioindustrial application often necessitates low plasmid copy number and inducible expression in order to balance cell growth against transgene product yield and quality. Both these aims have been met historically by empirical screening procedures, literature searches, word-of-mouth information exchange within communities and proprietary commercial product offerings.

In contrast to this *ad hoc* approach, a major aim of synthetic biology is to characterise the performance of genetic elements, such as promoters and ribosome binding sites, to a degree of rigour and reproducibility that will ultimately enable predictable gene design to fulfil a user-defined set of performance levels and functions. Standard units and measures of activity are central to this aim and are discussed in Chapter 3 and the excellent review by Arkin (2008). Also key to 'bespoke gene design' is abstraction, an approach common in engineering, where simplified graphical representations of complex underlying elements are used to aid the design process. Symbols used to represent elements such as promoters, open reading frames (ORFs) and transcription terminators (TTs) are an example of design abstraction with which most molecular life scientists will be familiar. In an ideal synthetic biology framework such symbols would represent underlying DNA sequences that are compatible and interchangeable as building blocks and also have accurate and stable performance metrics for transcription, translation, regulation and so on. Abstracted gene design could enable a computational process in which a user-defined wish list of objectives is translated algorithmically into a set of component assembly instructions. Critically, knowledge of the underlying assembly mechanisms would not be necessary for users to define a desired function, in much the same way that knowledge of computer programming or machine language is not necessary in order to use application programmes. A major step towards such an abstraction is the establishment and expansion of catalogued repositories of standardised gene components. Developing software and laboratory ('wetware') tools for generating such repositories is a key area of synthetic biology research, examples including computational tools for tuning codon usage and libraries of promoter and TT sequences, such as the Registry of Standard Biological Parts.

4.2.1 Controlling Transcription

Natural transcription control systems are based on the principle that specialised proteins, typically within heteromeric multi-protein complexes, bind to regions of DNA in a sequence-specific manner to initiate a level of transcription that is up- or downregulated by defined cellular triggers and/or sensor mechanisms. DNA sequences with affinity for these specialist 'transcription factors' proteins are generally termed 'promoter' regions. Successful promoter binding and transcription commencement, or 'initiation', can be influenced by a number of factors; in prokaryotes the presence of a second DNA sequence, often known as an 'operator', can prevent a transcription factor from gaining access to the promoter by recruiting distinct 'repressor' proteins into close proximity with the promoter site. This repression is often reversed by conformational changes to the repressor proteins caused by their binding certain small molecules (Hillen et al. 1983). Natural transcription factor and repressor pairings have been exploited extensively in basic research and also by synthetic biologists (Khalil and Collins 2010; Guazzaroni and Silva-Rocha 2014).

Mutational libraries of transcription factor or repressor variants can also be used as *de facto* dials that researchers can 'turn' to choose a desired strength of gene expression or repression with a high degree of accuracy. Examples include the *E. coli* 'Anderson Promoters' collection in the Registry of Standard Biological Parts (http://parts.igem.org/Catalog), *E. coli* promoters selected for their insensitivity to genome integration locus (Davis et al. 2010) and evolutionary instability (Yang et al. 2012) and a synthetic promoter library for the human commensal bacterium *Lactobacillus plantarum* (Rud et al. 2006).

Synthetic biologists frequently seek to construct novel, non-natural networks of genes either to effect new biological functions or to re-construct the logic gate functions that underlie digital computing. These tasks are distinct from typical biomolecular research in that they can potentially require ever greater numbers of 'orthogonal' transcriptional control elements. As discussed by Rao (2012), a truly orthogonal transcription factor '*x*' will ideally bind a specified promoter '*y*' for 100% of all transcription initiation events it is responsible for, and will have 0% activity for promoter sites '*a*', '*b*' or '*c*' and so on. However in reality, partly due to the fact genomes evolve over time rather than being constructed from a static blueprint, natural transcription factors can often 'cross-talk' with several promoters. The equivalent 'cross talk' can also occur for operators and repressors.

For synthetic gene circuit design, typically 'on' or 'off' are the ideal states of each individual component. Non-orthogonality could increase the likelihood of graduated levels of activity between 'on' or 'off' for a given switch and could therefore compromise overall circuit function. So the hunt is always on for orthogonal elements. One straightforward way to reduce the risk of cross-talk is to use exogenous elements from an organism sufficiently divergent from the host cell: from prokaryotes to yeasts, or from yeasts to mammalian cells.

Zaslaver et al. (2006) conducted a genome wide experimental search of *E. coli* for all DNA sequences, excluding ORFs, with the ability to recruit transcription factors and initiate transcription (Figure 4.1A). All regions between ORFs ('intergenic regions') of over 40 bp in size were identified *in silico* and then materialised *in vitro* using designed oligonucleotide primers for preparative polymerase chain reactions (PCR). Successfully amplified intergenic regions were sub-cloned into plasmids upstream of an ORF encoding a green fluorescent protein (GFP) reporter. These plasmids were transformed into *E. coli cells* and the ability of the intergenic regions to function as promoters was tested by subjecting the cells to different conditions followed by analysis of GFP expression. 1,820 potential promoters were screened and of these 78 were previously uncharacterised. The remainder

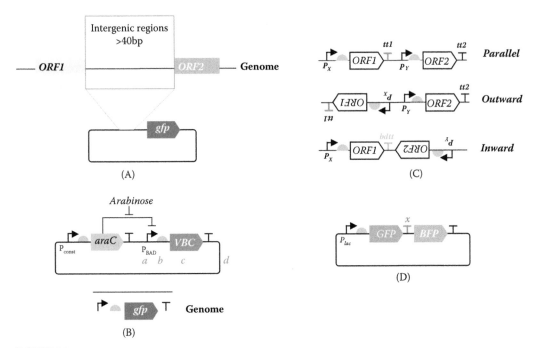

(A)

(B)

(C)

(D)

FIGURE 4.1
Designing synthetic genes. (A) Regions between ORFs greater than 40 bp in size were derived from the entire genome of *E. coli* and placed upstream of a GFP to assess their propensity to act as promoters by Zaslaver et al. (2006). (B) Four factors (grey lettering) were varied by Ceroni et al. (2015) and their effect on expression of plasmid-encoded VBC and chromosomally encoded GFP expression quantified as an indicator of transgenic and global expression capacity: (a) promoter sequence, (b) ribosome binding site sequence, (c) codon usage and (d) plasmid copy number. P_{const} = constitutive promoter, VBC = VioB mCherry, P_{BAD} = arabinose-inducible promoter, araC = inhibitor of P_{BAD}. (C) Inward orientation of expression cassettes enables the use of a single, bidirectional transcription terminator (bdtt) as opposed to the two transcription terminators (tt1, tt2) needed for parallel and outward-oriented expression cassettes. Px, Py = different promoters, grey semi-circle = ribosome binding site, ORF1, ORF2 = different open reading frames. (D) Plasmid designed by Nojima et al. (2005) to test the propensity of a given sequence (indicated as *x*) to function as a transcription terminator.

either had no activity, a previously characterised activity or an activity associated with a housekeeping function such as cell growth. Parallel characterisation of these 1,820 intergenic regions constitutes a *de facto* genome-wide promoter activity sensor in living cells, with significant advantages over array-based *in vitro* techniques for asking the same questions. In addition to constituting a valuable basic research tool, such libraries are valuable to synthetic biologists seeking access to a wider selection of potential promoter 'parts' to characterise in terms of their degree of orthogonality and responsiveness to inputs (Comba et al. 2012).

4.2.2 Controlling Translation

Translation of messenger RNA is another key control point to consider when designing genes, either as part of wider networks or pathways or simply to maximise yield of an economically valuable recombinant protein. There is a preponderance of bioinformatics tools for annotation of potential translation initiation sites within genome sequences, such as TiCO (Tech et al. 2006), and a similar diversity of tools for predicting RNA tertiary structure, from RNAfold (Zuker and Stiegler 1981) to RNAComposer (Popenda et al. 2012). However, there are fewer computational tools for predicting the influence of a given translation initiation sequence on translation rate. Recently developed software applications, the Ribosome Binding Site (RBS) Calculator (Salis et al. 2009, 2011), UTR Designer (Seo et al. 2013) and RBS Designer (Na and Lee 2010), attempt to accurately predict translation rate as a function of mRNA sequence, structure and engagement with ribosomes. Reeve et al. (2014) compare and contrast the three models and discuss the assumptions made by each. All three models have valuable capabilities and UTR Designer enables users to screen different codon usage biases (CUBs) for possible effects on translation rate.

Factors such as codon usage can clearly have a strong influence on translation rates, with aminoacyl-tRNA abundance frequently observed to be a limiting factor. In addition, gene copy number and transcription rates are also strong factors in determining total mRNA at any given point in time. Ceroni et al. (2015) demonstrated an experimental approach to quantifying the relative influence of such factors on cellular capacity to express a reference, reporter protein (Figure 4.1B). A reporter expression cassette featuring a synthetic, strongly constitutive promoter controlling expression of GFP was integrated into the host genome of a number of different *E. coli* strains. Expression of this gene functioned as a readout of the global capacity of the cell at any given time to transcribe DNA, translate RNA and fold polypeptides into proteins.

Plasmid-based expression of an exogenous VioB-mCherry (VBC) fusion protein was then used to monitor output from a variety of synthetic gene constructs with different architectures and CUBs but encoding the same VBC protein with the same amino acid sequence. In this way, synthetic gene performance and cellular capacity could be followed in parallel. Using this system, the authors were able to quantitate the relative influence of different molecular strategies for controlling synthetic gene transcription and translation and also the influence of gene dosage (plasmid copy number) and CUB. Interestingly, RNA-based regulation of transcription demanded less of the host cell expression capacity than control systems consisting of protein-based regulators. The binding strength of mRNA RBS was shown to have a key influence on host cell expression capacity. Rare codon usage only significantly influenced the output of synthetic genes with a strong RBS.

Mathematical modelling could then be used to simulate a larger portion of design space then would be experimentally practical. This approach predicted a weak-RBS, high-dosage, construct would be more efficient than a strong-RBS, medium-dosage construct despite greater requirements with respect to mRNA and DNA synthesis. The *in vivo* and *in silico* cellular capacity monitors of Ceroni et al. (2015) represent a powerful and elegant route by which synthetic constructs and networks can have their function remain the same, while the underlying architecture is reconfigured for optimal functional efficiency.

This process is analogous to the computer science practice of 'refactoring' (discussed further in Section 5.6) in which a given passage of computer code is rewritten to alter its internal structure while external behaviour is unchanged beyond an increase in underlying efficiency. Refactoring in biotechnology and synthetic biology most commonly takes the form of changing the CUB of a given ORF so that translation rates increase, while the amino acid sequence of the protein is unchanged. Codon refactoring, or 'optimisation', can be used to increase efficiency of bioindustrial processes such as microbial biotransformation (Elena et al. 2014) and production of biotherapeutics (Mauro and Chappell 2014). Gould et al. (2014) have performed a thorough review of computational approaches to codon optimisation for the interested reader.

4.2.3 Controlling Transcription Termination

Given the importance of 'off' switches in engineering, it is perhaps surprising that termination of transcription is not a more active area of synthetic biology research. This is likely due to the utility of most natural and synthetic TTs currently available, evidenced by the paucity of reports of significant challenges faced in preventing unwanted 'read through' beyond the stop codon into neighbouring genes. However, there is a well-established bioinformatic effort to computationally predict TT sites in sequence data (Nair et al. 1994) and libraries of TTs have since been developed, such as the WebGeSTer DB (Mitra et al. 2011). Bi-directional TTs (Postle and Good 1985) are particularly useful for conserving DNA use in SGNs by designing expression cassettes in which the direction of transcription is inwardly oriented, as opposed to outward or parallel orientation (Figure 4.1C). A potentially standardisable approach to assessing TTs experimentally was developed by Nojima et al. (2005). Figure 4.1D depicts how they positioned putative TTs between the stop codon of GFP optimised for fluorescence under ultraviolet (UV) light (GFPuv) and the start codon of a blue-shifted GFP variant (BFP). The fluorescence intensity ratio of the GFPuv to BFP can be used to directly infer the termination efficiency of a given sequence or putative TT. Importantly, when combined with degenerate PCR approaches, this setup also allows fine-tuning of a given TT by screening a library for variants with more or less read-through, depending on user requirements.

4.3 Bacterial Synthetic Gene Networks

Synthetic gene networks (SGNs) are perhaps the most characteristic achievement of synthetic biology as they illustrate vividly how analogues of a quintessential engineering technology, the electronic circuit, can be built with biological technology. Achievements in constructing SGNs with classical electronic circuit functions also run counter to the perception of predictable, controllable electronics and intractably complex biology as polar opposites. However, no gene is an island and naturally occurring gene networks have been recognised as important determinants of phenotype since the early days of genetics. Construction and application of SGNs has gained in research momentum since the emergence of synthetic biology. Below, we discuss a selection of SGNs, constructed in prokaryotes that exhibit classical circuit functions either as a proof of principle or through being coupled to useful biotechnological outputs.

E. coli is a commonly used chassis to house SGNs for the development of both basic and complex SGN architectures (Markson and Elowitz 2014). This is due to its rapid doubling time, physical robustness (Liu et al. 2013) and, critically, its position as the core tool for DNA manipulation (Voigt 2011), which brings a huge repository of troubleshooting know-how (Cameron et al. 2014). However, although *E. coli* provides the capabilities for SGN construction, among synthetic biology researchers standardisation of data capture and reporting remains a challenge (Nielsen et al. 2013).

Most *E. coli* SGNs are plasmid-based, which brings some uncertainty with respect to gene dosage per cell. Chromosomally integrated SGNs, such as the Perry and Ninfa oscillator (2012), allow greater confidence in gene dosage measurements. Cell-free SGNs have also been constructed in which RNA is cyclically produced *in vitro* from a DNA template by the action of T7 RNA polymerase, for RNA synthesis, and ribonuclease H, for RNA degradation (Kim and Winfree 2011).

4.3.1 Classical Gene Circuits

Awareness of biological cycles, such as diurnal behaviour patterns, dates back to antiquity. In modern science, the most widely studied cyclic biological system is the circadian rhythm, a dual-feedback, natural regulatory gene network with a 24-hour cycle (or 'period') that is responsible for diurnal cycles in mammals, bacteria and plants (McClung 2006). Circadian rhythms are of particular interest to synthetic biologists due to their ability to produce consistent behaviour despite the high degree of 'noise' from the external environment and the internal homeostatic mechanisms both of multicellular organisms and within cells themselves (Sayut et al. 2007). Oscillation has also been observed of human insulin secretion (Koseska et al. 2011), adenosine triphosphate (ATP) production during glycolysis (Novák and Tyson 2008) and of multiple proteins during mitosis. Dysregulation of natural biological oscillation is associated with health problems such as diabetes (Aziz et al. 2010), epileptic seizures (Danino et al. 2010) and sleep disorders (Weldemichael and Grossberg 2010).

Today, SGNs can be designed and implemented in cells to dictate periodic phenotypes as a result of cycles of expression and non-expression of a given protein or proteins. Theoretical exploration of SGNs started in the 1960s, when a simple oscillator circuit was proposed by Goodwin (1963). The 'Goodwin' oscillator serves to illustrate how a phenotype with relatively complex output and high degree of control can be encoded by a simple configuration of genetic elements. Figure 4.2 illustrates how placing a constitutive promoter sequence incorporating a repressor binding site (an 'operator' site, Op) upstream of an ORF that encodes a self-repressor protein (SR) that binds the Op will result in oscillating SR expression. The Goodwin oscillator architecture has remained relevant for theoretical studies of gene networks, including circadian rhythms (Ruoff et al. 2001; Gonze et al. 2005).

4.3.2 Expansion of SGN Design

After many years of *in silico* investigation of the Goodwin oscillator, *in vivo* construction of SGNs progressed to the development of auto-stabilising circuits (Becskei and Serrano 2000), a bi-stable toggle switch (Gardner et al. 2000) and several non-Goodwin oscillators

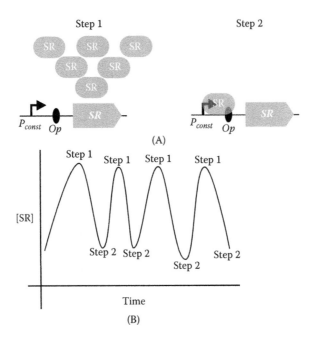

FIGURE 4.2
Minimal genetic oscillator scheme. (A) In Step 1 a constitutive promoter (P_{const}) drives expression of a downstream ORF encoding a self-repressor protein (SR). In Step 2 the SR protein, on reaching a certain level of abundance, will bind to and occupy the operator sequence (Op) positioned downstream of the P_{const}. This binding will prevent further SR expression. After a given time the SR protein will eventually become unstable or be degraded, leaving the Op unbound. Once the Op is unbound transcription will re-initiate at the P_{const} promoter, starting another round of Step 1. (B) Cartoon illustrating how levels of SR could oscillate as a result of the function of the minimal genetic oscillator scheme.

based on control of transcription initiation at promoter sites. Since 2000 the field has rapidly expanded to encompass evolved networks (Yokobayashi et al. 2002; Peisajovich 2012), a mammalian oscillator (Tigges et al. 2009) and calculator networks (Friedland et al. 2009) to name but a few examples. The interested reader should consult excellent discussions by Heinemann and Panke (2006), Weber and Fussenegger (2007), Tigges and Fussenegger (2009), Chen et al. (2012), Khalil and Collins (2010) and Singh (2014) to gain further understanding of the rich variety of SGNs that have been demonstrated to approximate logic gates, time-delay circuits, hysteresis and communication.

A major application of SGN design and function is to shed light on natural biological phenomena. The Elowitz and Leibler (2000) 'repressilator' is based on three genes and can be thought of as a molecular clock and as so has interesting parallels with circadian rhythms that are also influenced by three time-keeping genes (Ahmad et al. 2010). The drive for forward-engineering of synthetic oscillators as a means of gaining understanding of natural systems is echoed by Fung et al. (2005) in their development of the first metabolic oscillator. See Table 4.1 for a list of proof-of-principle synthetic oscillators that have been made and measured *in vivo*.

Although synthetic biology as a field aspires to universal standards, most SGNs to date have not been constructed using standardised parts or measurements. For example, few authors report quantitative dynamics of network behaviour such as the amplitude of oscillation (Danino et al. 2010; Tigges et al. 2010) and even when such metrics are provided no standard unit measure has been agreed, making comparison between SGNs difficult.

TABLE 4.1

Synthetic Oscillatory Gene Network Designs, Characterisation and Mathematical Modelling

	Organism	Percentage Cells with Oscillation	Oscillation Period	Oscillation Amplitude	Oscillation Duration	Oscillation Regulation	Mathematical Model	Notes
1	*E. coli*	40%	160 min ± 40	Not quantified.	c. 10 hrs	Negative feedback loop using LacI, λ and TetR.	Deterministic and stochastic. Parameters estimated.	Oscillatory period triple the duration of cell cycle; Cells remain correlated for 95 ± 10 min; significant cell-to-cell variation observed; proteins oscillate in anti-phase. Elowitz and Liebler 2000.
2	*E. coli*	None reported.	11–20 hrs	None reported.	40–70 hrs	Transcriptional, amplified negative feedback loop using Lac and Ntr.	Deterministic.	Synchronous dampened oscillations. Atkinson et al. 2003.
3	*E. coli*	60%	45 min ± 10	Variable	+4 hrs	Transcriptional, amplified negative feedback loop with metabolism controlled by acetyl phosphate.	Deterministic and stochastic.	The 'metabolator' is a metabolic synthetic oscillator featuring tunable response to carbon source; model sensitive to glycolytic rates and acetate concentration. Fung et al. 2005.
4	*E. coli*	90%	40 min	None reported.	+4 hrs	Transcriptional, amplified negative feedback loop based on Arabinose and Lactose operons, araC positively regulates network and lacI inhibits transcription.	Deterministic and stochastic.	Application of a microfluidic device to control cell environment; robust to arabinose levels; period tunable by IPTG, temperature, doubling time. Stricker et al. 2008.

(Continued)

TABLE 4.1

(*Continued*) Synthetic Oscillatory Gene Network Designs, Characterisation and Mathematical Modelling

	Organism	Percentage Cells with Oscillation	Oscillation Period	Oscillation Amplitude	Oscillation Duration	Oscillation Regulation	Mathematical Model	Notes
5	CHO	None reported.	170 min ± 71	1.8 ± 2 f.u.	None reported.	See Section Chapter 5, Section 5.5.1.	Deterministic. Parameters estimated.	Mammalian synthetic oscillator; tunable via gene dosage & reporter stability; robust to protein and mRNA degradation variations; 20% cell-cell variation of GFP expression; f.u. = fluorescence units. Tigges et al. 2009.
6	CHO	12%	25.8 hrs ± 7.6	34.8 ± 22.5 r.a	+100 hrs	Transcriptional, positive regulation via auto-induction of tTA. Post-transcriptional, negative feedback via siRNA mediated silencing of tTA.	Deterministic. Parameters estimated.	Oscillations have low frequency and are repressed by antibiotics; oscillation robust to changes in gene dosage; r.a. = relative amplitude (%). Tigges et al. 2010.
7	CHO	None reported.	55–90 min	30 ± 54 a.u.	None reported.	Transcriptional, positive feedback loop using quorum sensing.	Deterministic.	First bacterial synthetic oscillator incorporating quorum sensing; use of microfluidic device for characterisation; synchronised oscillations; period is roughly proportional to enzyme decay time; a.u. = arbitrary units. Danino et al. 2010.

Useful abstraction and modularity in SGN design are unlikely to be achieved without standardisation and Canton et al. (2008) proposed a minimum data information sheet format to try and address this issue.

4.3.3 Mathematical Modelling of SGN Behaviour

Mathematics is regarded as an indispensable part of synthetic biology and readers interested in the mathematics most commonly applied to synthetic biology should consult the exploration of the topic provided by Zheng and Sriram (2010). Mathematical modelling can enable rational design of complex SGNs and reveal counter-intuitive conditions in which oscillation should occur. Gramelsberger (2013) and Stricker et al. (2008) both performed computational simulations of their oscillatory SGNs that revealed unexpected sets of parameter conditions predicted, and experimentally confirmed, to favour oscillation. The necessity of mathematical models in synthetic biology is also evidenced by studies that show complex and unexpected dynamics can emerge from even simple SGNs due to non-linear behaviours (Borg et al. 2014). Many such *in silico* studies have been able to establish SGN design algorithms (Chang et al. 2013, Chen and Chen 2010) and to simulate SGN impact on cell division (Gonze 2013), growth rate (Osella and Lagomarsino 2013), nutrient availability, cellular homeostasis (O'Brien et al. 2012) and quorum sensing (Garcia-Ojalvo et al. 2004; Lang et al. 2011; Chen and Hsu 2012). Figure 4.3 briefly sets out some of the basic mathematical considerations that come into play when seeking to model and predict gene expression and repression processes within a simple oscillator.

Synthetic biology aims to manipulate biological systems for which the information available is often very limited. Transcription, translation and protein turnover rates may be unknown, as may optimum cell growth conditions and the compatibility or toxicity of different genetic elements. Filling these gaps in knowledge typically requires collaboration across multiple fields including biologists, chemists, biochemists, mathematicians, computer scientists and engineers (Vinson and Pennisi 2011). Mathematics, in particular, is utilised to develop realistic theoretical models of the SGNs and perform time series simulations, steady state analysis and sensitivity analysis. Computation makes it possible to test an orders of magnitude greater range of different hypotheses and experimental conditions than would ever be possible to perform at the laboratory bench (Chandran et al. 2008). Endler et al. (2009) suggest that iterative rounds of a 'design-test-validate' cycle for SGN refinement will lead to an economically feasible procedure for designing and implementing SGNs to address bioindustrial challenges.

Mathematical modelling tends to fulfil three purposes in synthetic biology: quantitative analysis, qualitative analysis and parameter estimation. Quantitative analysis leads to the development of a deterministic or stochastic mathematical model from which profiles of protein dynamics can be simulated. Qualitative analysis can be used to analyse the innate properties of an SGN such as stability and robustness of its intended behaviour. Parameter estimation is used in the deduction of biochemical rates, which are often experimentally unknown.

Typically in the analysis of SGN behaviour, a quantitative mathematical model is used to describe protein turnover due to feedback signals under different environmental conditions (Chandran et al. 2008; de Jong 2002; Smolen et al. 2000). This reduces the need for time-consuming and costly experiments and provides what Bailey (1998) describes as a *'rigorous, systematic, and quantitative linkage between molecular and microscopic phenomena on one hand and macroscopic process performance on the other'*.

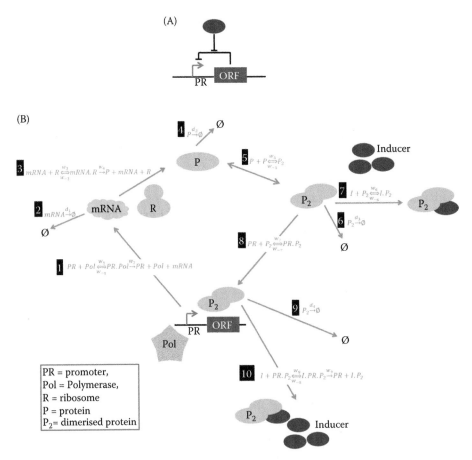

FIGURE 4.3

Mathematical modelling of a simple genetic oscillator. (A) Simplified diagram of a basic circuit in which self-repression is derepressed by an inducer. The promoter (PR) drives expression of the ORF (purple box) and the resultant protein binds to and prevents expression from the promoter (see Figure 4.2). The promoter can be de-repressed by the presence of an inducer (blue oval). (B) Detailed mathematical analysis of factors involved in a simple oscillator. Starting from the purple box in the centre of the diagram, moving clockwise, the polymerase (Pol.) binds to the promoter in a reversible reaction. This drives transcription of the gene, resulting in mRNA (Equation 1). The mRNA can either degrade (Equation 2) or else bind to the ribosome, R, in a reversible reaction to drive translation and the synthesis of protein P (Equation 3). The protein P can either degrade (Equation 4) or else bind to a second protein P in a reversible dimerisation reaction (Equation 5) to form P_2. P_2 can either degrade (Equation 6), bind to an inducer (Equation 7) or bind to the promoter (Equation 8). P_2 has greater affinity for the inducer than the promoter and therefore will bind to the inducer if it is present at sufficient concentration. In the absence of inducer, P_2 will bind to the promoter (Equation 8) and inhibit expression. This inhibition by P_2 can be reversed through natural degradation of P_2 (Equation 9) or any further binding of inducer to P_2 (Equation 10), either of which will result in unbinding of the promoter, allowing the polymerase access again to drive expression. w_1 is the rate of binding of PR to Pol, w_{-1} is the inverse reaction, w_2 is the rate of transcription of the gene, w_3 is the rate of binding of mRNA to R, w_{-3} is the reverse reaction, w_4 is the rate of translation of mRNA, w_5 is the rate of dimerisation of P, w_{-5} is the reverse reaction, w_6 is the rate of binding of P_2 to I, w_{-6} is the reverse reaction, w_7 is the rate of binding of P_2 to PR, w_{-7} is the reverse reaction, w_8 is the rate of binding of PR.P_2 to I, w_{-8} is the reverse reaction, w_9 is the rate of unbinding of PR from I. Degradation rates d_1, d_2, d_3 and d_4 describe, respectively, degration of mRNA, monomeric P, dimeric P_2 and P_2-bound-to-PR.

4.3.4 Deterministic and Stochastic Modelling of SGNs

A model can be either deterministic or stochastic. Deterministic models of SGNs ignore all sources of statistical noise and randomness within and surrounding the gene network, assuming that the initial conditions of an SGN within a cell faithfully dictate, and therefore predict, future states of the SGN. Non-linear, first-order ordinary differential equations (ODEs) are often used to represent SGNs, although alternative representations can be used such as directed graphs (Yongling and Su-Shing 2006) and Bayesian and Boolean networks (Saadatpour and Albert 2013), which make use of logic networks to represent network topology and dynamics. Crucially, deterministic models predict protein concentration levels over time and are reproducible.

Stochastic models, although more biologically relevant, are typically more computationally demanding to solve and analyse. Stochastic modelling is used to represent the intracellular and extracellular noise that in reality SGNs must function within (Zheng and Sriram 2010). Stochastic models represent biological systems more accurately, especially when the SGN is present at low dosage within the host chassis and natural biological fluctuations are felt more strongly than in high gene dosages when the impact of noise is more likely to be averaged out (El Samad et al. 2005). This is known as the 'finite-number effect' and is considered to be the main source of intrinsic stochasticity within genetic systems (Kaern et al. 2003, 2005). Stochastic models consider the probability that a protein (or any other molecule, such as mRNA) will have a given abundance (number of molecules) at a given time, for multiple time points. Stochastic models are unlikely to follow the same trajectory twice even when starting from the same initial starting point (de Jong 2003).

In simulations using stochastic models, the output (taken to be the protein expression dynamics) of each simulation is unique due to the element of stochasticity which shifts the output marginally each time. Within a stable environment, the output of a deterministic model can equal the average output of a stochastic model (Kepler and Elston 2001; Stricker et al. 2008). The outputs of stochastic models tend to be equivalent to the output of their deterministic counterpart but at a theoretical thermodynamic limit or in systems larger in size (Gillespie 1976). This does not imply that stochastic models can generally be bypassed in favour of deterministic ones. In fact, with their stochastic model Elowitz and Leibler (2000) observed variable oscillation amplitudes and a finite autocorrelation time between oscillations, features which were not observed in deterministic simulations. Similarly, Tigges et al. (2009) found that deterministic and stochastic simulations yielded trajectories with the same trends but different values.

For many SGNs, the development of a simple deterministic model is followed by the development of a more complex and more realistic stochastic model. Examples exist for both a bacterial (Purcell et al. 2011) and a mammalian (Tigges et al. 2010) oscillatory SGN in which stochastic models predicted *in vivo* dynamics more faithfully than deterministic models.

Depth and scope of mathematical analysis varies significantly between studies. Elowitz and Leibler (2000), Danino et al. (2010), Tigges et al. (2009) and Fung et al. (2005) used mathematical modelling to carry out parameter space analysis whereas Atkinson et al. (2003) use modelling to scan multiple SGN designs and select those networks predicted to function as oscillators. Mathematical models have also been employed in an iterative cycle of designing and testing SGNs. For example, Stricker et al. (2008) used mathematical analysis to locate a second region in the parameter space where oscillations occur to develop a second oscillatory SGN based on a singular dominant negative feedback loop.

4.3.5 Importance of Biological Noise for SGN Design and Function

Biological phenomena at the cellular scale exhibit variability, or 'noise', in the numbers of the different molecules from which they arise. This is despite the relative rigidity of the genomic sequences that encode these phenomena, both within an individual cell and across a clonal population of cells. As an example of intrinsic noise, two copies of an identical gene within a single cell may be observed to have different expression levels. Identical genes in two genetically identical cells may also exhibit different expression levels, known as extrinsic noise (Strelkowa and Barahona 2010). It is believed that biological noise can have a detrimental effect on SGNs: varying output amplitude (Drubin et al. 2007), increasing stochasticity (Elowitz and Leibler 2000), dampening dynamics (Fung et al. 2005; Atkinson et al. 2003) and even a resulting loss of uniform distribution of genetic material during cell division (Elowitz and Leibler 2000). These biological noise effects are felt more strongly when SGN dosage per cell is low (Yoda et al. 2007; Tigges et al. 2009). Larger numbers of gene and mRNA copies can buffer the effect of noise. Andrianantoandro et al. (2006) suggest extrinsic biological noise has a larger impact as it causes fluctuations that are not synchronous with the SGN within a given cell.

Tigges et al. (2010) propose that maintaining the robustness of SGN-encoded phenotypes in the presence of biological noise is a major challenge to synthetic biology. This is partly because biological noise is an unintuitive phenomenon (Guantes and Poyatos 2006) given the seemingly ordered state of biology as observed in everyday life. Raser and O'Shea (2004; 2005) and Elowitz et al. (2002) caution that biological noise cannot be ignored when it comes to SGN design. Danino et al. (2010) suggest that quorum sensing has the potential to buffer the effects of internal stochasticity and Barkai and Leibler (2000) propose that the effects of noise can be minimised by parameter rate combination choices made at the design and modelling stage of SGN construction.

Despite the earlier observations, the level of biological noise within the 3.7×10^{13} cells of the human body still permits the level of order necessary for the reader to understand the words on this page. As such, the role of biological noise is not well understood, particularly as to whether it has a constructive or destructive role on gene network dynamics. Purcell et al. (2011), Bates et al. (2014) and Borg et al. (2014) showed that different levels of noise could have a varied effect on SGN dynamics, with optimal noise levels leading to stochastic resonance in certain SGN architectures. Bratsun et al. (2005) performed deterministic and stochastic analysis on a time-delayed negative feedback network and showed that while no limit cycles were observed in the deterministic case, the instabilities caused by biological noise resulted in oscillations. Guantes and Poyatos (2006) suggest that the stochasticity of the biological process within a cell can increase survival chances within a constantly changing environment.

4.3.6 Probing the Relationship between SGN Architecture, Biological Noise and SGN Output

The role of biological noise in confounding or supporting natural and SGNs remains to be fully understood. Çağatay et al. (2009) sought to investigate the extent to which the performance of a given gene network can be influenced by the choice of two different network architectures, each predicted to bring about the same function, although with possibly different sensitivities to biological noise. As a test case, they selected a native *Bacillus subtilis* (*B. subtilis*) gene network that controls the onset of 'competence', the ability of cells to take up DNA from the external milieu (Dubnau 1999; Grossman 1995).

When subjected to environmental stress, *B. subtilis* cells can differentiate into a state of competence, allowing cells to uptake extracellular DNA to increase the likelihood of acquiring advantageous traits. Sufficiently high ComK protein levels trigger the cell to adopt a competent state. When ComK protein falls below a given threshold, the cell returns to a non-competent state (Figure 4.4A). ComK directly stimulates its own expression, but this auto-stimulation alone does not result in sufficient intracellular ComK concentration to trigger a competent state. Environmental stress initiates a series of cellular events that lead to induction of the P_{hyp} promoter that drives expression of ComS protein. ComS expression also initiates a series of cellular events that, in this case, lead to increased expression of ComK, ultimately to a level sufficient to trigger cellular transition to a competent state. In addition to auto-activating its own expression, ComK also triggers events that lead to inhibition of the P_{hyp} promoter. The level of ComK abundance required to initiate competence is also sufficient to effect downregulation of ComS and, as a result, eventual downregulation of ComK back to the starting levels where competence is no longer being stimulated.

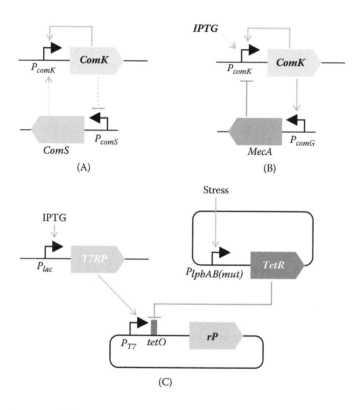

FIGURE 4.4

Exploring SGN architectures. (A) Natural gene network present in *B. subtilis* that controls onset of a phenotype where cells become 'competent' for DNA uptake in response to cell stress. ComK protein stimulates its own expression by inducing its own promoter P_{comK}. Although auto-induced, this level of expression is not sufficient for ComK to stimulate competence. ComK also indirectly represses expression of the protein ComS. When cells experience stress, ComS expression is induced to an extent that overrides ComK-mediated inhibition and indirectly increases ComK expression to a level sufficient to stimulate competence. (B) SGN used in *B. subtilis* to oscillate onset of the competence phenotype. An optimal level of IPTG addition indirectly induces the P_{ComK} promoter, which is also auto-induced by ComK protein. These joint effects can result in sufficient ComK protein concentration to stimulate competence. ComK protein also induces expression of MecA protein by direct induction of the P_{comG} promoter. MecA protein directly represses ComK expression by binding the P_{comK} promoter, bringing cells out of competence. In the absence of ComK expression, MecA expression will itself eventually also decrease, allowing another cycle of ComK expression to begin. (C) An SGN designed to improve performance of *E. coli* host strains in biomanufacturing. Production of a recombinant protein (rP) is driven by the strong P_{T7} promoter which is induced by T7 RNA polymerase (T7RP) controlled by addition of IPTG at a given point in high cell density large scale cultivation. If rP expression is driven to levels that induce cell stress, stress-sensitive promoters such as $P_{lpbAB(mut)}$ will be induced, directing expression of the Tet Repressor protein (TetR) which will bind a tet operator (tetO) sequence positioned downstream of the P_{T7} promoter, reducing rP expression back to sub-stressful levels.

The involvement of external stimuli and multi-step intracellular pathways make this network what Çağatay et al. (2009) describe as a *'noise-driven excitable system'*. Small perturbations in the balance of stress detection and the levels of ComS and ComK can generate transient but large-amplitude responses.

To dissect some of the factors at play in the natural ComK/ComS network, Çağatay et al. (2009) designed an oscillatory SGN (Figure 4.4B) predicted to periodically increase and decrease the abundance of ComK in a manner uncoupled from stress detection. In the native ComK/ComS network, ComK auto-stimulation is a direct process but all other relationships are indirect and transition through multiple steps. Growing *B. subtilis* cells in a nutrient-deficient medium was shown by Sterlini et al. (1969) to cause cell stress so was used by Çağatay et al. (2009) to initiate a natural cycle of ComK production and therefore competence. For the SGN, expression of the ComS gene (not depicted in Figure 4.4B) is uncoupled from cell stress and placed under the control of an isopropyl β-D-1-thiogalactopyran (IPTG)-inducible promoter. This enables ComK to be indirectly induced by IPTG to achieve concentrations sufficient to trigger competence in the absence of a cell stress input. ComK in the SGN also directly induces expression from a P_{comG} promoter placed upstream of an ORF encoding the protein MecA. MecA protein in turn directly inhibits the P_{comK} promoter, downregulating expression of ComK. As such, the SGN directs cycles of ComK concentration sufficient to trigger competence but in a direct manner, largely uncoupled from pathways external to the SGN and hence far less exposed to biological noise.

Mathematical simulation showed that the SGN is able to autonomously achieve ComK pulses that initiate transition to competence and this was confirmed experimentally. In physical experiments, there was, however, a clear difference observed between the natural and synthetic gene networks over one cycle of ComK accumulation. In terms of the median duration of the period for which cells were in a competent state, the native network achieved 15 hours of competence compared with 7 hours for the SGN. The level of stochasticity (coefficient of variation) observed was 0.26 for the SGN and 0.51 for the native network. Stochastic simulations indicated that the source of the variability within the native network was due to low concentration of ComS proteins, causing the effects of fluctuations to be magnified and reflected in the duration of the competence pulse.

The tighter regulation of competence period observed in the SGN also reduced the amount of DNA these cells were able to uptake. Cells with the native network, on the other hand, were able to take up DNA from the external milieu even when very low DNA concentrations were present during the state of competence. These observations illustrate a possible role of biological noise in enhancing a function encoded within a network of genes, a design feature that could perhaps be exploited in future SGN design.

4.3.7 Bioindustrial Application of SGNs: Optimising Recombinant Protein Quality and Yield

In the biopharmaceutical industry, recombinant proteins produced in *E. coli* are often complex, antibody-based therapeutics (Huang et al. 2012) representing multi-million pound revenues. As such, the absolute yield of a recombinant protein from these cells can be a higher priority than the efficiency with which protein quality (native folding) is achieved. Hypothetically, even if 90% of the translated recombinant proteins were to form unusably denatured material (Surinder et al. 2005), an industrial process could still be deemed acceptable if the remaining 10% represents the required mass of product.

E. coli cells have been shown to experience stress (Valdez-Cruz et al. 2010) during over-expression of exogenous proteins and Dragosits et al. (2012) sought to exploit this fact to optimise folding, and therefore the quality, of recombinant protein production. One possible cause, or consequence, of stress in *E. coli* cells overexpressing recombinant proteins is that inclusion bodies (IBs) my form in the cytosol. IBs are typically comprised of misfolded recombinant proteins that have become bound together as insoluble homoaggregates. Dragosits et al. (2012) designed an SGN to couple the onset of cell stress with downregulation of recombinant protein expression, the intention being that recombinant protein expression would only occur when the cellular protein expression and folding machinery are not overloaded – increasing correctly folded recombinant protein as a percentage of total recombinant protein expressed.

Figure 4.4B illustrates the SGN architecture used. The *E. coli* C41 strain was used as a host due to the presence of a chromosomally inserted expression cassette encoding the T7 RNA polymerase downstream of a P_{lac} promoter, inducible by the lactose analogue, IPTG. Two compatible plasmids were then introduced into the cell, one encoding GFP protein, as a proxy for a recombinant biotherapeutic, downstream of a T7 promoter incorporating Tet operator sequence, which could be bound by the Tet Repressor to abolish transcription. High levels of GFP expression result in insoluble, IB-like, aggrgates of GFP. The second plasmid encodes the Tet Repressor under the control of the stress-sensitive P_{IbpAB} promoter.

The authors first used a selection of control SGNs to confirm functionality of the Tet Repressor protein and operator sequence and the increased activity of the stress-sensitive P_{IbpAB} promoter in response to increasing levels of IPTG-induced GFP expression. In the absence of the P_{IbpAB}-TetR plasmid, overall levels of IPTG-induced GFP expression were high, as was the ratio of insoluble, IB-like, GFP to soluble GFP. The presence of the P_{IbpAB}-TetR plasmid increased the ratio of soluble GFP but also reduced the overall level of GFP produced. Mutation of the P_{IbpAB} to weaken its activity, and therefore its sensitivity to stress, did result in higher yield of GFP but also a higher ratio of insoluble GFP.

The authors coupled experiments with the development of a deterministic mathematical model of the SGN based in part on the work of Yildirim and Mackey (2003). A set of differential equations were used to represent the network as three main sets of events: the expression of T7 polymerase via induction, the expression of the GFP via induction by T7 plus inhibition by TetR and third, the expression of TetR as induced by cell stress caused by GFP overexpression. The model incorporates growth rates, the movement of inducer molecules into the cell from the external milieu differentiates between soluble and insoluble GFP uses standard Michaelis–Menten equations to represent transcription reactions and uses least-square methods in order to estimate unknown parameter values. The equations were not simulated via stochastic simulation algorithms but were instead solved using Runge–Kutta methods (Abramowitz and Stegun 1964). Simulated trajectories of GFP quantities over time agreed with observed results. Sensitivity analysis, which served to assess robustness of the system, showed that the model is sensitive to GFP misfolding and TetR production but otherwise robust to parameter and rate perturbations.

4.3.8 Bioindustrial Application of SGNs: Optimising Recombinant Protein Recovery from Host Chassis

Although *E. coli* is long established as an industrial host chassis for recombinant protein production it also poses a long-established challenge: how to efficiently liberate the valuable recombinant protein either from the cell interior (cytoplasm) or the periplasmic space (Chen et al. 2012). One option to release recombinant protein is lysis of the host cell by co-expression of the exogenous, bacteriophage proteins holin and lysozyme. Holin punctures the inner membrane that separates the cytosol from the periplasm, allowing lysozyme to access and degrade the outer cell wall leading to total cell lysis and release of intracellular proteins (Young et al. 2000). Although the holin/lysozyme (HL) method is effective at achieving cell disassembly the challenge is triggering the HL activity at precisely the right juncture within an industrial process that ensures acceptable levels of both cell growth and recombinant protein production. Leaky expression of holin or lysozyme will compromise these two performance metrics. Typically, strategies for triggering expression of the biotherapeutic protein product are cheap and effective, so triggering of HL activity must be achieved by a separate, orthogonal method that is also cheap and tightly controlled.

Pasotti et al. (2011) developed two SGNs for triggering HL activity, defined as 'output ORFs' in Figure 4.5. In one SGN (Figure 4.5A), HL activity is under the control of the P_{Lux} promoter, which is induced by N-3-oxohexanoyl-L-homoserine lactone (HSL). HSL is the small signal molecule component of a quorum-sensing system compatible with *E. coli* (Conway and Greenberg 2002). Although not attempted by Pasotti et al. (2011), this P_{Lux} promoter-based SGN could potentially be combined with a further SGN encoding HSL production in *E. coli* to automate cell lysis. In this way, recombinant therapeutic protein release would be triggered by a P_{Lux} promoter variant optimised for induction only by that HSL concentration characteristic of the cell density and growth duration needed to accumulate the desired mass of recombinant protein. The second SGN (Figure 4.5B) featured the P_R promoter to control HL activity. The P_R promoter is induced by a 30°C–42°C temperature shift, a relatively inexpensive, and certainly non-invasive, triggering mechanism.

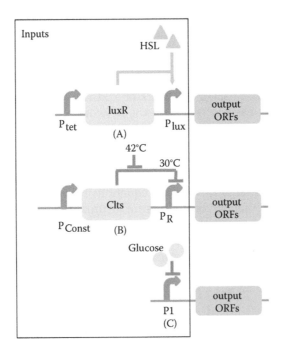

FIGURE 4.5
Input control options within SGNs. (A) When sufficient levels of the chemical HSL accumulate as a result of large numbers of HSL-producing cells, the luxR protein becomes HSL-bound and in this configuration can bind to and induce expression from the P_{lux} promoter. When an inducible promoter, such as P_{tet}, is used to control luxR levels, the sensitivity of the input sensor can be varied. (B) The Clts protein loses the conformation necessary to repress the P_R promoter at 42°C so switching cell growth to this temperature is able to trigger expression of output ORFs without the need for chemical additions to cell growth media. (C) When present in a multicopy plasmid the P1 promoter is inhibited by the presence of glucose.

An alternative SGN for triggering HL activity was proposed by Yun et al. (2007) who used the P1 promoter, which is activated in the absence of glucose when present on a multicopy plasmid, and GFP as proxy for a recombinant biotherapeutic (Figure 4.5C). Thus, as glucose levels in the surrounding growth media decrease, induction of HL activity is upregulated. Yun et al. (2007) discuss the potential use of their SGN to automate cell lysis by following the progress of cell lysis as a function of glucose concentration, which decreases with cell growth.

All three systems (Figure 4.5) showed successful induction of the HL-mediated cell lysis in response to the designed input method, with concomitant release of recombinant protein (GFP or RFP). HSL-induced output first occurred 15 minutes after induction whereas temperature shift induced cell lysis only after a 55-minute lag. For the temperature-shift SGN, the highest percentage of lysed cells occurred when the system was triggered during exponential growth phase, with up to 76% of cells lysed. If the system was triggered at late stationary phase, only 50% of cells were lysed. In the HSL-based system, the concentration of HSL could also be used to tune the released protein output, with a dosage of over 10 nM giving optimal results. Cell lysis by the temperature-shift and HSL systems caused release of 95% of RFP compared with 25% in non-triggered systems.

A peculiarity of the temperature-shift and HSL systems was that, following cell lysis, revival of a sub-population was possible. Analysis showed the presence of mutations within the sub-population which prevent the triggering system from being effective. In the glucose-depletion SGN, 0.5 mg/mL glucose inhibited HL induction at the start of cell cultivation but cell lysis, and concomitant GFP release, proceeded in proportion to the decrease in glucose levels to <0.1 mg/mL over 12 hours.

All three systems are encouraging first steps towards development of industrially robust SGNs. The next generation of SGNs will ideally address the challenge of tightly controlling the HL trigger so that accumulation of sufficient cellular biomass is achieved as rapidly as with conventional approaches. Sufficient recombinant protein accumulation must also be preserved and the SGN must retain function during large scale, high cell density cultivation in bioreactors.

4.3.9 Expanding Design Space for SGNs

The networks described earlier serve to introduce some of the basic themes and recurrent challenges in SGN design but are just a small sample from a rapidly increasing repertoire of SGNs being developed *in silico* and *in vivo* by the synthetic biology research community, including SGNs of multiple functions (Purcell et al. 2011). Boyle and Silver (2012) and Stephanopoulos (2012) provide a thorough discussion of how SGNs can be applied to challenges in metabolic engineering while Silva-Rocha and de Lorenzo (2008), Densmore and Anderson (2009) and Seaone and Sole (2013) provide excellent explorations of how SGN design can be mapped over conventional logic gates. In addition to transcriptional regulation, Bonnet et al. (2013) and Siuti et al. (2013) have developed systems in which concentric recombination target sites are used to rearrange SGN components into different configurations, the outputs of which can be interpreted in the context of logic gates. A particularly exciting advance in SGN design is the effort to build 'classifier' circuits (Tan et al. 2007; Didovyk et al. 2014) which in some examples (Macia and Sole 2014) are designed to be distributed throughout multiple cell types to achieve computation.

4.4 Genome-Scale Re-Engineering of Bacterial Cells

The post-genomic era is well underway and beginning to benefit from developments in the fields of 'big data' (advanced analytics of large data sets) and the industrial internet – both of which were in their relative infancy when the draft human genome sequence was first made available in 2000. In the last decade, the number of whole genomes that have been sequenced has risen rapidly as the cost of genome-scale sequencing projects has steadily decreased (Carlson 2011). Over this time, the many tools and techniques used to investigate the organisation and function of naturally occurring genomes, particularly large DNA fragment assembly, have been put to use by synthetic biologists seeking to build genomes from the bottom up (Esvelt and Wang 2013), for instance using chemically synthesised DNA as starting material. Synthetic biologists are attempting to design and build novel genomes while the interplay of factors that dictate the organisation, size, evolution and functions of naturally occurring genomes are still not fully understood. The inevitable failures and setbacks in this process will no doubt improve our knowledge of genome-scale function, both natural and synthetic.

4.4.1 A Genome of Synthetic Provenance Can Control Living, Replicating Cells

In a landmark proof of concept experiment, Gibson et al. (2010) designed and assembled an entire prokaryotic genome 'off line', both *in vitro* and *in vivo* using *E. coli* and yeast as DNA assembly tool organisms (ATOs), before using the assembled genome to transform and ultimately control cells originating from *Mycoplasma capricolum* (*M. capricolum*). The synthetic genome was a rewritten version of the comparatively small (Klasson and Andersson 2010), ≈1 Mb, natural genome of the bacterium *Mycoplasma mycoides* (*M. mycoides*). It was possible to 're-write' the genome in this way because the initial, pre-assembled version was composed entirely of designed oligonucleotides (Figure 4.6) in solution, synthesised by the company Blue Heron Biotechnology, Inc.

These oligonucleotides were first annealed *in vitro* to assemble ≈1 kb fragments, each of which was designed to have a degree of overlap with its consecutive partner to encompass the entire genomic sequence (Figure 4.6A). Batches of 10 of these ≈1 kb fragments were used to transform yeast to generate ≈10 kb plasmids by recombination *in vivo* (Figure 4.6B). The plasmid backbones used to achieve this featured elements that enabled its replication as circular DNA in both yeast and *E. coli* cells. These ≈10 kb plasmids were purified from yeast and used to transform *E. coli* to aid confirmation of their size and identity. One hundred nine ≈10 kb plasmids were generated in total this way and confirmed to be correct. Groups of 10 of these plasmids were then used to transform yeast for *in vivo* recombination to generate ≈100 kb plasmids, each again incorporating a degree of overlap with its upstream and downstream partners (Figure 4.6C). Ten of these ≈100 kb plasmids were purified from yeast, pooled *in vitro* and used to transform yeast cells one final time. This final yeast transformation yielded a clone, designated sMmYCp235, confirmed to house the ≈1 Mb plasmid encoding the entire, re-written genome of *M. mycoides* (Figure 4.6D).

Lartigue et al. (2009) had previously shown that the natural *M. mycoides* genome could be modified to feature tetracycline resistance and β-galactosidase genes plus genetic elements necessary for replication in yeast cells. Once modified in this way, the *M. mycoides* genome could be purified from the host organism and used to transform yeast cells in which it was then stably maintained. The entire circular *M. mycoides* genome could then be

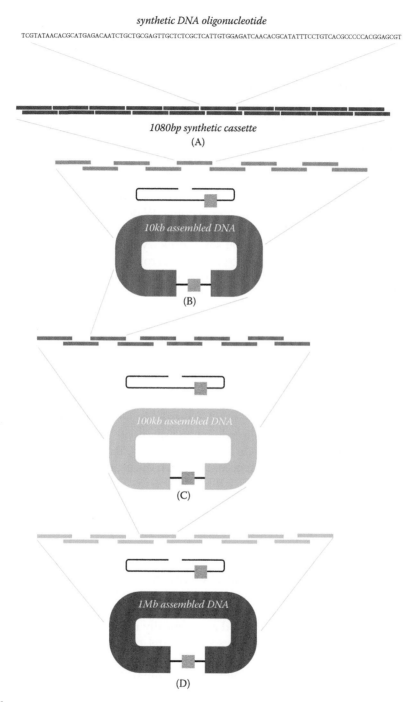

FIGURE 4.6

Assembly of a genome-scale circular DNA molecule. (A) Synthetic oligonucleotides are annealed to form ≈1 kb fragments. (B) Groups of ten ≈1 kb fragments and a plasmid backbone are used to co-transform yeast cells in which they recombine to form a 10 kb region of the synthetic genome. (C) Groups of ten 10 kb fragments and a plasmid backbone are used to co-transform yeast cells in which they recombine to form a 100 kb region of the synthetic genome. (D) Groups of ten 100 kb fragments and a plasmid backbone are used to co-transform yeast cells in which they recombine to form a complete 1 Mb synthetic genome.

purified from yeast cells and used to transform *M. capricolum*. Growing the transformed bacterial cells in tetracycline lead to emergence of β-galactosidase-positive (indicated by blue colonies in the presence of X-gal substrate), tetracycline-resistant colonies harbouring the *M. mycoides* genome and in which the original *M. capricolum* genome was undetectable by sequencing analysis. Karas et al. (2013) also demonstrated transfer of bacterial genomes of up to 1.8 Mb in size to yeast cells where they persisted as circular DNA.

The genome transfer method developed by Lartigue et al. (2009) using modified *M. mycoides* genomes was successfully applied to the synthetic, sMmYCp235 version of the *M. mycoides* genome. The circular, 1 Mb synthetic sMmYCp235 genome was purified from yeast cells and used to transform *M. capricolum* cells. Post-transformation, a β-galactosidase-positive, tetracycline resistant clone was isolated, designated 'JCVI-syn1.0' and also known informally as 'Synthia'. Synthia cells grow and replicate normally and contain none of their original *M. capricolum* genome. Wider ethical discussion of the significance of Synthia is examined in Chapter 1. In the context of bench research, Synthia is an important proof of the principle that small prokaryotic genomes of wholly synthetic origin constitute sufficient information to control a surrounding small prokaryotic cellular structure and coordinate its sustained growth, viability and replication.

4.4.2 DNA Editing at Genomic Scale

The proof of principle experiment by Gibson et al. (2010) involved a final *en bloc* transfer of the synthetic genome to the intended cellular environment. However, the modular steps in assembly of the 1-Mb DNA molecule were performed 'off line': *in vitro* and in ATOs. Re-writing larger, more complex genomes is likely to require that, during the assembly process, sub-genomic segments of a synthetic genome are also tested 'live' within their intended recipient cellular environment in order to detect deleterious or lethal errors. This 'testing' could be of *ab initio* designed genome segments, perhaps based on naturally occurring clusters of genes that encode a given function (Fischbach and Voigt 2010), or a high-throughput, parallel screening process by which the best tolerated and best performing sequences are identified (Cobb et al. 2012).

Lajoie et al. (2013) applied such a screening approach in their use of a three-stage strategy to ultimately construct a new and fully functional version of the 4.6 Mb *E. coli* genome in which (a) all natural UAG stop codons have been replaced with UAA stop codons and (b) the UAG codon has been repurposed to encode a non-standard amino acid (NSAA). The three stages of the process were: (i) precise editing of modular segments of the genome in order to substitute all UAG codons with UAA, (ii) testing of strains which contained these modified genome segments for growth and viability and (iii) stepwise, hierarchical assembly of these modified genome segments into an entirely modified whole genome. The final strain was called a 'genetically recoded' organism (GRO) and its gradual construction involved multiple research teams, referred to collectively here as the 'GRO group'.

A novel technique, known as multiplexed, automated genome engineering (MAGE), was used for the first two steps of genome editing and phenotype testing. MAGE builds on an approach developed by Ellis et al. (2009) who showed that oligonucleotides of 30–70 bp in length can be used to efficiently define edits of host DNA via their association with the lagging strand of the DNA replication fork during cell division (Figure 4.7A) and the action of recombinant λ bacteriophage protein Beta. The procedure is sufficiently well tolerated by host cells that multiple sites in a genome can be targeted simultaneously with no significantly deleterious effects beyond those encoded

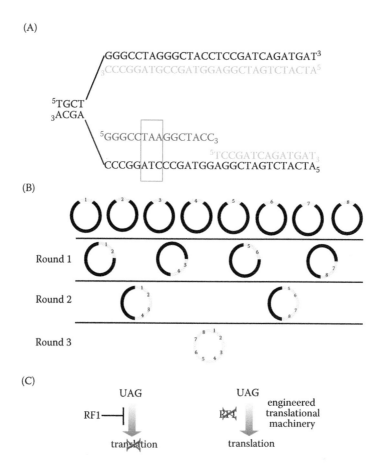

FIGURE 4.7

Recoding the *E. coli* genome. (A) Central to multiplexed, automated genome engineering (MAGE) is the observation that, during transient periods of expression of the λ bacteriophage Beta protein, synthetic oligonucleotides of 30–70 bp in length can be used to efficiently make point mutations across a genome by sequence-specific associating with the lagging strand (lower, green sequence) during DNA replication. *Nota bene*, for graphical brevity the synthetic oligonucleotide depicted (red sequence) is shorter than the 30–70 bp that would in reality be needed for Beta-mediated mutagenesis. In the example, the oligonucleotide would effect a TAG-to-TAA substitution (blue box). (B) The CAGE method enables stepwise accumulation of a larger and larger proportion of MAGE-edited genome by controlled conjugation of selected donor and recipient cells. In the example shown, eight strains, each of which have accumulated 575 kb of MAGE-edited genome, undergo three rounds of conjugation until a 4.6 Mb, entirely MAGE-edited genome is assembled. (C) Leftmost diagram illustrates how the RF1 machinery recognises UAG codons as a signal to stop translation. Right diagram: when UAG codons have been removed from the entire genome it is safe to delete the RF1 machinery. This also enables UAG codons to be used in transgenes alongside engineered translation machinery enabling translation of UAG codons into NSAAs.

by the inserted edit. The GRO group first used MAGE to improve performance of an industrially valuable metabolic pathway in *E. coli* (Wang et al. 2009) by parallel editing of 24 genomic targets. For genome-scale recoding, they identified 312 UAG stop codons in the *E. coli* genome and then used MAGE to convert them to UAA in batches of 10 or less across 32 different strains (Isaacs et al. 2011). Each strain was assessed in terms of maintenance of wild type growth phenotype and to confirm whether the codon substitutions had been made or not. Across the 32 strains a total of 255 mutations were also detected in addition to the UAG codon substitutions.

4.4.3 DNA Assembly at Genomic Scale

Assembly of the 32 regions of re-coded genome into one single, entirely recoded genome involved a further novel technique, termed conjugative assembly genome engineering (CAGE). An exquisitely clear and straightforward description of the fine detail of the CAGE procedure is provided in the excellent method paper by Ma et al. (2014). Briefly, CAGE is predicated on a streamlined version of conventional *E. coli* conjugation tools (Smith 1991) in which donor cells are modified to harbour either an F plasmid or, within their genome, recombinant genetic elements for high frequency recombination (Hfr). The F plasmid and the hfr elements both effect transfer of genetic material from a defined donor to a defined recipient cell type.

In CAGE, all the genes necessary for conjugation are placed within a plasmid pRK4, which also encodes an origin of transfer (oriT) sequence. The pRK4 plasmid is transformed into 'donor' cells that contain MAGE-edited genome regions. The genome of donor cells is also modified to include an oriT element adjacent to the MAGE-edited region intended for transfer to a recipient. MAGE-edited genome regions in donor cells are also flanked with inserted selectable markers to confirm their successful transfer to recipient cells. Induction of the pRK4-encoded conjugative machinery results in complete or near-complete duplicative transfer of all genetic material containing an oriT signal to recipient cells via mechanisms that have yet to be fully characterised. By this process a complete copy of the pRK4 plasmid is transferred from donor to recipient cells. Critically, up to 50% of the donor genome, starting with the region closest to the oriT, is also transferred. Once present in the recipient cells, the transferred portion of MAGE-edited donor genome replaces the homologous region of the recipient genome by recombination. Figure 4.7B provides an overview of how hierarchical rounds of CAGE can be used to combine the MAGE-edited genome segments in a stepwise fashion, until an entirely MAGE-edited genome is assembled.

Successful genome-wide replacement of UAG stop codons with UAA stop codons had no major viability cost to host cells (Lajoie et al. 2013). Against this UAG-free background, the host cell release factor 1 (RF1) machinery, required for recognition of UAG as a stop codon, could also be knocked out with no deleterious effects. This was achieved by knockout of the prfA gene. The RF1-minus GRO was then further modified by introduction of a plasmid encoding an evolved amino acyl tRNA synthetase (aaRS) plus its cognate substrates: a unique tRNA and NSAA (p-acetylphenylalanine) pair. These additional factors should enable UAG codons to be translated into the NSAA within a protein sequence, instead of functioning as a stop signal (Figure 4.7C). A modified GFP ORF was introduced to test this, featuring three UAG codons within its coding sequence. The NSAA-GFP was successfully expressed in its complete form, demonstrating incorporation of p-acetylphenylalanine residues encoded by the repurposed UAG codon.

Furthermore, Lajoie et al. (2013) showed that the RF1-minus GRO *E. coli* strain was relatively resistant to infection by the T7 bacteriophage compared with the unmodified MG1655 strain. Ten percent of T7 bacteriophage genes feature UAG stop codons, including Gene 6 which is critical to the biological activity of the virus. UAG codons in the invading T7 genome would not be recognised by the RF1-negative translation machinery of the GRO strain. It is therefore likely that this refactored gene expression machinery of the host cell is responsible for the bacteriophage resistance.

4.4.4 Genome-Scale Perspectives

In addition to physical assembly of genomes, computational genome design is also a very active area of synthetic biology and fundamental genomics research. Carrera et al. (2012)

used computational simulation to demonstrate that a refactored sub-genomic region could retain important elements of its behaviour while undergoing a 73% reduction in the degree to which its constituent genes are transcriptionally regulated. Chan et al. (2005) modelled refactoring of the genome of bacteriophage T7 before testing *in vivo* the function of a virus in which ≈25% of the genome sequence was refactored. In addition to genome refactoring, genome reduction also holds the promise of a more efficient 'industrial cell' controlled by only that minimum quantity of genetic material required to encode a given list of functions (Gao et al. 2010; Juhas et al. 2012). A cell designed to this level of stringency could represent a considerable 'de-risking' of biomanufacturing due to its resistance to pathogens and consistency of performance.

4.5 Engineering Periplasmic Space

Typically Gram negative bacteria such as *E. coli* are formed of two concentric membrane boundaries: the inner and outer membranes, which define the innermost, cytoplasmic chamber and a space up to 70 nm thick (Sochacki et al. 2011) between the inner and outer membranes known as the periplasm. A major function of the periplasm is to form a chamber in which host proteins that engage the external milieu are correctly folded, monitored and, when no longer functional, degraded. The periplasm is also the first cellular compartment encountered by incoming material from outside the cell.

Oxidising conditions within the periplasmic space favour disulphide bridge formation between the cysteine residues of periplasm-resident proteins. Conversely the oxidising environment of the cytosol inhibits disulphide bridge formation. The periplasm has been thoroughly exploited by biotechnologists for *E. coli*-hosted expression of recombinant biotherapeutics, particularly antibody-based proteins for which disulphide bonding is functionally critical. To boost recombinant protein quantity, quality and ease of purification at industrial scale, numerous genetic modifications have been used to remodel the periplasm and the inner and outer membranes that form its boundaries. Many conventional biotechnological modifications of the periplasm are predicated only on the peculiarities of a specific recombinant protein of interest, on a protein-by-protein basis. Synthetic biology platforms would ideally involve generic remodelling of the periplasm in ways that improve bioprocessing across a broad spectrum of recombinant protein types.

4.5.1 Periplasm as Secretion Conduit

The principle motivation for biotechnological exploitation of the periplasm is production of biotherapeutics whose correct folding and pattern of disulphide bond formation is favoured by the oxidising environment within the periplasmic space (Skerra and Pluckthun 1991). Many recombinant proteins also auto-aggregate into structures known as IBs when expressed in the *E. coli* cytoplasm. This IB formation can occur for a number of reasons, including exposure of hydrophobic patches that might otherwise be buried during translation in the native host cell during protein folding and/or glycosylation. Redirecting recombinant protein expression to the periplasmic space is one route to addressing the IB formation that would otherwise occur in the cytoplasm (Boström et al. 2005). In addition to correct folding, the periplasm also represents a desirable subcellular location in terms of recombinant protein recovery from the host cell.

Achieving isolation of recombinant protein in a manner that is as rapid and simple as possible is a key consideration at the laboratory bench and, to an order of magnitude greater degree, at industrial scales for production of biotherapeutics. There are three principle means by which recombinant proteins can be liberated from *E. coli* host cells: biological, chemical and mechanical. All these routes tend to be more efficient when the recombinant proteins are expressed at the cell periphery, in the periplasmic space, rather than within the cell interior, in the cytoplasm.

One chemical approach (Pierce et al. 1997) relies on osmotic disruption of only the outer membrane of cells, releasing the contents of the periplasm to the external milieu. This leaves behind a 'spheroplast', which is bounded only by the inner membrane and contains all the cytosolic proteins and small molecule solutes and all macromolecular DNA. Preservation of intact spheroplasts is critical as it enables their efficient removal by centrifugation at both bench and industrial scales. Another chemical route is to increase the rate at which food substrate for energy and growth is supplied to cells within bioreactors. This has been shown to stimulate release of recombinant proteins from the periplasm (Bäcklund et al. 2008), by as yet uncharacterised mechanisms. By contrast, total cell disruption by mechanical means greatly increases the complexity and, due to the presence of genomic DNA (Balasundaram et al. 2009), the viscosity of the material from which recombinant protein must be purified. Addressing such a complex mixture is more costly in terms of recombinant protein purification compared with a dilute solution consisting of only periplasm contents and outer membrane fragments. A proposed biological solution, of triggering outer membrane disruption by concerted action of recombinant holin and lysozyme, is discussed in Section 4.3 of this chapter.

4.5.2 Periplasmic and Trans-Periplasmic Routing of Recombinant Proteins

Figure 4.8 details biological routes for directing recombinant protein expression to the periplasmic space or for secretion to the cell exterior. In broad terms, recombinant proteins directed for periplasmic expression by the SEC route (du Plessis et al. 2011) are translocated in an unfolded state into the periplasm in which they subsequently fold. Recombinant proteins directed by the SRP route (Bornemann et al. 2008) are co-translationally translocated into the periplasm as nascent polypeptides by a process involving inner-membrane associated ribosomes. By contrast, the TAT pathway (Cline 2015) is capable of transporting fully folded proteins from the cytosol into the periplasm. Branston et al. (2012) showed that overexpression of the host protein components of the TAT pathway greatly increased periplasmic levels of a recombinant protein. Most biological steps taken to transport recombinant proteins to the periplasm subsequently employ a chemical method to effect release from the periplasm or for total cell disruption. Biological routes to recombinant protein release tend to involve downregulation of outer membrane proteins, such as lpp (Ni et al. 2007). Downregulation of lpp has been shown to effect secretion of 90% of recombinant periplasmic proteins while leaving behind intact spheroplasts (Shin and Chen 2008).

Further pathways exist by which recombinant proteins can be transported directly from the cytoplasm to the cell exterior, effectively bypassing the periplasmic space. The haemolysin (Hly) system occurs naturally in *E. coli* for secretion of the host protein HlyA (Figure 4.8D). A number of recombinant proteins have been modified for secretion by the Hly system. Ni and Chen (2009) provide a detailed discussion of the Hly route for the interested reader and also consider several further biological routes for possible secretion of recombinant proteins from *E. coli*, as well as certain protein sequence motifs that direct secretion by unknown mechanisms.

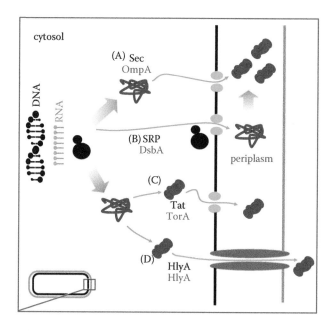

FIGURE 4.8
Periplasmic and trans-periplasmic routing of recombinant proteins. Scheme of trafficking routes within *E. coli* cells, starting with DNA transcription then translation of RNA by ribosomes (joined small and large black circles) and translocation through pore complexes (green ovals). (A) Recombinant proteins (red lines = unfolded, red blob = folded) can be modified with an N-terminal signal peptide, typically from the host protein, OmpA, to direct their translocation to the periplasmic space via the SEC route. In this route translation takes place in the cytosol prior to translocation to the periplasm where correct folding can occur. (B) Recombinant proteins can be modified with an N-terminal signal peptide from the host protein, DsbA, to direct their translocation to the periplasmic space via the SRP route. In this route translation of nascent polypeptides takes place simultaneously with their translocation to the periplasm where correct folding can then occur (grey arrow). (C) Recombinant proteins modified with an N-terminal signal peptide from the host protein, TorA, can achieve a fully folded state in the cytosol and then be translocated to the periplasm by action of the TAT translocation route. (D) Recombinant proteins modified with motifs from the host protein, HlyA, can achieve a fully folded state in the cytosol then be translocated directly to the cell exterior by action of the Hly route.

4.5.3 Periplasm Engineering for Improved Biomanufacturing

At industrial scale, a process of high-pressure homogenisation is typically used to release recombinant therapeutic proteins expressed in the *E. coli* cytosol or periplasm. Typically this disruption results in significant release of the 4.6 Mb *E. coli* genome into the 'process stream', causing a level of viscosity that reduces the efficiency with which the material is handled at large scales. Balasundaram et al. (2009) sought to address the presence of high molecular weight genomic DNA in cell homogenates by engineering the periplasm contents (Figure 4.9). They designed a recombinant, broad-spectrum nuclease targeted to the periplasm by the SEC pathway. Cytosolic expression of the nuclease would be lethal due to the recombinant nuclease acting on the host genome as substrate. As such, periplasmic localisation of the nuclease is essential to preserve cell viability. The periplasmic nuclease, although active, was non-toxic when expressed on its own and when co-expressed alongside a periplasmic Fab' fragment biotherapeutic (Figure 4.9). Upon homogenisation the liberated nuclease successfully degraded the released genomic DNA and reduced the viscosity of the process stream.

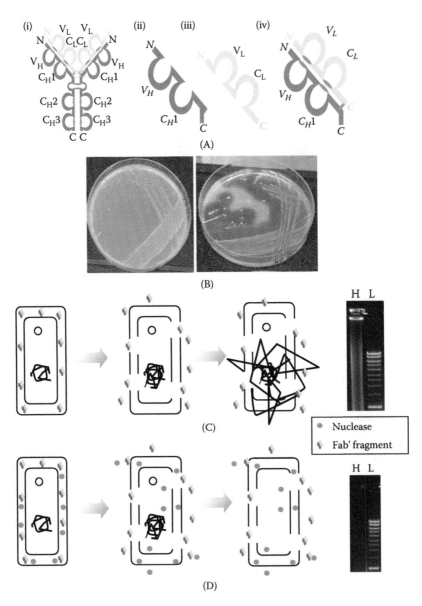

FIGURE 4.9
Periplasm engineering for improved biomanufacturing. (A) (i) A typical whole antibody structure with two light chains (light grey) and two heavy chains (dark grey) and disulphide bonds (flat, grey ovals). Constant domains of the heavy chain are denoted C_H1, C_H2 and C_H3, variable regions as V_H. In the light chain there are also constant (C_L) and variable (V_L) domains. (ii) The heavy chain region of a Fab' fragment. (iii) The light chain region of a Fab' fragment. (iv) An assembled Fab' fragment, requiring four intra- and one inter-molecular disulphide bonds. (B) Nuclease activity detected on DNAse agar plates. Left plate is wild type strain and the right plate is a strain expressing a periplasmic nuclease. Cloudy regions are precipitated DNA. (C) Illustration of cell disruption by homogenisation for release of periplasmic Fab' causing parallel release of high molecular weight genomic DNA. Rightmost image is of homogenate (H lane) run on agarose gel with a DNA ladder (L lane) and stained with ethidium to visualise DNA present. (D) Cell disruption with parallel release of high molecular weight genomic DNA and periplasmic nuclease, resulting in DNA hydrolysis. Ethidium staining of agarose gel reveals absence of DNA in homogenates (H) with DNA ladder (L) as control.

In the context of mass production of recombinant proteins, the periplasm can be thought of as a nano-scale biological foundry in which proteins are ideally folded and stored at high concentrations. A number of studies have shown that cell growth conditions strongly influence the performance of the periplasm in this regard, and several groups have used computational statistical approaches to optimise growth conditions with periplasm performance as the objective (Boström et al. 2005; Azaman et al. 2010). Co-overexpression of chaperones in the periplasm can also boost the capacity of the compartment to accommodate and fold recombinant proteins, a recent example being the work of O'Reilly et al. (2014) for periplasmic production of a recombinant β-scorpion toxin.

4.5.4 Engineered Periplasm as Gene Moat

In the 2012 document 'Principles for the Oversight of Synthetic Biology' over 100 organisations, including Friends of the Earth (Hoffman et al. 2012), discussed concerns regarding the environmental release of organisms designed by synthetic biologists. One possibility raised was of post-mortem gene transfer from a released synthetic biology organism to naturally occurring organisms. This scenario was highlighted due to the fact that many of the proposed biological routes to 'gene containment', 'firewalls' or 'quarantine' rely on a living, metabolically active cell for their enactment (Mandell et al. 2015). For example, cell death due to the absence of an essential, non-natural nutrient may still result in contamination of the natural environment with dead cells, complete with their engineered genomes. Rigorous studies into these factors are of course essential for accurate assessment of the risks and solutions going forward.

The 2012 iGEM team, Plastic Republic, proposed re-purposing engineered periplasmic nuclease activity to act as a 'gene moat' within a putative bacterial synthetic biology chassis that may be released into the environment at some point in mankind's future for purposes of bioremediation or biogeoengineering. Several groups have shown that periplasmic nuclease expression is well tolerated by didermal bacteria (Takahara et al. 1985; Boynton et al. 1999; Cooke et al. 2003; Song and Zhang 2008; Nesbeth et al. 2012). A classic indicator of periplasmic nuclease activity is halo formation around colonies on DNA-impregnated agar plates as shown in Figure 4.9. This assay is predicated on the level of cell death that occurs within agar plate colonies after overnight growth, resulting in release of cell contents. In theory, the gene moat (BioBrick™ part number BBa_K729019) would safely surround host cell DNA in living cells and prevent post-mortem horizontal gene transfer by releasing a genetically cataclysmic wave of nucleic acid hydrolysis upon cell death.

In indicative experiments (personal communications and data reported on the BBa_K729019 BioBrick registry entry), plasmid-bearing 'gene moat' cells and plasmid-bearing unmodified cells were first disrupted and then the resultant cell debris was used to attempt plasmid transformation of competent 'target' cells optimised for plasmid uptake. Any plasmids that survived the cell disruption procedure should be able to transform the healthy 'target' cells. Debris from unmodified cells contain sufficient intact plasmid to successfully transform the target competent cells. However, debris from the gene moat cells was completely unable to transform competent 'target' cells. This indicates the post-mortem wave of nuclease activity in the 'gene moat' cells successfully cleared plasmid DNA from the cell debris. However, further research would be needed to validate this approach as a bona fide route to genetic containment for environmentally released synthetic biology organisms.

4.6 Re-Imagining the Bacterial Surface and Glycocalyx

The proteins and lipopolysaccharides that form the outer membrane of *E. coli* mediate all interactions with the external environment, including movement (Zhong et al. 2014), attack, defence (Bonsor et al. 2007), nutrient uptake, communication, waste egress and intake of both advantageous and deleterious genetic material. Biotechnological strategies to direct secretion of recombinant proteins from *E. coli* cells often involve downregulating expression of genes required for the structural integrity of the outer membrane. Surface-bound effectors also mediate many of the pathogenic effects of virulent *E. coli* strains in the mammalian gut and as a result are often highly immunogenic. Removal of immunogenic lipopolysaccharides and endotoxins from industrial bioprocess streams is an absolute regulatory requirement due to the potential for their presence to cause life-threatening immune reactions in patients. Therefore, when appraising *E. coli* cell as an engineerable chassis, remodelling the outer surface is clearly an important consideration for bringing about effector functions.

4.6.1 Cell Surface for Biocatalyst Immobilisation

Figure 4.10 describes the two compartments and membranes of most Gram negative, didermal bacteria and some possible configurations of membrane-bound proteins. Surface display of membrane-bound proteins requires transcription and translation in the relatively reducing environment of the inner, cytosolic compartment. Proteins targeted for surface display must then be translocated from the cytosol cross the inner membrane into the periplasmic space, an oxidising environment where they will be inserted into the outer membrane.

As hosts for collections of soluble enzymes that catalyse steps in a metabolic pathway, *E. coli* cells are often considered as factories that house multiple interconnected production lines. However, when recombinant enzymes are embedded in membranes it can also be useful to regard *E. coli* cells as nano-scale catalytic reactors with architectures that can be designed to maximise factors such as catalytic surface area, substrate feeding and recycling, product removal and catalyst regeneration. In studies with the mammalian enzyme adrenal ferredoxin, Jose et al. (2006) estimated that approaching 200,000 copies of a membrane-anchored version of the enzyme could be displayed at the *E. coli* surface for biocatalysis. Hwang et al. (2011) expressed a fusion protein consisting of a recombinant transaminase fused to the transmembrane domain of the cotG outer membrane protein and demonstrated activity when expressed on the outer membrane of *Bacillus subtilis* cells. Henriques et al. (2013) adopted the opposite approach of expressing putative phosphorylation substrates on *E. coli* cell surfaces as part of a screen to characterise kinases and small molecule kinase inhibitors. Saffar et al. (2007) also applied biological functionalization of the bacterial outer surface for bioremediation of cadmium and nickel from water by heavy metal adsorption to surface expressed hexahistidine motifs.

4.6.2 Designing Vaccines by Bacterial Cell Surface Modification

Disease states can become established when the immune system fails to recognise antigens associated with cancerous cells or pathogens as aberrant, and instead tolerates them as 'self' antigens. A longstanding therapeutic challenge has been to stimulate the immune system by re-presenting these antigens in a more immunogenic context, in order to elicit an immune response. This can be achieved by co-expressing a tolerated antigen juxtaposed

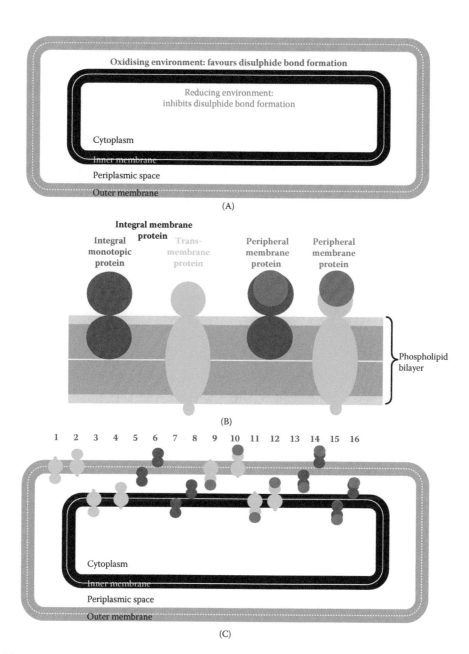

FIGURE 4.10
E. coli cell as nano-scale catalyst. (A) The two chambers that comprise an *E. coli* cell. (B) Four configuration options for design of a membrane-anchored biocatalyst. For the transmembrane protein, two domains are soluble and a middle domain is membrane-embedded. Of the soluble domains, the larger green circle indicates the domain where the catalytic function would reside. (C) Sixteen options for subcellular location of a membrane-anchored biocatalyst.

with highly immunogenic proteins. Ingram et al. (2009) demonstrated this strategy by expressing immunogenic or immunomodulatory proteins on the surface of an otherwise unrecognised cancer cell. Another approach is to genetically fuse a tolerated antigen to an immunogenic protein.

The human protein endoglin is specifically associated with angiogenesis within solid tumours but is often not recognised as an aberrant antigen by the immune system. Huang et al. (2014) exploited an immunogenic *E. coli* transmembrane protein, Ag43, which is abundant at the *E. coli* cell surface to 'de-tolerise' endoglin. Their strategy relies on the observation that Ag43 can be fused to other protein domains at its amino surface and still be correctly folded and trafficked to the *E. coli* outer membrane (Figure 4.11A). They constructed an Ag43 variant that incorporates the extracellular domain of mouse endoglin. Upon expression the endoglin portion of the fusion protein is successfully displayed at the cell surface and is recognised by anti-endoglin antibodies in Western blot experiments. Heating to 60°C was sufficient to cause the host *E. coli* cells to shed the Ag43-endoglin fusion protein. The fusion protein was subsequently used to immunise mice and successfully raise an immune response against native endoglin. Immune response to the Ag43-endoglin fusion protein also resulted in reduced tumour sizes in an animal model, with few side effects.

A more conventional application of bacterial surface expression for vaccine development is to display genetic libraries of possible antigens (Klemm and Schembri 2000; Verhoeven et al. 2009; Hudson et al. 2012) or antibodies (Daugherty et al. 1999; Wang et al. 2015). The level of native folding achieved by recombinant proteins displayed on the surface of *E. coli*

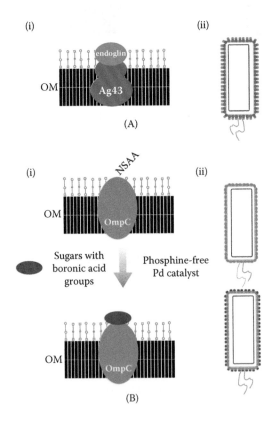

FIGURE 4.11
E. coli surface design. (A) (i) Fusing endoglin to Ag43 enables its display on the outer membrane (OM), (ii) from which it can subsequently be purified. (B) (i) An exposed NSAA on the otherwise native OpmC protein can be functionalised using the biocompatible Suzuki–Miyaura cross-coupling process involving the materials indicated, to (ii) control surface glycosylation.

surfaces can even be sufficient to screen for binding partners of intracellular mammalian proteins, such as Bcl-2, for basic or cancer research (Zhang and Link 2011).

4.6.3 Cell Surface Engineering for Whole Cell Bacterio-Therapeutics

Whole cell therapeutic approaches often involve using dead or attenuated cells for vaccination and immunotherapy or living cells that can populate organs to restore function, for example in regenerative medicine (Rao et al. 2015). Piñero-Lambea et al. (2015) have utilised a whole cell therapy approach using surface engineering of a non-pathogenic *E. coli* strain for tumour treatment. Populations of commensal, non-pathogenic strain *E. coli* cells have been shown to have a natural propensity to grow at the surfaces of solid tumours upon being systemically administered in mouse animal models. Piñero-Lambea et al. (2015) redesigned the surface of such a non-pathogenic *E. coli* strain known to populate tumour surfaces. First they designed a fusion protein incorporating three functional elements. The first element spans the outer membrane (Figure 4.12D) and comprises an anchor region from the protein LysM and a beta barrel. The next element is a surface-exposed carboxy terminal region, D0, which features a stable, immunoglobulin-like fold and promotes adhesion to surfaces (Figure 4.12C). The final element is an antigen-binding domain, VHH, fused to the D0 (Figure 4.12B). The VHH domain normally forms the antigen-binding region of heavy chain only antibodies (HCAb), as depicted in Figure 4.12 (compare A and B). The final fusion protein, referred to as a synthetic adhesion (SA), exposes the VHH domain at the cell surface (Figure 4.12E and F).

The next step was to demonstrate the efficacy of cells coated with an adhesion-promoting protein that also has antigen specificity. To do this, Piñero-Lambea et al. (2015) had chosen a GFP-specific VHH domain to incorporate into the SA. A mammalian tumorigenic cell line, HeLa, modified to express GFP on its surface was then used as a model for antigen-based binding by the SA-displaying *E. coli* cells. When GFP-expressing HeLa cells were mixed with SA-display *E. coli* cells, the bacteria were shown to preferentially colonise the HeLa cell surfaces in a manner dependent on binding GFP. GFP-HeLa cells were then used to grow tumours that were grafted subcutaneously into a mouse animal model. Wild type,

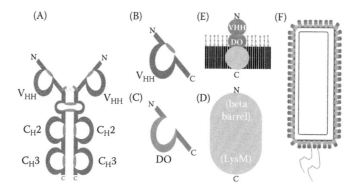

FIGURE 4.12
Targeting whole-cell bacteriotherapy. (A) The antigen-specific VHH domain in its original context within a HCAB. Constant domains are denoted C_H2 and C_H3, variable regions as V_{HH}. Simple graphics of (B) a VHH domain, (C) a DO domain, (D) the membrane-spanning domain and (E) the complete synthetic adhesion (SA) *in situ* in the *E. coli* outer membrane. (F) Successful SA-coating of an *E. coli* cell.

non-pathogenic *E. coli* cells were introduced systematically into the animals and the presence of *E. coli* cells measured in the tumour and in normal tissues. When non-pathogenic, anti-GFP SA-VHH *E. coli* cells were also used in this way, they colonised tumours to a greater degree than wild type non-pathogenic *E. coli* cells and were also less prevalent in non-tumour tissues.

4.6.4 Remodelling the Glycocalyx

Glycosylation of native bacterial proteins, as opposed to their lipids, was discovered for the first time in 2003 when Upreti et al. reported characterising glycoproteins in an archaeon. Since then many bacterial glycoproteins have been described along with the glycosyl transferase enzymes that direct their glycosylation. For example, Sánchez-Rodríguez et al. (2014) report *in silico* identification of over 40 glycosyltransferases in the genome of *Lactobacillus rhamnosus GG*. This growing knowledge is beginning to give synthetic biologists the ability to control prokaryote surface protein glycosylation by genetic means (Merritt et al. 2013). Designing surface sugars in prokaryotes could be a powerful route to immune evasion for whole cell bacterio-therapeutics. Further possibilities include improved tolerance of cryopreservation or extreme osmotic conditions.

Spicer and Davis (2013) developed a semi-genetic approach to rewrite the *E. coli* glycocalyx using a combination of unnatural amino acids, the technology for which is discussed further in Chapter 9, and a mild, biocompatible chemical treatment called Suzuki–Miyaura cross-coupling (Figure 4.11B). An *E. coli* variant strain was developed in which UAG stop codons can be used to encode an NSAA, p-iodophenyl alanine (pIPhe), using plasmids encoding evolved aminoacyl-tRNA synthetase and transfer RNA molecules (Young et al. 2010). The outer membrane protein OmpC was also modified to incorporate a surface-exposed pIPhe residue. Suzuki–Miyaura chemical cross-coupling was then performed with minimal cytotoxicity to covalently link the OmpC to glucose, galactose and mannose moieties (Figure 4.11B). All chemically linked carbohydrate moieties were shown to be biologically recognizable by successful binding with their cognate leptins.

References

Andrianantoandro, E., S. Basu, D. K. Karig, and R. Weiss. 2006. Synthetic biology: New engineering rules for an emerging discipline. *Molecular Systems Biology* 2: 2006.0028. doi:10.1038/msb4100073.

Arkin, A. 2008. Setting the standard in synthetic biology. *Nature Biotechnology* 26 (7): 771–74. doi:10.1038/nbt0708-771.

Atkinson, M. R., M. A. Savageau, J. T. Myers, and A. J. Ninfa. 2003. Development of genetic circuitry exhibiting toggle switch or oscillatory behavior in *Escherichia Coli*. *Cell* 113 (5): 597–607. doi:10.1016/S0092-8674(03)00346-5.

Azaman, S. N., N. R. Ramakrishnan, J. S. Tan, R. A. Rahim, M. P. Abdullah, and. A. B. Ariff. 2010. Optimization of an induction strategy for improving interferon-α2b production in the periplasm of *Escherichia Coli* using response surface methodology. *Biotechnology and Applied Biochemistry* 56 (4): 141–50. doi:10.1042/BA20100104.

Aziz, N. A., G. V. Anguelova, J. Marinus, G. J. Lammers, and R. A. Roos. 2010. Sleep and circadian rhythm alterations correlate with depression and cognitive impairment in Huntington's disease. *Parkinsonism and Related Disorders* 16 (5): 345–50. doi:10.1016/j.parkreldis.2010.02.009.

Bäcklund, E., D. Reeks, K. Markland, N. Weir, L. Bowering, and G. Larsson. 2008. Fedbatch design for periplasmic product retention in *Escherichia Coli*. *Journal of Biotechnology* 135 (4): 358–65. doi:10.1016/j.jbiotec.2008.05.002.

Bailey, J. E. 1998. Mathematical modeling and analysis in biochemical engineering: Past accomplishments and future opportunities. *Biotechnology Progress* 14 (1): 8–20. doi:10.1021/bp9701269.

Balasundaram, B., D. Nesbeth, J. M. Ward, E. Keshavarz-Moore, and D. G. Bracewell. 2009. Step change in the efficiency of centrifugation through cell engineering: Co-expression of Staphylococcal nuclease to reduce the viscosity of the bioprocess feedstock. *Biotechnology and Bioengineering* 104 (1): 134–42. doi:10.1002/bit.22369.

Barkai, N. and L. Serrano. 2000. Biological rhythms: Circadian clocks limited by noise. *Nature* 403 (6767): 267–68. doi:10.1038/35002258.

Bates, R., O. Blyuss, and A. Zaikin. 2014. Stochastic resonance in an intracellular genetic perceptron. *Physical Review E* 89 (3): 032716. doi:10.1103/PhysRevE.89.032716.

Becskei, A. and L. Serrano. 2000. Engineering stability in gene networks by autoregulation. *Nature* 405 (6786): 590–93. doi:10.1038/35014651.

Bonnet, J., P. Yin, M. E. Ortiz, P. Subsoontorn, and D. Endy. 2013. Amplifying genetic logic gates. *Science (New York, N.Y.)* 340 (6132): 599–603. doi:10.1126/science.1232758.

Bonsor, D. A., I. Grishkovskaya, E. J. Dodson, and C. Kleanthous. 2007. Molecular mimicry enables competitive recruitment by a natively disordered protein. *Journal of the American Chemical Society* 129 (15): 4800–4807. doi:10.1021/ja070153n.

Bornemann, T., J. Jöckel, M. V Rodnina, and W. Wintermeyer. 2008. Signal sequence–independent membrane targeting of ribosomes containing short nascent peptides within the exit tunnel. *Nature Structural Molecular Biology* 15 (5): 494–99. doi:10.1038/nsmb.1402.

Borg, Y., E. Ullner, A. Alagha, A. Alsaedi, D. Nesbeth, and A. Zaikin. 2014. Complex and unexpected dynamics in simple genetic regulatory networks. *International Journal of Modern Physics B* 28 (14): 1430006. doi:10.1142/S0217979214300060.

Boström, M., K. Markland, A. M. Sandén, M. Hedhammar, S. Hober, and G. Larsson. 2005. Effect of substrate feed rate on recombinant protein secretion, degradation and inclusion body formation in *Escherichia Coli*. *Applied Microbiology and Biotechnology* 68 (1): 82–90. doi:10.1007/s00253-004-1855-4.

Boyle, P. M., and P. A. Silver. 2012. Parts plus pipes: Synthetic biology approaches to metabolic engineering. *Metabolic Engineering* 14 (3): 223–32. doi:10.1016/j.ymben.2011.10.003.

Boynton, Z. L., J. J. Koon, E. M. Brennan, J. D. Clouart, D. M. Horowitz, T. U. Gerngross, and G. W. Huisman. 1999. Reduction of cell lysate viscosity during processing of poly(3-hydroxyalkanoates) by chromosomal integration of the staphylococcal nuclease gene in pseudomonas putida. *Applied and Environmental Microbiology* 65 (4): 1524–29.

Branston, S. D., C. F. Matos, R. B. Freedman, C. Robinson, and E. Keshavarz-Moore. 2012. Investigation of the impact of tat export pathway enhancement on E. coli culture, protein production and early stage recovery. *Biotechnology and Bioengineering* 109 (4): 983–91. doi:10.1002/bit.24384.

Bratsun, D., D. Volfson, L. S. Tsimring, and J. Hasty. 2005. Delay-induced stochastic oscillations in gene regulation. *Proceedings of the National Academy of Sciences of the United States of America* 102 (41): 14593–98. doi:10.1073/pnas.0503858102.

Çağatay, T., M. Turcotte, M. B. Elowitz, J. Garcia-Ojalvo, and G. M. Süel. 2009. Architecture-dependent noise discriminates functionally analogous differentiation circuits. *Cell* 139 (3): 512–22. doi:10.1016/j.cell.2009.07.046.

Cameron, D. E., C. J. Bashor, and J. J. Collins. A brief history of synthetic biology. 2014. *Nature Reviews Microbiology* 12 (5): 381–90. doi:10.1038/nrmicro3239.

Canton, B., A. Labno, and D. Endy. 2008. Refinement and standardization of synthetic biological parts and devices. *Nature Biotechnology* 26 (7): 787–93. doi:10.1038/nbt1413.

Carlson, R. H. 2011. *Biology Is Technology: The Promise, Peril, and New Business of Engineering Life.* Cambridge, MA: Harvard University Press.

Carrera, J., S. F. Elena, and A. Jaramillo. 2012. Computational design of genomic transcriptional networks with adaptation to varying environments. *Proceedings of the National Academy of Sciences* 109 (38): 15277–82.

Ceroni, F., R. Algar, G. B. Stan, and T. Ellis. 2015. Quantifying cellular capacity identifies gene expression designs with reduced burden. *Nature Methods* 12 (5): 415–18. doi:10.1038/nmeth.3339.

Chan, L. Y., S. Kosuri, and D. Endy. 2005. Refactoring bacteriophage T7. *Molecular Systems Biology* 1 (1): E1–10. doi:10.1038/msb4100025.

Chandran, D., W. B. Copeland, S. C. Sleight, and H. M. Sauro. 2008. Mathematical modeling and synthetic biology. *Drug Discovery Today: Disease Models* 5 (4): 299–309. doi:10.1016/j.ddmod.2009.07.002.

Chang, Y. C., C. L. Lin, and T. Jennawasin. 2013. Design of synthetic genetic oscillators using evolutionary optimization. *Evolutionary Bioinformatics Online* 9: 137–50. doi:10.4137/EBO.S11225.

Chen, B. S., and C. Y. Hsu. 2012. Robust synchronization control scheme of a population of nonlinear stochastic synthetic genetic oscillators under intrinsic and extrinsic molecular noise via quorum sensing. *BMC Systems Biology* 6: 136. doi:10.1186/1752-0509-6-136.

Chen, N., F. L. Hong, H. H. Wang, Q. H. Yuan, W. Y. Ma, X. N. Gao, R. Shi, R. J. Zhang, C. S. Sun, and S. B. Wang. 2012. Modified recombinant proteins can be exported via the Sec pathway in *Escherichia Coli*. *PLoS ONE* 7 (8): e42519. doi:10.1371/journal.pone.0042519.

Chen, Y. Y., K. E. Galloway, and C. D. Smolke. 2012. Synthetic biology: Advancing biological frontiers by building synthetic systems. *Genome Biology* 13 (2): 240. doi:10.1186/gb-2012-13-2-240.

Cline, K. 2015. Mechanistic aspects of folded protein transport by the twin arginine translocase (Tat). *The Journal of Biological Chemistry* 290 (27): 16530–38. doi:10.1074/jbc.R114.626820.

Cobb, R. E., T. Si, and H. Zhao. 2012. Directed evolution: An evolving and enabling synthetic biology tool. *Current Opinion in Chemical Biology* 16 (3–4): 285–91. doi:10.1016/j.cbpa.2012.05.186.

Comba, S., A. Arabolaza, and H. Gramajo. 2012. Emerging engineering principles for yield improvement in microbial cell design. *Computational and Structural Biotechnology Journal* 3 (4): 1–6. doi:10.5936/csbj.201210016.

Conway, B. A., and E. P. Greenberg. 2002. Quorum-sensing signals and quorum-sensing genes in *Burkholderia vietnamiensis*. *Journal of Bacteriology* 184 (4): 1187–91. doi:10.1128/jb.184.4.1187-1191.2002.

Cooke, G. D., R. M. Cranenburgh, J. A. Hanak, and J. M. Ward. 2003. A modified *Escherichia Coli* protein production strain expressing staphylococcal nuclease, capable of auto-hydrolysing host nucleic acid. *Journal of Biotechnology* 101 (3): 229–39. doi:10.1016/S0168-1656(02)00339-5.

Danino, T., O. Mondragón-Palomino, L. Tsimring, and J. Hasty. 2010. A synchronized quorum of genetic clocks. *Nature* 463 (7279): 326–30. doi:10.1038/nature08753.

Daugherty, P. S., M. J. Olsen, B. L. Iverson, and G. Georgiou. 1999. Development of an optimized expression system for the screening of antibody libraries displayed on the *Escherichia Coli* surface. *Protein Engineering* 12 (7): 613–21. doi:10.1093/protein/12.7.613.

Davis, J. H., A. J. Rubin, and R. T. Sauer. 2011. Design, construction and characterization of a set of insulated bacterial promoters. *Nucleic Acids Research* 39 (3): 1131–41. doi:10.1093/nar/gkq810.

De Jong, H. 2002. Modeling and simulation of genetic regulatory systems: A literature review. *Journal of Computational Biology: A Journal of Computational Molecular Cell Biology* 9 (1): 67–103. doi:10.1089/10665270252833208.

De Jong, H. 2003. *Mathematical Modeling of Genetic Regulatory Networks*. Rhone-Alps: Inria.

Densmore, D., and J. C. Anderson. 2009. Combinational logic design in synthetic biology. In *Circuits and Systems, 2009. ISCAS 2009. IEEE International Symposium on*, 301–4. IEEE. http://ieeexplore.ieee.org/xpls/abs_all.jsp?arnumber=5117745.

Didovyk, A., O. I. Kanakov, M. V. Ivanchenko, J. Hasty, R. Huerta, and L. Tsimring. 2015. Distributed classifier based on genetically engineered bacterial cell cultures. *ACS Synthetic Biology* 4 (1): 72–82. doi:10.1021/sb500235p.

Dragosits, M., D. Nicklas, and I. Tagkopoulos. 2012. A synthetic biology approach to self-regulatory recombinant protein production in *Escherichia Coli*. *Journal of Biological Engineering* 6 (2). http://www.biomedcentral.com/content/pdf/1754-1611-6-2.pdf.

Drubin, D. A., J. C. Way, and P. A. Silver. 2007. Designing biological systems. *Genes and Development* 21 (3): 242–54. doi:10.1101/gad.1507207.

Dubnau, D. 1999. DNA uptake in bacteria. *Annual Review of Microbiology* 53: 217–44. doi:10.1146/annurev.micro.53.1.217.

Elena, C., P. Ravasi, M. E. Castelli, S. Peirú, and H. G. Menzella. 2014. Expression of codon optimized genes in microbial systems: Current industrial applications and perspectives. *Frontiers in Microbiology* 5. doi:10.3389/fmicb.2014.00021.

Ellis, H. M., D. Yu, T. DiTizio, and D. L. Court. 2001. High efficiency mutagenesis, repair, and engineering of chromosomal DNA using single-stranded oligonucleotides. *Proceedings of the National Academy of Sciences of the United States of America* 98 (12): 6742–46. doi:10.1073/pnas.121164898.

Elowitz, M. B., and S. Leibler. 2000. A synthetic oscillatory network of transcriptional regulators. *Nature* 403 (6767): 335–38. doi:10.1038/35002125.

Elowitz, M. B., A. J. Levine, E. D. Siggia, and P. S. Swain. 2002. Stochastic gene expression in a single cell. *Science (New York, N.Y.)* 297 (5584): 1183–86. doi:10.1126/science.1070919.

El Samad, H., M. Khammash, L. Petzold, and D. Gillespie. 2005. Stochastic modelling of gene regulatory networks. *International Journal of Robust and Nonlinear Control* 15 (15): 691–711. doi:10.1002/rnc.1018.

Endler, L., N. Rodriguez, N. Juty, V. Chelliah, C. Laibe, C. Li, and N. Le Novère. 2009. Designing and encoding models for synthetic biology. *Journal of the Royal Society Interface* 6 (Suppl 4): S405–17. doi:10.1098/rsif.2009.0035.focus.

Esvelt, K. M., and H. H. Wang. 2013. Genome-scale engineering for systems and synthetic biology. *Molecular Systems Biology* 9 (1): 641. doi:10.1038/msb.2012.66.

Fischbach, M., and C. A. Voigt. 2010. Prokaryotic Gene Clusters: A rich toolbox for synthetic biology. *Biotechnology Journal* 5 (12): 1277–96. doi:10.1002/biot.201000181.

Friedland, A. E., T. K. Lu, X. Wang, D. Shi, G. Church, and J. J. Collins. 2009. Synthetic gene networks that count. *Science* 324 (5931): 1199–1202. doi:10.1126/science.1172005.

Fung, E., W. W. Wong, J. K. Suen, T. Bulter, S. G. Lee, and J. C. Liao. 2005. A synthetic gene–metabolic oscillator. *Nature* 435 (7038): 118–22. doi:10.1038/nature03508.

Gao, H., Y. Zhuo, E. Ashforth, and L. Zhang. 2010. Engineering of a genome-reduced host: Practical application of synthetic biology in the overproduction of desired secondary metabolites. *Protein and Cell* 1 (7): 621–26. doi:10.1007/s13238-010-0073-3.

Gibson, D. G., J. I. Glass, C. Lartigue, V. N. Noskov, R.-Y. Chuang, M. A. Algire, G. A. Benders, et al. 2010. Creation of a bacterial cell controlled by a chemically synthesized genome. *Science* 329 (5987): 52–56. doi:10.1126/science.1190719.

Gillespie, D. T. 1976. A general method for numerically simulating the stochastic time evolution of coupled chemical reactions. *Journal of Computational Physics* 22 (4): 403–34. doi:10.1016/0021-9991(76)90041-3.

Gonze, D. 2013. Modeling the effect of cell division on genetic oscillators. *Journal of Theoretical Biology* 325: 22–33. doi:10.1016/j.jtbi.2013.02.001.

Gonze, D., S. Bernard, C. Waltermann, A. Kramer, and H. Herzel. 2005. Spontaneous synchronization of coupled circadian oscillators. *Biophysical Journal* 89 (1): 120–29. doi:10.1529/biophysj.104.058388.

Goodwin, B. C. 1963. *Temporal organization in cells: A dynamic theory of cellular control processes.* London: Academic Press. Retrieved from http://www.archive.org/details/temporalorganiza00good.

Gould, N., O. Hendy, and D. Papamichail. 2014. Computational tools and algorithms for designing customized synthetic genes. *Frontiers in Bioengineering and Biotechnology* 2 (October). doi:10.3389/fbioe.2014.00041.

Gramelsberger, G. 2013. The simulation approach in synthetic biology. *Studies in History and Philosophy of Science Part C: Studies in History and Philosophy of Biological and Biomedical Sciences,* Philosophical Perspectives on Synthetic Biology, 44 (2): 150–57. doi:10.1016/j.shpsc.2013.03.010.

Grossman, A. D. 1995. Genetic networks controlling the initiation of sporulation and the development of genetic competence in *Bacillus subtilis. Annual Review of Genetics* 29: 477–508. doi:10.1146/annurev.ge.29.120195.002401.

Guantes, R., and J. F Poyatos. 2006. Dynamical principles of two-component genetic oscillators. *PLoS Comput Biol* 2 (3): e30. doi:10.1371/journal.pcbi.0020030.

Guazzaroni, M. E., and R. Silva-Rocha. 2014. Expanding the logic of bacterial promoters using engineered overlapping operators for global regulators. *ACS Synthetic Biology* 3 (9): 666–75. doi:10.1021/sb500084f.

Heinemann, M., and S. Panke. 2006. Synthetic biology—putting engineering into biology. *Bioinformatics* 22 (22): 2790–99. doi:10.1093/bioinformatics/btl469.

Henriques, S. T., L. Thorstholm, Y. H. Huang, J. A. Getz, P. S. Daugherty, and D. J. Craik. 2013. A novel quantitative kinase assay using bacterial surface display and flow cytometry. *PLoS ONE* 8 (11): e80474. doi:10.1371/journal.pone.0080474.

Hillen, W., C. Gatz, L. Altschmied, K. Schollmeier, and I. Meier. 1983. Control of expression of the Tn10-encoded tetracycline resistance genes. Equilibrium and kinetic investigation of the regulatory reactions. *Journal of Molecular Biology* 169 (3): 707–21.

Huang, C. J., H. Lin, and X. Yang. 2012. Industrial production of recombinant therapeutics in *Escherichia Coli* and its recent advancements. *Journal of Industrial Microbiology & Biotechnology* 39 (3): 383–99. doi:10.1007/s10295-011-1082-9.

Huang, F. Y., L. Li, Q. Liu, Y. N. Li, R. Z. Bai, Y. H. Huang, H. G. Zhao, et al. 2014. Bacterial surface display of endoglin by antigen 43 induces antitumor effectiveness *via* bypassing immunotolerance and inhibition of angiogenesis: Anticancer effects by Ag43'/END^e chimeric protein. *International Journal of Cancer* 134 (8): 1981–90. doi:10.1002/ijc.28511.

Hoffman, E., J. Hanson, and J. Thomas. 2012. Principles for the oversight of synthetic biology. *Friends of the Earth*. http://libcloud.s3.amazonaws.com/93/11/7/1204/1/Principles_for_the_oversight_of_synthetic_biology.pdf.

Hudson, E. P., M. Uhlen, and J. Rockberg. 2012. Multiplex epitope mapping using bacterial surface display reveals both linear and conformational epitopes. *Scientific Reports* 2 (October). doi:10.1038/srep00706.

Hwang, B. Y., B. G. Kim, and J. H. Kim. 2011. Bacterial surface display of a co-factor containing enzyme, Ω-transaminase from *Vibrio fluvialis* using the *Bacillus subtilis* spore display system. *Bioscience, Biotechnology, and Biochemistry* 75 (9): 1862–65. doi:10.1271/bbb.110307.

Ingram, W., L. Chan, H. Guven, D. Darling, S. Kordasti, N. Hardwick, L. Barber, G. J. Mufti, and F. Farzaneh. 2009. Human CD80/IL2 lentivirus-transduced acute myeloid leukaemia (AML) cells promote natural killer (NK) cell activation and cytolytic activity: Implications for a phase I clinical study. *British Journal of Haematology* 145 (6): 749–60. doi:10.1111/j.1365-2141.2009.07684.x.

Isaacs, F. J., Carr, P. A., Wang, H. H., Lajoie, M. J., Sterling, B., Kraal L., et al. 2011. Precise manipulation of chromosomes in vivo enables genome-wide codon replacement. *Science* 333 (6040): 348–53. doi:10.1126/science.1205822.

Jose et al. 2006. Autodisplay: Efficient bacterial surface display of recombinant proteins. *Applied Microbiology and Biotechnology* 69 (6): 607–14. doi:10.1007/s00253-005-0227-z.

Juhas, M., L. Eberl, and G. M. Church. 2012. Essential genes as antimicrobial targets and cornerstones of synthetic biology. *Trends in Biotechnology* 30 (11): 601–7. doi:10.1016/j.tibtech.2012.08.002.

Kaern, M., W. J. Blake, and J. J. Collins. 2003. The engineering of gene regulatory networks. *Annual Review of Biomedical Engineering* 5: 179–206. doi:10.1146/annurev.bioeng.5.040202.121553.

Kaern, M., T. C. Elston, W. J. Blake, and J. J. Collins. 2005. Stochasticity in gene expression: From theories to phenotypes. *Nature Reviews Genetics* 6 (6): 451–64. doi:10.1038/nrg1615.

Karas, B. J., J. Jablanovic, L. Sun, L. Ma, G. M Goldgof, J. Stam, A. Ramon, et al. 2013. Direct transfer of whole genomes from bacteria to yeast. *Nature Methods* 10 (5): 410–12. doi:10.1038/nmeth.2433.

Khalil, A. S., and J. J. Collins. 2010. Synthetic biology: Applications come of age. *Nature Reviews Genetics* 11 (5): 367–79. doi:10.1038/nrg2775.

Kim, J., and E. Winfree. 2011. Synthetic in vitro transcriptional oscillators. *Molecular Systems Biology* 7 (1): 465. doi:10.1038/msb.2010.119.

Klasson, L., and S. G. Andersson. 2010. Research on small genomes: Implications for synthetic biology. *BioEssays* 32 (4): 288–95. doi:10.1002/bies.200900165.

Klemm, P., and M. A. Schembri. 2000. Fimbriae-assisted bacterial surface display of heterologous peptides. *International Journal of Medical Microbiology: IJMM* 290 (3): 215–21. doi:10.1016/S1438-4221(00)80118-6.

Koseska, A., E. Volkov, and J. Kurths. 2011. Synthetic multicellular oscillatory systems: Controlling protein dynamics with genetic circuits. *Physica Scripta* 84 (4): 045007. doi:10.1088/0031-8949/84/04/045007.

Lajoie, M. J., A. J. Rovner, D. B. Goodman, H.-R. Aerni, A. D. Haimovich, G. Kuznetsov, J. A. Mercer, et al. 2013. Genomically recoded organisms expand biological functions. *Science* 342 (6156): 357–60. doi:10.1126/science.1241459.

Lang, M., T. T. Marquez-Lago, J. Stelling, and S. Waldherr. 2011. Autonomous synchronization of chemically coupled synthetic oscillators. *Bulletin of Mathematical Biology* 73 (11): 2678–2706. doi:10.1007/s11538-011-9642-8.

Lartigue, C., S. Vashee, M. A. Algire, R.-Y. Chuang, G. A. Benders, L. Ma, V. N. Noskov, et al. 2009. Creating bacterial strains from genomes that have been cloned and engineered in yeast. *Science* 325 (5948): 1693–96. doi:10.1126/science.1173759.

Liu, L., H. Redden, and H. S. Alper. 2013. Frontiers of yeast metabolic engineering: diversifying beyond ethanol and Saccharomyces. *Current Opinion in Biotechnology* 24(6): 1023–30. doi:10.1016/j.copbio.2013.03.005.

Ma, N. J., D. W. Moonan, and F. J. Isaacs. 2014. Precise manipulation of bacterial chromosomes by conjugative assembly genome engineering. *Nature Protocols* 9 (10): 2285–2300. doi:10.1038/nprot.2014.081.

Mandell, D. J., M. J. Lajoie, M. T. Mee, R. Takeuchi, G. Kuznetsov, J. E. Norville, C. J. Gregg, B. L. Stoddard, and G. M. Church. 2015. Biocontainment of genetically modified organisms by synthetic protein design. *Nature* 518 (7537): 55–60. doi:10.1038/nature14121.

Markson, J. S., and M. B. Elowitz. 2014. Synthetic biology of multicellular systems: New platforms and applications for animal cells and organisms. *ACS Synthetic Biology* 3(12): 875–76. doi:10.1021/sb500358y.

Mauro, V. P., and S. A. Chappell. 2014. A critical analysis of codon optimization in human therapeutics. *Trends in Molecular Medicine* 20 (11): 604–13. doi:10.1016/j.molmed.2014.09.003.

McClung, C. R. 2006. Plant circadian rhythms. *The Plant Cell* 18 (4): 792–803. doi:10.1105/tpc.106.040980.

Merritt, J. H., A. A. Ollis, A. C. Fisher, and M. P. DeLisa. 2013. Glycans-by-design: Engineering bacteria for the biosynthesis of complex glycans and glycoconjugates. *Biotechnology and Bioengineering* 110 (6): 1550–64.

Mitra, A., A. K. Kesarwani, D. Pal, and V. Nagaraja. 2011. WebGeSTer DB—A transcription terminator database. *Nucleic Acids Research* 39 (Database): D129–35. doi:10.1093/nar/gkq971.

Na, D., S. Lee, and D. Lee. 2010. Mathematical modeling of translation initiation for the estimation of its efficiency to computationally design mRNA sequences with desired expression levels in prokaryotes. *BMC Systems Biology* 4: 71. doi:10.1186/1752-0509-4-71.

Nair, M. T., S. S. Tambe, and B. D. Kulkarni. 1994. Application of artificial neural networks for prokaryotic transcription terminator prediction. *FEBS Letters* 346 (2–3): 273–77.

Nesbeth, D. N., M. A. Perez-Pardo, S. Ali, J. Ward, and E. Keshavarz-Moore. 2012. Growth and productivity impacts of periplasmic nuclease expression in an *Escherichia Coli* fab' fragment production strain. *Biotechnology and Bioengineering*, 109(2): 517–27. doi:10.1002/bit.23316.

Ni, Y., and R. Chen. 2009. Extracellular recombinant protein production from *Escherichia Coli*. *Biotechnology Letters* 31 (11): 1661–70. doi:10.1007/s10529-009-0077-3.

Ni, Y., J. Reye, and R. R. Chen. 2007. Lpp deletion as a permeabilization method. *Biotechnology and Bioengineering* 97 (6): 1347–56. doi:10.1002/bit.21375.

Nielsen, A. A., T. H Segall-Shapiro, and C. A. Voigt. 2013. Advances in genetic circuit design: novel biochemistries, deep part mining, and precision gene expression. *Current Opinion in Chemical Biology.* 17 (6): 878–92. doi:10.1016/j.cbpa.2013.10.003.

Nojima, T., A. C. Lin, T. Fujii, and I. Endo. 2005. Determination of the termination efficiency of the transcription terminator using different fluorescent profiles in green fluorescent protein mutants. *Analytical Sciences: The International Journal of the Japan Society for Analytical Chemistry* 21 (12): 1479–81.

Novák, B., and J. J. Tyson. 2008. Design principles of biochemical oscillators. *Nature Reviews Molecular Cell Biology* 9 (12): 981–91. doi:10.1038/nrm2530.

O'Brien, E. L., E. V. Itallie, and M. R. Bennett. 2012. Modeling synthetic gene oscillators. *Mathematical Biosciences,* 236(1): 1–15.

O'Brien, T. F., B. K. Gorentla, D. Xie, S. Srivatsan, I. X. McLeod, Y. W. He, and X. P. Zhong. 2011. Regulation of T-cell survival and mitochondrial homeostasis by TSC1. *European Journal of Immunology* 41 (11): 3361–70. doi:10.1002/eji.201141411.

O'Reilly, A. O., A. R. Cole, J. L. S. Lopes, A. Lampert, and B. A. Wallace. 2014. Chaperone-mediated native folding of a β-scorpion toxin in the periplasm of *Escherichia Coli. Biochimica et Biophysica Acta (BBA)–General Subjects* 1840 (1): 10–15. doi:10.1016/j.bbagen.2013.08.021.

Osella, M., and M. C. Lagomarsino. 2013. Growth-rate-dependent dynamics of a bacterial genetic oscillator. *Physical Review E* 87 (1). doi:10.1103/PhysRevE.87.012726.

Pasotti, L., S. Zucca, M. Lupotto, M. G. Cusella De Angelis, and P. Magni. 2011. Characterization of a synthetic bacterial self-destruction device for programmed cell death and for recombinant proteins release. *Journal of Biological Engineering* 5 (8). http://www.biomedcentral.com/content/pdf/1754-1611-5-8.pdf.

Peisajovich, S. G. 2012. Evolutionary synthetic biology. *ACS Synthetic Biology* 1 (6): 199–210. doi:10.1021/sb300012g.

Perry, N., and A. J. Ninfa. 2012. Synthetic networks: Oscillators and toggle switches for *Escherichia Coli. Methods in Molecular Biology (Clifton, N.J.)* 813: 287–300. doi:10.1007/978-1-61779-412-4_17.

Pierce, J. J., C. Turner, E. Keshavarz-Moore, and P. Dunnill. 1997. Factors determining more efficient large-scale release of a periplasmic enzyme from *E. coli* using lysozyme. *Journal of Biotechnology* 58 (1): 1–11.

Piñero-Lambea, C., G. Bodelón, R. Fernández-Periáñez, A. M. Cuesta, L. Álvarez-Vallina, and L. Á. Fernández. 2015. Programming controlled adhesion of *E. Coli* to target surfaces, cells, and tumors with synthetic adhesins. *ACS Synthetic Biology* 4 (4): 463–73. doi:10.1021/sb500252a.

Popenda, M., M. Szachniuk, M. Antczak, K. J. Purzycka, P. Lukasiak, N. Bartol, J. Blazewicz, and R. W. Adamiak. 2012. Automated 3D structure composition for large RNAs. *Nucleic Acids Research* 40 (14): e112. doi:10.1093/nar/gks339.

Postle, K., and R. F. Good. 1985. A bidirectional rho-independent transcription terminator between the *E. coli* tonB gene and an opposing gene. *Cell* 41 (2): 577–85.

Purcell, O., M. di Bernardo, C. S. Grierson, and N. J. Savery. 2011. A multi-functional synthetic gene network: A frequency multiplier, oscillator and switch. *PLoS ONE* 6 (2): e16140. doi:10.1371/journal.pone.0016140.

Rao, C. V. 2012. Expanding the synthetic biology toolbox: Engineering orthogonal regulators of gene expression. *Current Opinion in Biotechnology* 23: 689–94.

Rao, M., C. Mason, and S. Solomon. 2015. Cell therapy worldwide: An incipient revolution. *Regenerative Medicine* 10 (2): 181–91. doi:10.2217/rme.14.80.

Raser, J. M. 2005. Noise in gene expression: Origins, consequences, and control. *Science (New York, N.Y.)* 309 (5743): 2010–13. doi:10.1126/science.1105891.

Raser, J. M., and E. K. O'Shea. 2004. Control of stochasticity in eukaryotic gene expression. *Science* 304 (5678): 1811–14. doi:10.1126/science.1098641.

Reeve, B., T. Hargest, C. Gilbert, and T. Ellis. 2014. Predicting translation initiation rates for designing synthetic biology. *Frontiers in Bioengineering and Biotechnology* 2. doi:10.3389/fbioe.2014.00001.

Rud, I. 2006. A synthetic promoter library for constitutive gene expression in lactobacillus plantarum. *Microbiology* 152 (4): 1011–19. doi:10.1099/mic.0.28599-0.

Ruoff, P., M. Vinsjevik, C. Monnerjahn, and L. Rensing. 2001. The Goodwin model: simulating the effect of light pulses on the circadian sporulation rhythm of *Neurospora crassa*. *Journal of Theoretical Biology* 209 (1): 29–42. doi:10.1006/jtbi.2000.2239.

Saadatpour, A., and R. Albert. 2013. Boolean modeling of biological regulatory networks: A methodology tutorial. *Methods (San Diego, Calif.)* 62 (1): 3–12. doi:10.1016/j.ymeth.2012.10.012.

Saffar, B., B. Yakhchali, and M. Arbabi. 2007. Development of a bacterial surface display of hexahistidine peptide using CS3 pili for bioaccumulation of heavy metals. *Current Microbiology* 55 (4): 273–77. doi:10.1007/s00284-005-0511-2.

Sánchez-Rodríguez, A., H. L. Tytgat, J. Winderickx, J. Vanderleyden, S. Lebeer, and K. Marchal. 2014. A network-based approach to identify substrate classes of bacterial glycosyltransferases. *BMC Genomics* 15 (1): 349.

Salis, H. M. 2011. The ribosome binding site calculator. *Methods in Enzymology* 498: 19–42. doi:10.1016/B978-0-12-385120-8.00002-4.

Salis, H. M., E. A. Mirsky, and C. A. Voigt. 2009. Automated design of synthetic ribosome binding sites to control protein expression. *Nature Biotechnology* 27 (10): 946–50. doi:10.1038/nbt.1568.

Sayut, D. J., P. K. Kambam, and L. Sun. 2007. Engineering and applications of genetic circuits. *Molecular BioSystems* 3 (12): 835–40. doi:10.1039/b700547d.

Seo, S. W., J. Yang, B. E. Min, S. Jang, J. H. Lim, H. G. Lim, S. C. Kim, S. Y. Kim, J. H. Jeong, and G. Y. Jung. 2013. Synthetic biology: Tools to design microbes for the production of chemicals and fuels. *Biotechnology Advances* 31 (6): 811–17. doi:10.1016/j.biotechadv.2013.03.012.

Shin, H. D., and R. R Chen. 2008. Extracellular recombinant protein production from an *Escherichia Coli* Lpp deletion mutant. *Biotechnology and Bioengineering* 101 (6): 1288–96. doi:10.1002/bit.22013.

Silva-Rocha, R., and V. de Lorenzo. 2008. Mining logic gates in prokaryotic transcriptional regulation networks. *FEBS Letters* 582 (8): 1237–44. doi:10.1016/j.febslet.2008.01.060.

Singh, V. 2014. Recent advancements in synthetic biology: Current status and challenges. *Gene* 535 (1): 1–11. doi:10.1016/j.gene.2013.11.025.

Siuti, P., J. Yazbek, and T. K Lu. 2013. Synthetic circuits integrating logic and memory in living cells. *Nature Biotechnology* 31 (5): 448–52. doi:10.1038/nbt.2510.

Skerra, A., and A. Plückthun. 1991. Secretion and in vivo folding of the fab fragment of the antibody McPC603 in *Escherichia Coli*: Influence of disulphides and cis-prolines. *Protein Engineering* 4 (8): 971–79.

Smith, G. R. 1991. Conjugational recombination in *E. coli*: Myths and mechanisms. *Cell* 64 (1): 19–27.

Smolen, P., D. A. Baxter, and J. H. Byrne. 2000. Mathematical modeling of gene networks. *Neuron* 26 (3): 567–80.

Sochacki, K. A., I. A. Shkel, M. T. Record, and J. C. Weisshaar. 2011. Protein diffusion in the periplasm of *E. coli* under osmotic stress. *Biophysical Journal* 100 (1): 22–31. doi:10.1016/j.bpj.2010.11.044.

Song, Q., and X. Zhang. 2008. Characterization of a novel non-specific nuclease from thermophilic bacteriophage GBSV1. *BMC Biotechnology* 8 (1): 43. doi:10.1186/1472-6750-8-43.

Spicer, C. D., and B. G. Davis. 2013. Rewriting the bacterial glycocalyx via Suzuki–Miyaura cross-coupling. *Chemical Communications* 49 (27): 2747. doi:10.1039/c3cc38824g.

Stegun, I. A., and P. M. Morse. 1964. *Handbook of Mathematical Functions with Formulas, Graphs, and Mathematical Tables*. Edited by Milton Abramowitz. Corrected edition. National Bureau of Standards.

Stephanopoulos, G. 2012. Synthetic biology and metabolic engineering. *ACS Synthetic Biology* 1 (11): 514–25. doi:10.1021/sb300094q.

Sterlini, J. M., and J. Mandelstam. 1969. Commitment to sporulation in *Bacillus subtilis* and its relationship to development of actinomycin resistance. *Biochem. J* 113: 29–37.

Strelkowa, N., and M. Barahona. 2010. Switchable genetic oscillator operating in quasi-stable mode. *Journal of the Royal Society Interface* 7 (48): 1071–82. doi:10.1098/rsif.2009.0487.

Stricker, J., S. Cookson, M. R. Bennett, W. H. Mather, L. S. Tsimring, and J. Hasty. 2008. A fast, robust and tunable synthetic gene oscillator. *Nature* 456 (7221): 516–19. doi:10.1038/nature07389.

Tan, C., H. Song, J. Niemi, and L. You. 2007. A synthetic biology challenge: Making cells compute. *Molecular BioSystems* 3 (5): 343. doi:10.1039/b618473c.

Takahara, M., D. W. Hibler, P. J. Barr, J. A. Gerlt, and M. Inouye. 1985. The ompA signal peptide directed secretion of Staphylococcal nuclease A by *Escherichia Coli*. *The Journal of Biological Chemistry* 260 (5): 2670–74.

Tech, M., B. Morgenstern, and P. Meinicke. 2006. TICO: A tool for postprocessing the predictions of prokaryotic translation initiation sites. *Nucleic Acids Research* 34 (Web Server issue): W588–90. doi:10.1093/nar/gkl313.

Tigges, M., N. Dénervaud, D. Greber, J. Stelling, and M. Fussenegger. 2010. A synthetic low-frequency mammalian oscillator. *Nucleic Acids Research* 38 (8): 2702–11. doi:10.1093/nar/gkq121.

Tigges, M., and M. Fussenegger. 2009. Recent advances in mammalian synthetic biology—design of synthetic transgene control networks. *Current Opinion in Biotechnology* 20 (4): 449–60. doi:10.1016/j.copbio.2009.07.009.

Tigges, M., T. T. Marquez-Lago, J. Stelling, and M. Fussenegger. 2009. A tunable synthetic mammalian oscillator. *Nature* 457 (7227): 309–12. doi:10.1038/nature07616.

Upreti, R. K., M. Kumar, and V. Shankar. 2003. Bacterial glycoproteins: Functions, biosynthesis and applications. *Proteomics* 3 (4): 363–79.

Valdez-Cruz, N. A., L. Caspeta, N. O. Pérez, O. T. Ramírez, and M. A. Trujillo-Roldán. 2010. Production of recombinant proteins in *E. coli* by the heat inducible expression system based on the phage lambda pL And/or pR promoters. *Microbial Cell Factories* 9: 18. doi:10.1186/1475-2859-9-18.

Verhoeven, G. S., S. Alexeeva, M. Dogterom, and T. den Blaauwen. 2009. Differential bacterial surface display of peptides by the transmembrane domain of OmpA. *PLoS ONE* 4 (8): e6739. doi:10.1371/journal.pone.0006739.

Vinson, V., and E. Pennisi. 2011. The allure of synthetic biology. *Science* 333 (6047): 1235–1235. doi:10.1126/science.333.6047.1235.

Voigt, C. 2011. *Synthetic Biology: Methods for Part/Device Characterization and Chassis Engineering*. San Diego, CA: Academic Press.

Wang, H. H., F. J. Isaacs, P. A. Carr, Z. Z. Sun, G. Xu, C. R. Forest and G. M. Church. 2009. Programming cells by multiplex genome engineering and accelerated evolution. *Nature* 460 (7257): 894–98. doi:10.1038/nature08187.

Wang, L. X., M. Mellon, D. Bowder, M. Quinn, D. Shea, C. Wood, and S. H. Xiang. 2015. *Escherichia Coli* surface display of single-chain antibody VRC01 against HIV-1 infection. *Virology* 475: 179–86. doi:10.1016/j.virol.2014.11.018.

Weber, W., and M. Fussenegger. 2007. Inducible product gene expression technology tailored to bioprocess engineering. *Current Opinion in Biotechnology* 18: 399–410.

Weber, W., W. Bacchus, F. Gruber, M. Hamberger, and M. Fussenegger. 2007. A novel vector platform for vitamin H-inducible transgene expression in mammalian cells. *Journal of Biotechnology* 131 (2): 150–58. doi:10.1016/j.jbiotec.2007.06.008.

Weldemichael, D. A., and G. T. Grossberg. 2010. Circadian rhythm disturbances in patients with Alzheimer's disease: A review. *International Journal of Alzheimer's Disease* 2010. doi:10.4061/2010/716453.

Yang, Z., T. Zhou, P. M. Hui, and J. H. Ke. 2012. Instability in evolutionary games. *PLoS ONE* 7 (11). doi:10.1371/journal.pone.0049663.

Yildirim, N., and M. C. Mackey. 2003. Feedback regulation in the lactose operon: A mathematical modeling study and comparison with experimental data. *Biophysical Journal* 84 (5): 2841–51. doi:10.1016/S0006-3495(03)70013-7.

Yoda, M., T. Ushikubo, W. Inoue, and M. Sasai. 2007. Roles of noise in single and coupled multiple genetic oscillators. *The Journal of Chemical Physics* 126 (11): 115101. doi:10.1063/1.2539037.

Young, T. S., I. Ahmad, J. A. Yin, and P. G. Schultz. 2010. An enhanced system for unnatural amino acid mutagenesis in *E. coli*. *Journal of Molecular Biology* 395 (2): 361–74. doi:10.1016/j.jmb.2009.10.030.

Young, I., I. Wang, and W. D. Roof. 2000. Phages will out: Strategies of host cell lysis. *Trends in Microbiology* 8 (3): 120–28.

Yokobayashi, Y., R. Weiss, and F. H. Arnold. 2002. Directed evolution of a genetic circuit. *Proceedings of the National Academy of Sciences* 99 (26): 16587–91.

Yun, J., J. Park, N. Park, S. Kang, and S. Ryu. 2007. Development of a novel vector system for programmed cell lysis in *Escherichia Coli*. *Journal of Microbiology and Biotechnology* 17 (7): 1162–68.

Zaslaver, A., A. Bren, M. Ronen, S. Itzkovitz, I. Kikoin, S. Shavit, W. Liebermeister, M. G. Surette, and U. Alon. 2006. A comprehensive library of fluorescent transcriptional reporters for *Escherichia Coli*. *Nature Methods* 3 (8): 623–28. doi:10.1038/nmeth895.

Zhang, S., and A. J. Link. 2011. Bcl-2 family interactome analysis using bacterial surface display. *Integrative Biology* 3 (8): 823. doi:10.1039/c1ib00023c.

Zheng, Y., and G. Sriram. 2010. Mathematical modeling: Bridging the gap between concept and realization in synthetic biology. *Journal of Biomedicine and Biotechnology* 2010: 1–16. doi:10.1155/2010/541609.

Zuker, M., and P. Stiegler. 1981. Optimal computer folding of large RNA sequences using thermodynamics and auxiliary information. *Nucleic Acids Research* 9 (1): 133–48.

5

Eukaryotae Synthetica: Synthetic Biology in Yeast, Microalgae and Mammalian Cells

Desmond Schofield, Alexander Templar, Yanika Borg,
René Daer, Karmella Haynes and Darren N. Nesbeth

CONTENTS

5.1. Introduction

Synthetic biology aims to engineer novel cellular functions by assembling well-characterised molecular parts (i.e. nucleic acids and proteins) into biological 'devices' that exhibit predictable behaviour. For instance, a genetic toggle switch that responds to transient input signals, similar to how a machine might respond to an 'on/off' switch, would allow long-term maintenance of protein expression without the need for constant drug administration. This goal was first demonstrated by seminal work in bacteria where a simple bi-stable switch was built using two mutually repressive genes. Each gene encoded a transcriptional repressor of the other and each repressor was blocked by a chemical input. This system could be switched between two stable states, that is, 'gene 1 on, gene 2 off' versus 'gene 1 off, gene 2 on'. Gardner et al. (2000) used mathematical modelling to predict that such bi-stability is feasible for any set of promoters or repressors as long as they fulfil certain conditions, such as balanced promoter strengths. Kramer et al. (2004) demonstrated that such genetic toggle switches could also be established in Chinese hamster ovary (CHO) cells. Designing synthetic circuits to operate reliably in the context of differentiating and morphologically complex cells still presents unique challenges and opportunities for progress in mammalian synthetic biology.

Mammalian cells are hugely important to fundamental research into the elements of human cellular biology that can determine disease states and broader health issues such as aging and fertility. Unicellular eukaryotes, such as yeast, can also provide invaluable insight into the biomolecular machinery of human cells. In the applied field of biotechnology, mammalian cells, particularly CHO cells and yeasts, such as *Saccharomyces cerevisiae*, have emerged as predominant platforms for the manufacture of the large, complex therapeutic proteins that have become established in the last two decades as mainstays of the pharmaceutical industry. Mammalian and yeast cells have also been used extensively as screening tools for identification of small molecule drug candidates.

Synthetic biologists have begun in earnest to establish systems of designed gene circuitry, modularity and standardised measurement in eukaryotes. Since the inception of International Genetically Engineered Machine (iGEM) Foundation, the establishment of rigorous engineering approaches in eukaryotes has been a steady, if minority, pursuit of the synthetic biology community. Synthetic biology research to a varying degree is now ongoing across most of the canon of eukaryotic model organisms such as *Caenorhabditis elegans*, *Drosophila melanogaster*, *Xenopus laevis*, zebrafish, algae and the protozoa. Genome design and integration are more critical to establishing synthetic biology in eukaryotes due to the fact that, unlike in prokaryotes, plasmids are not the principle vehicle for transgene expression.

Yeast species, such as *Saccharomyces cerevisiae* and *Pichia pastoris*, have key properties that enable them to be realistically regarded as 'chassis' that can be re-designed, controlled and measured using the standardised approaches of synthetic biology. Both grow relatively rapidly and have a well-established repertoire of recombinant DNA tools, open source and commercial (David and Siewers 2014). Yeasts are also resistant to bacteriophage infection, grow at relatively low (<37°C) temperature and tolerate a broader pH range than *E. coli* (Liu et al. 2013). A significant limitation of yeasts is that, unless engineered specifically to do so (Nett et al. 2011), they cannot confer higher-eukaryote post-translational modifications, typically glycosylation, on the exogenous proteins they produce (Kelwick et al. 2014).

Establishing synthetic biology devices in mammalian cells has the most exciting potential in the context of medical applications (Aubel and Fussenegger 2010; Bacchus et al. 2013; Lienert et al. 2014). Engineered mammalian chassis could also represent considerable gains as cellular factories producing correctly formatted, complex human therapeutic proteins (Wuest et al. 2012). Challenges for mammalian cell factories include expensive growth media (Hirokawa et al. 2013), and a more technically demanding landscape of DNA techniques (Markson and Elowitz 2014) with many important cell lines, and primary cells, posing real difficulties with respect to transfection. Even when elegant and robust synthetic gene networks (SGNs) are demonstrated in mammalian cells, standardisation of DNA componentry and output measurement still lags somewhat behind what has been achieved in *E. coli* (Auslander and Fussenegger 2013a).

The complex intracellular architecture and intercellular communication systems of many eukaryotes represent both constraints and opportunities when compared with prokaryotic systems such as *E. coli*. The organisms covered in this chapter illustrate a diversity of eukaryotic synthetic biology research – plus notable achievements to date. Standardised measurement and genome-scale redesign of eukaryotes will also be discussed and the chapter concludes with an examination of how synthetic biology has been applied in the design of SGNs to address disease states such as cancer and diabetes.

5.2 Remodelling Outer Cell Membranes as Engineered Chassis Surfaces

Surfaces and the exterior of chassis are important to engineers. Conventionally, chassis serve to support, encase and protect processes within from thermal, physical or chemical assault. Beyond serving as a simple barrier structure, cellular surfaces can also function as sensors, filters, communication media, limbs and motility devices. When appraising a biological cell as an engineered chassis, the outer membrane is critically important as it can effect two-way communication between the internal controlling genome and the external environment, relaying sensory information on external conditions, transducing chemical signals to and from other cells and releasing effector molecules.

5.2.1 Functionalising the Yeast Cell Surface

One of the most direct approaches to re-designing a chassis surface is simply to paint useful molecules onto it. Jin et al. (2012) have taken such an approach by expressing a fusion protein in *P. pastoris* consisting of the soluble, catalytic domain of the enzyme *Candida antarctica* lipase B (CALB), fused to the glycosylphosphatidylinositol (GPI) anchor signal domain of the native cell wall protein GCW21p to direct membrane-anchored expression on the cell surface. This mode of expression removes any requirement for enzyme substrates to diffuse across yeast cell membranes into the cytosol and also any reliance on resulting products diffusing outward from the cell interior. When mixed with a second strain expressing *Rhizomucor miehei* lipase (RML) on its surface by the same route, the dual-mode, whole cell bioreactor system was able to successfully achieve biodiesel yields of 85%–90% from vegetable oil.

Hong Lim et al. (2012) devised a more flexible system by successfully expressing active streptavidin on the surface of yeast cells, stabilised by supplementing growth media with biotin. The avidin:biotin binding pair is one of the strongest non-covalent bonds in biology (Kd ≈ 10^{-14} M) and recombinant proteins are frequently functionalised with biotin to enable their binding to avidin, often for immobilisation within diagnostic assay kits. Hong Lim et al. (2012) showed that surface-displayed streptavidin, although stabilised by the presence of biotin, still retained the ability to bind a biotinylated antibody. This system potentially enables surface display of any biotinylated protein, without the need for its transgene-based expression within the same cell.

A more nuanced approach to yeast surface engineering has been explored by Fukuda and Honda (2014) who expressed the cytosolic domain of a human epidermal growth factor receptor (EGFR) in yeast cells and showed that its autophosphorylation can trigger the native yeast Gγ signal transduction system to stimulate cell growth. Erroneous EGFR auto-phosphorylation is a significant mechanism of oncogenesis in human cells and thus compounds that suppress this auto-phosphorylation may prove to be effective anticancer agents. Within the Fukuda and Honda (2014) system, inhibitors of EGFR autophosphorylation will also suppress yeast cell growth and as such the system can be used to screen compounds as potential anticancer drugs at low cost and with high throughput.

5.2.2 Engineering the Immune Synapse

The cell surface is a critical means of intercellular communication in the mammalian immune system. The field of immuno-oncology is rapidly developing, in part due to developments in understanding of the 'immune synapse' that forms between the surfaces of immune cells and antigen presenting cells, and holds great promise in enabling synthetic biology-based cancer therapies (Lienert et al. 2014). Two exciting areas of research within T cell therapy are chimeric antigen receptor-based T cells (CAR-Ts) and T cell receptor (TCR) therapeutics (Kershaw et al. 2013).

CAR-T therapy involves extracting patient T cells, in a process known as leukapheresis, engineering them *ex vivo* to express a chimeric antigen receptor (CAR) then re-administering to the patient the engineered T cells, which are now capable of CAR-mediated cancer cell recognition and deletion. CARs typically feature single-chain variable domains (scFv) that recognise cancer-specific antigens. T cells armed with a surface-expressed CAR are able to bind any cells presenting the cancer-specific antigen and target them for destruction (Figure 5.1). Engineering T cells outside the body is typically achieved using lentivirus particles containing a 'payload' in their RNA genome encoding a designed CAR expression cassette. The engineered lentiviral particles are able to efficiently infect the T cells where their delivered RNA genome is reverse-transcribed to DNA in the cytosol then integrated into the T cell host genome to ensure long lived expression of transgenes. Alternative transfection and genome-editing techniques (Kalos and June 2013) are being developed using tools such as TALEs discussed in Section 5.5 and clustered regulatory short palindromic repeats (CRISPR) discussed in Section 5.6.

So-called 'on-target, off-tumour' effects have been observed with CAR-T therapy, where similar antigens in non-tumour locations trigger an unwanted response (Tey 2014). Synthetic biology can play a role within T cell therapy by ablating this unwanted toxicity. Logic gates (discussed in Section 5.5) can be introduced into engineered T cells to enhance

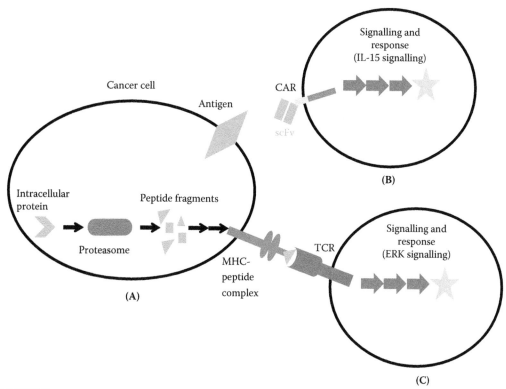

FIGURE 5.1

Engineered cell surface proteins for design of whole cell immunotherapeutics. Mammalian cell surface proteins designed to recognise cancer-specific antigens and trigger expression of immune effector genes upon antigen binding. Cell surface antigens of a cancer cell (A) are recognised by a T cell (B) engineered to express a recombinant chimeric antigen receptor (CAR) consisting of a single-chain Fv (scFv) domain fused to a transmembrane domain (lighter grey) and endodomain (darker grey) derived from other T cell immunomodulatory proteins. Antigen binding by the scFv domain induces conformational changes in the CAR that trigger the IL-15 signal transduction pathway (thick arrows) that concludes in expression of immune response effector genes (grey star). Intracellular cancer cell proteins (C) are processed into short peptide fragments by the proteasome for presentation at the cell surface by the major histocompatibility complex (MHC). The peptide-bound MHC is recognised by an engineered T cell receptor (TCR). Upon binding to the MHC-peptide, conformational changes in the TCR trigger the extracellular signal-regulated kinases (ERK) transduction pathway (thick arrows) which concludes with immune effector gene expression (grey star).

specificity, moderate response level or regulate the cell proliferation response. An AND gate functionality is depicted in Figure 5.2A which prevents a response from being elicited unless two CARs bind antigen to co-stimulate immune effector activation (Kloss et al. 2013; Lanitis et al. 2013). The CARs have been engineered with scFvs that only weakly bind antigen, requiring multiple binding events to effect an internal response. In this way, the AND gate could enhance the specificity of the treatment and reduce off-tumour effects.

However, due to the heterogeneous nature of tumours (Edwards 1985), using weak scFvs may also limit the efficacy of the therapy. Engineering transcriptional and translational logic circuits into effector cascades within the CAR-expressing T cells may enable response regulation without decreasing the efficacy of the therapy. The interleukin 15 (IL-15) signal transduction pathway ultimately stimulates cell proliferation and other immune effector

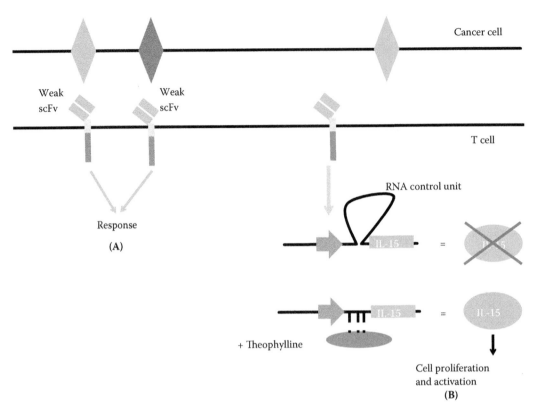

FIGURE 5.2
Modulation of chimeric antigen receptor (CAR) immunotherapeutic output. The scale of response elicited by immune recognition of antigens can be controlled by (A) deliberately reducing the strength of antigen binding by the scFv region within a CAR. This reduction in antigen binding strength also reduces the degree of signal transduction so increasing the number of antigen binding events required to trigger an output. Activation of only one CAR would be insufficient. An alternative strategy (B) is to insert a genetic control element into the output signal transduction pathway. An RNA control unit prevents expression of a signal transduction pathway step protein (IL-15) in the absence of a small molecule (theophylline). Providing the small molecule at a given concentration therefore controls the activity of the signal transduction pathway and the scale of the immune effector response.

responses. Chen et al. (2010) designed an RNA-based AND gate to control the level of IL-15 expression triggered as an output of CAR binding to antigen (Figure 5.2B). IL-15 translation is prevented unless the drug theophylline is present (Chen et al. 2010), allowing greater control of the magnitude of the CAR-T response.

TCR therapy is a method of cancer treatment where patient T cells are engineered to express a designed version of the TCR in order to better recognise and destroy cancerous cells. Whereas CARs recognise antigens that are components of the target cell exterior, TCRs recognise intracellular peptide fragments when presented on the cell surface by major histocompatibility complexes (MHC), as depicted in Figure 5.1. Briefly, intracellular proteins are constantly and randomly degraded into short peptide fragments at a background level by a large, cytosolic multi-protein complex known as the proteasome. These fragments are then processed and presented at the exterior of the cell by the MHC. The fragments can then be engaged by the immune system in order to assess the biological status of the cell. For instance, if a virus infects a cell, intracellular viral peptides, generated by the proteasome, will be surface-presented by MHC for immune surveillance and the cell targeted for destruction as a result. For cancerous cells, these fragments are often less likely to be recognised as a sign of danger and more likely to be recognised as a benign 'self' antigen. As such, engineered TCRs can be used to recognise antigens known to be cancer-specific (Kalos and June 2013).

Wei et al. (2012) engineered T cells by introducing an SGN encoding a negative feedback loop designed to control the strength of immune response elicited by engagement of TCRs with foreign, MHC-presented antigen. Activation of TCRs normally leads to a signalling cascade within T cells that triggers immune responses via the extracellular signal-regulated kinases (ERK) pathway. Wei et al. (2012) designed a genetic TCR control circuit featuring a negative feedback loop to dampen the magnitude of the ERK-based immune response. They placed expression of the bacterial virulence protein, OspF (Figure 5.3A), under the control of an ERK sensitive promoter. When the ERK pathway is activated by TCR binding to antigen, subsequent expression of OspF represses the ERK pathway and serves as an amplitude dampener *in vivo*. Furthermore, use of a doxycycline-activated promoter (Figure 5.3B) instead of an ERK-responsive promoter allows an 'off-switch' to be introduced into the system, providing a method of stopping activity should toxic side-effects be observed.

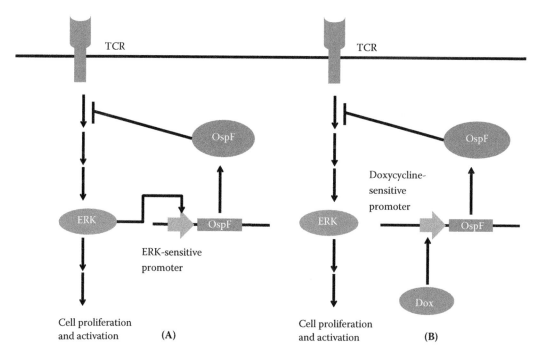

FIGURE 5.3
Modulation of engineered T cell receptor (TCR) immunotherapeutic output. Magnitude of immune response elicited by engineered TCRs can be controlled by (A) a feedback loop that automatically dampens the strength of the TCR output signal. An extracellular signal-regulated kinases (ERK)-sensitive promoter for a downstream effector gene induction step is engineered to also control expression of OspF, which inhibits ERK synthesis. Alternatively (B), OspF expression is controlled by administration of the antibiotic doxycycline that stimulates activity of a doxycycline-sensitive promoter. Providing doxycycline at a given concentration can be used to control the intensity of the immune effector response.

5.2.3 Surface Auto-Biotinylation by Mammalian Cells

Mammalian cell lines are important tools for dissecting aspects of basic cell biology and disease states. They can also be workhorses for industrial scale production of recombinant proteins, virus-based vaccines and gene therapy vectors. Tannous et al. (2006) sought to improve monitoring of an immortalised glioma cancer cell line as it was introduced into a mouse model to investigate tumour formation. To do this, they first fused the 75 amino acid biotinylation target domain from *Propionibacterium shermanii* 1.3S transcarboxylase (PSTCD) to the transmembrane domain of platelet-derived growth factor receptor (PDGFR). Endogenous mammalian biotin ligases are then understood to ligate biotin to the PSTCD domain resulting in a biotinylated cell surface protein (Figure 5.4). The surface-biotinylated cells were amenable to further labelling with a variety of streptavidin-coated magnetic reagents and, after being introduced into animal mouse models of disease, were readily detected within tissues by magnetic resonance imaging.

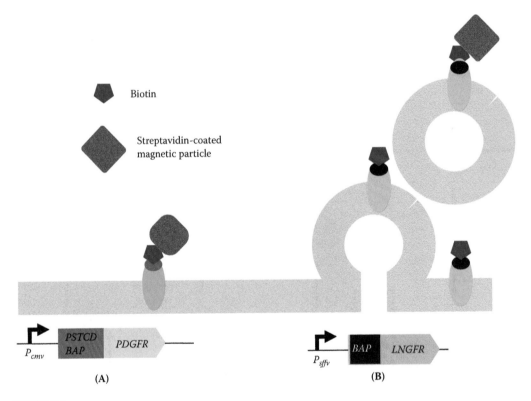

FIGURE 5.4
Engineered cell surface proteins for preparative and analytical applications. (A) The strong P_{CMV} promoter was used to overexpress a recombinant surface protein, platelet-derived growth factor receptor (PDGFR), featuring the biotinylation acceptor domain of *Propionibacterium shermanii* 1.3S transcarboxylase (PSTCD). The protein was biotinylated, most likely by host biotin ligases, and displayed on the cell surface. Streptavidin-functionalised paramagnetic particles were then used to detect biotin-coated cells *in vitro* and *in vivo*. (B) The strong P_{sffv} promoter was used to overexpress an engineered cell surface fusion protein with no cytosolic domain, but transmembrane and extracellular domains from low-affinity nerve growth factor receptor (LNGFR), plus a biotinylation acceptor peptide (BAP), on the surface of a lentiviral packaging cell. A co-expressed, bacterial biotin ligase, BirA, was directed to the lumen of the endoplasmic reticulum to biotinylate the LNGFR-BAP (not shown). Biotinylated LNGFR-BAP was incorporated into the outer envelope of budding lentiviruses, enabling their capture by streptavidin-coated paramagnetic particles.

Production of lentivirus particles for applications such as T cell transduction, discussed earlier, is challenging in terms of industrial scale up. Conventionally, lentivirus particles are allowed to accumulate in the growth medium of infected, adherent cells after which this media is removed and virus particles concentrated by ultracentrifugation. Lentivirus particles bud from the plasma membrane surface, so they always capture a random portion of the cell surface as their own outer envelope. As such, surface-expressed transmembrane proteins are incorporated into progeny lentiviral envelopes in proportion to their abundance at the cell surface.

Nesbeth et al. (2006) engineered a recombinant cell surface protein, low-affinity nerve growth factor receptor (LNGFR), first, to have no intracellular domain, rendering the receptor inert with respect to signal transduction, and, second, to feature a short, 13 amino acid biotin acceptor peptide (BAP) signal at the amino terminal region of the extracellular domain (Figure 5.4). The LNGFR-BAP was coexpressed with a version of the bacterial biotin ligase, BirA, engineered for transport to, and retention in, the lumen of the endoplasmic reticulum (ER) (not shown in Figure 5.4). When these engineered cells are grown in media supplemented with biotin, the ER-resident BirA enzyme ligates biotin to the BAP within the ER lumen. The biotinylated membrane protein then progresses through the secretory pathway to be displayed on the outer, plasma membrane, where it is incorporated into the surface of budding lentiviral particles. These metabolically biotinylated virus particles can then be concentrated onto streptavidin-coated paramagnetic beads and achieve dramatically enhanced infection titres *ex vivo*.

5.2.4 Engineering Mammalian Cell Surfaces

The examples discussed in Sections 5.2.1–5.2.3 show that eukaryotic cell surfaces are a highly effective setting for synthetic biologists seeking to engineer useful 'handles' for capturing or monitoring cells and their products. The cell surface is also a sophisticated information exchange junction by which natural intercellular communication can be exploited, mastered and perhaps eventually replicated in non-natural synthetic cellular systems. Daringer et al. (2014) present a framework in which conditional dimerisation of the external domains of engineered transmembrane proteins is used to control proximity-based interactions between internal proteins which in turn effect a given user-defined output. Nagaraj et al. (2013), on the other hand, constructed a mammalian cell SGN that directs cells to express vesicular stomatitis virus glycoprotein (VSV-G) on their surface in response to blue light. Cell surface VSV-G expression drives formation of membrane–membrane fusion events, termed syncytia. When targeted to tumour cells, syncytia formation was able to cause previously sequestered, immunogenic tumour-specific antigens to be revealed, or cause previously tolerated tumour antigens to become immunogenic by bringing about their juxtaposition with immunogenic antigens. Both effects would result in an increased immunogenicity of the tumour, targeting it for immune clearance.

Finally, Majerle et al. (2014) use an elegant approach to enable screening for inhibitor-resistant proteases of human immunodeficiency virus (HIV). In their method, a transcription factor (TF) was fused to the intracellular domain of a transmembrane cell surface protein by a peptide sequence known to be cleaved by HIV protease. Inhibitor-sensitive protease variants fail to release the TF, resulting in zero reporter gene expression. Inhibitor-resistant HIV protease variants cleave the TF causing reporter gene expression. The diversity of approaches presented in this section serve to illustrate how cell surfaces are a key area of synthetic biology research as a versatile and effective output device.

5.3 Programmable Organelles

Organelles are arguably the most pivotal development in life on Earth due to the key role they are believed to have played in providing sufficient energy to enable aggregation of cells into multicellular organisms. Understanding the processes involved in formation, maintenance and division of organelles such as the Golgi apparatus remains a major goal of fundamental life science research (Ronchi et al. 2014). Synthetic biologists are also beginning to exploit intracellular trafficking routes as a modular sequence of nanoscale unit operations for protein assembly (Nagarajan and Elowitz 2011; Vogl et al. 2013). A useful toolbox of approaches such as co-overexpression of native chaperones (Jossé et al. 2010), or use of peptide signals shown to boost secretion performance (Lee et al. 1999), continues to emerge from biotechnology and synthetic biology, while establishment of standardisation and parts repositories still lags to a degree.

Plastids are organelles encompassed by double membranes such as mitochondria and chloroplasts. They are believed to have developed due to endosymbiosis and typically possess minimal genomes and a specialised set of functions with respect to their surrounding host cell. Plastids divide by binary fission in a process that is topologically distinct from the rest of their host cell. As nanoscale biological machines that fulfil a specific sub-role within a larger, more complex structure, plastids are perhaps an ideal system for synthetic biologists to study, control and reconstruct. First attempts have been made to develop synthetic chloroplasts in the form of photosynthetic bacteria engineered to invade mammalian cells (Agapakis et al. 2011).

5.3.1 Editing a Chloroplast Genome

Chlamydomonas reinhardtii is a green microalga that, via its chloroplasts, uses energy from the sun to drive cell growth. Up to 50% of *C. reinhardtii* total biomass can be soluble protein (Figure 5.5A) and as such the organism has attracted interest as a potential industrial host platform for production of recombinant proteins. The genome of the *C. reinhardtii* chloroplast has been sequenced and shown to alternate between linear and circular forms (Maul 2002). Insertion of foreign genes into the chloroplast genome can be a relatively reliable strategy as gene insertion can be targeted to a specific locus and transgene expression can be sustained with minimal gene silencing effects. A diverse range of chaperones and disulphide isomerases are naturally present within the *C. reinhardtii* chloroplast, making it an ideal chamber for efficient folding of complex, multi-component exogenous proteins from higher eukaryotes.

Braun-Galleani et al. (2015) developed a plasmid-based, chloroplast genome targeting strategy (Figure 5.5B) for insertion of a codon-optimised open reading frame (ORF) encoding vivid verde fluorescent protein (VFP) as reporter protein. The VFP ORF was under the control of the constitutive promoter of the native atpA gene and the target strain lacked an active version of the psbH gene, so it was unable to grow in phototrophic condition. Braun-Galleani et al. (2015) used introduction of an active copy of the psbH gene and subsequent incubation in phototrophic conditions as a successful method of selecting for transformants. The VFP and psbH genes were positioned between two genes of the chloroplast genome, trnE2 and psbN, within a plasmid. Transfection resulted in successful homologous recombination between plasmid and chloroplast genome (Figure 5.5B). VFP expression was then used as an indicator of likely productivity for other recombinant proteins. Using VFP expression as a guide, a set of

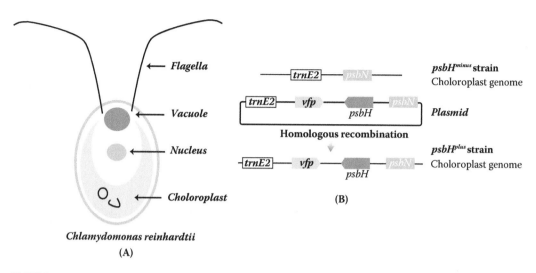

FIGURE 5.5

Editing a chloroplast genome. (A) Schematic overview of the green microalga *Chlamydomonas reinhardtii* (*C. reinhardtii*). Note the large proportion of cell volume occupied by the chloroplast and the depiction of the genome in both linear and circular conformations (black circle and line within chloroplast). (B) Plasmid-based editing of the *C. reinhardtii* genome as performed by Braun-Galleani et al. (2015). Transgenes intended for insertion are flanked by copies of trnE2 upstream and psbN downstream. Against a psbH-negative background, psbH expression can be used to select stable transformants. Screening of transformants for codon-optimised vivid verde fluorescent protein (VFP) expression levels indicates the likely performance of a given clone when expressing other recombinant proteins.

growth conditions was developed to favour maximum transgene expression. However the optimised VFP expression conditions did not predict the yield of a commercially relevant recombinant protein, indicating further steps will be necessary to make the *C. reinhardtii* chloroplast a robust and predictable industrial route to recombinant protein production.

5.3.2 Quantitating Intracellular Events

The ultrastructure of eukaryotic cells is complex (Figure 5.6A) and the interactions of different subcellular structures add another level of complexity to the events occurring within a eukaryotic cell at any given time. A long-term goal of synthetic and conventional biology is to have a clear and detailed an understanding of these complex phenomena. A step toward this is the ability to monitor transitory intracellular events such as protein–protein interactions, synthesis and conversion of short-lived metabolites and the assembly and disassembly of the multi-protein complexes that direct vesicle budding, migration and docking. Forster resonance energy transfer (FRET) is an effective and versatile tool for probing such dynamic intracellular events. FRET exploits the phenomenon whereby a non-radiative energy transfer can take place between two fluorophores in a manner defined by the physical distance between the fluorophores. The excitation spectrum of the acceptor fluorophore must overlap with the emission spectrum of the donor fluorophore for the non-radiative energy transfer to occur. The closer the proximity of the two fluorophores the higher the likelihood that transferred energy will lead to excitation within the acceptor.

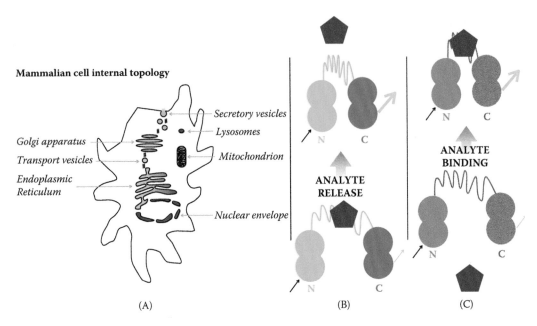

FIGURE 5.6

FRET sensors to probe within eukaryote cells. (A) Diagram of the ultrastructure of mammalian cell. Note that the lumen of the Golgi (green) and endoplasmic reticulum (light blue) and the interior of the nuclear envelope (dark blue) are topologically equivalent to the cell exterior. (B) Fusion protein configuration in which analyte binding suppresses (thin blue line) the FRET signal which it at its strongest (thick blue line) when analytes are absent. (C) Fusion protein configuration in which analyte binding stimulates (thick blue line) the FRET signal which it at its weakest (thin blue line) when analytes are absent.

Most FRET sensors are constructed from recombinant a fusion protein that incorporates two fluorescent proteins with overlapping spectra (Figure 5.6B and C). A third region links the two fluorescent protein domains within the fusion protein and determines the configuration of the FRET sensor. Upon binding to an analyte, typically a small molecule, this linker region will either bring the two fluorescent protein domains into close proximity (Figure 5.6A) or push them further apart (Figure 5.6B). In this way, FRET sensors can be configured to signal the appearance (Figure 5.6B) or loss (Figure 5.6A) of a given analyte. Bespoke design of FRET biosensors based on fusion proteins can require significant genetic screening of variants with different three-dimensional structures to optimise the expansion and contraction of fluorophore distance in response to analyte presence (Ha et al. 2007).

Fusion protein-based FRET biosensors have been developed to monitor cellular pH (Gjetting et al. 2012), oxygen levels (Potzkei et al. 2012) and metabolites such as amino acids and carbohydrates (Bogner and Ludewig 2007; Behjousiar et al. 2012). The redox status of the lumen of the ER can have a major influence on the productivity of CHO cells for production of biotherapeutics. Lin et al. (2011) developed FRET biosensors influenced by the oxidative formation of cystine disulphide bridges to monitor ER redox state.

5.3.3 Portable Mitochondria

As the principle actuating motor of biological chassis, the mitochondrion represents a tantalising challenge to engineers and synthetic biologists seeking to master the performance of biological machines (Tang 2015). The mitochondrion of *Ashbya* (*Eremothecium*) *gossypii*, a filamentous-growing plant pathogen, exhibits some unique codon reassignments (Ling et al. 2014), suggesting mitochondrial genomes may be more flexible than nuclear genomes with respect to their genetic code usage. Alford et al. (2012) successfully used dimerisation-dependent fluorescent proteins (ddFP) in a manner similar to FRET to monitor physical association between mitochondrial and ER membranes.

In the field of biomedicine, several genetic conditions are associated with the mitochondrial genome, such as Leigh disease. As yet it is not possible to edit mitochondrial genomes *in situ* in a way that would address mitochondrial disease. However, it is possible to transfer the nucleus of a human egg to an enucleated acceptor egg. It follows that it is also possible to transfer a nucleus from an egg cell containing mitochondria with a genetic disease to an enucleated egg cell with mitochondria free of genetic disease. Subsequent fertilisation of the egg with the 'donor mitochondrion' should result in disease-free individuals. This process has been made legal in the United Kingdom (UK Parliament 2015) but remains illegal in many countries. A major driver of acceptance of this procedure was the demonstration of its safety, reliability and robustness. Similarly safe, robust and reliable methods of mitochondrial genome editing could in the future achieve the same results as mitochondrial donation, while also preserving the remaining maternal genetic heritage within the mitochondria that have been genome-edited to be disease free.

Efforts to design, synthesise and implement a synthetic mitochondrial genome are still at an early stage. Yoon et al. (2010) successfully ligated the entire 16.3 kb mouse mitochondrion genome into a bacterial plasmid backbone. The resultant plasmid-encoded mitochondrial genome was transferred into genome-free recipient mouse mitochondria by electroporation. The transferred genome was shown to have transcriptional activity, but the hybrid mitochondria did not replicate. Yang and Koob (2012) are also developing techniques to stably transfer isolated recombinant mitochondria into recipient cells.

5.4 Eukaryotic Genes and Their Synthetic Counterparts

Model structures of mammalian and viral genes inform the design of artificial genes. Natural genes often contain multiple genetic elements. Life science fields such as molecular genetics, molecular evolutionary biology and biochemistry have elucidated which genetic elements are necessary for gene expression, and even many that can be swapped into different genes and between organisms. Synthetic biology owes much of its success to the models produced by data from comparative genomics, genetic disruption analyses, protein crystallography and reconstitution assays. Synthetic genes are often much more streamlined than their natural counterparts. For instance, synthetic biologists express proteins from ORFs derived from cDNA libraries, which therefore lack the introns common in natural genes. Such minimal gene designs help to overcome the practical constraints of constructing complex circuits.

Manipulation of SGNs is a powerful synthetic biology strategy for both probing gene function and gaining precise temporal control over phenotype. Regulation of the genes comprising an SGN is typically controlled via sequences upstream of ORFs known as promoters. Promoters serve as a binding site for transcription machinery and due to the central role they play in gene regulation and transcription they have become a primary target for synthetic biology-based engineering (Hasty et al. 2001). The eukaryotic DNA sequence elements that encode a given protein expression cassette are typically more complex than those of prokaryotes, especially when considering organisms commonly exploited by biotechnologists such as budding yeasts and mammalian cell lines. The genetic elements commonly found in eukaryotic expression cassette, or 'genes', are illustrated in Figure 5.7 alongside typical prokaryotic gene elements.

Some of the toughest engineering challenges posed by genetic elements are standardisation and absolute measurement of transcription and translation initiation events and rates of transcription and translation. Attempts by synthetic biologists to establish such measurements are discussed in Chapter 3. A more realistic challenge in the short term is to establish inventories of well-characterised, interoperable genetic components such as promoters that can be easily assembled to fabricate designed genes. The scaling of such schemes could then enable a higher order of complexity in biological design in which multiple genes are used to build networks with emergent or sophisticated properties. Moreover, parallel, as opposed to serial, modes of gene assembly would accelerate this approach, as discussed in Chapter 2.

Figure 5.7 depicts a typical prokaryote promoter (Figure 5.7A and B), which is clearly distinct from a typical eukaryotic promoter (Figure 5.7C and D). In typical natural eukaryote promoters, TFs are recruited to secondary sequences known as cis-regions from where they recruit RNA polymerase to the main promoter sequence and initiate transcription at the transcription start site (TSS). The cis elements and main promoter sequence form a region called the core promoter. Synthetic eukaryotic promoters have been designed in yeast (Hartner et al. 2008; Ruth et al. 2010) and mammalian (Tornøe et al. 2002; Hartenbach

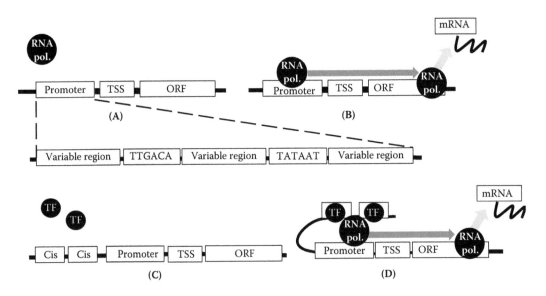

FIGURE 5.7
Natural promoters. (A) Structure of a prokaryotic gene, a promoter region, a transcription start site (TSS) and an open reading frame (ORF) consisting of codons in a given reading frame and sequence to direct expression of a unique protein. Zooming in to the promoter sequence (dashed lines) reveals a common structure of three variable regions separated by the sequences indicated. (B) Collectively, the sequences comprising the promoter have an affinity for RNA polymerase (RNA pol.) which binds the promoter then moves in one dimension (darker grey arrow) to transcribe DNA to RNA (lighter grey arrow). (C) Structure of a eukaryotic gene. In addition to core promoter, TSS and ORF sequences, *cis* elements exist upstream of the promoter. (D) Transcription factor binding of cis elements is often necessary for promoter activity. This binding tends to cause the *cis* elements to loop toward the promoter and activate the RNA polymerase *in situ*. The transcription complex then moves in one dimension (darker grey arrow) to transcribe DNA to RNA (lighter grey arrow).

and Fussenegger 2006) platforms to achieve a variety of goals, such as to explore and understand SGN performance, optimise productivity of industrial host chassis (Ruth and Glieder 2010) and enable precise control over the phenotype of a whole cell therapeutic (Purnick and Weiss 2009; Schlabach 2009).

5.4.1 Synthetic *Pichia Pastoris* Promoters

P. pastoris is an increasingly popular industrial host chassis for production of a variety of useful chemicals and recombinant proteins. Most strategies for engineering *P. pastoris* expression focus on the promoter of the Alcohol Oxidase 1 gene, P_{AOX1}, as it is well-characterised, allows for strong expression and can be induced by switching to methanol as carbon source (Goodman 1984). P_{AOX1} has been used as the starting point for the design of synthetic promoters that direct a range of expression levels and modes of transcriptional regulation, often by subjecting the core promoter and 5′ untranslated region (UTR) to random mutagenesis and selecting useful genotypes. Vogl et al. (2014) designed a fully synthetic core promoter and 5′ UTR and created a library of diverse P_{AOX1} promoters with a variety of properties. They then aligned core sequences of four natural promoters with different regulation profiles: P_{AOX1}, P_{GAP}, P_{HIS4} and P_{ScADH2}. The consensus sequence that emerged from this alignment (Figure 5.8) contributed to the design of a selection of synthetic promoter variants. Analysis of these variants with a green fluorescent protein (GFP) reporter showed productivity ranging from 10% to 117% of wild type expression levels, providing a useful 'dial' synthetic biologists can use to fine-tune gene expression levels within the wider context of an SGN or desired phenotype.

5.4.2 CRISPR/Cas9-Mediated Transcriptional Control

Farzadfard et al. (2013) developed a programmable system of eukaryotic transcriptional regulation (Figure 5.9) based on the 'clustered regulatory short palindromic repeats' (CRISPR) system, discussed in depth in Section 5.6, and synthetic TFs. In nature, CRISPR forms part of the bacterial immune system that deals with invading genetic material from bacteriophages. A key feature of the CRISPR system is the ability of small, 'guide' RNA (gRNA) molecules to bind an endonuclease, Cas9, and target it to invading bacteriophage DNA in a sequence-specific manner to effect its destruction. The CRISPR/Cas9 system can be exploited for targeted gene editing in eukaryotes by transgenic expression of user-defined gRNA molecules and engineered Cas9. The Cas9 is then targeted to a locus of choice and catalyses a double-strand break (DSB), the repair of which will then typically be error prone and cause gene knockdown if an ORF is targeted (Chakraborty 2014).

pCoreAOX1	CTAACCCCTACTTGACAGCAATATAAAACAGAAGG
pCoreGAP	ATTGGAAACCACCAGAATAGAATAAAAGGCGAACA
pCoreHIS4	CTTGAGTACGAACTATGTATCTATAAAATCGCAG
pCoreADH2	GTATAGCATGCCTATCACAACTATAAAAGAGACCG
Synthetic core	CTTGAG–ACGACC–A–ATAACTATAAAAGCGACCG

FIGURE 5.8

Synthetic promoter design. Alignment of regions of a selection of natural promoter sequences used by Vogl et al. (2014) to design a novel synthetic promoter sequence, the core of which is indicated lowermost in the diagram.

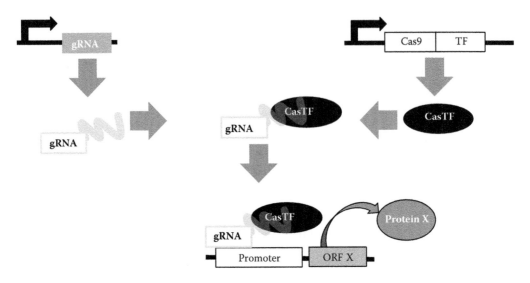

FIGURE 5.9
Synthetic transcription factor design. CRISPR-based promoters feature a fusion protein, 'CasTF', comprising a Cas9 nuclease (see Section 5.6) domain, modified to remove DNA cleavage activity, plus a domain with transcription factor (TF) activity. In parallel a short, guide RNA (gRNA) sequence must be designed with specificity for a promoter of interest and expressed. When co-expressed the gRNA binds to the Cas9 domain of the CasTF fusion protein and guides the complex to the promoter of interest whereupon transcription of downstream open reading frames (depicted here as 'ORF X') is coordinated by the TF domain of the CasTF fusion protein and no longer by a regulated host-cell TF.

To further exploit CRISPR/Cas9 as a programmable transcriptional regulatory system Farzafard et al. (2013) developed the CRISPR interference (CRISPRi) approach. Rather than modifying the often complex, natural TF protein machineries, CRISPRi replaces them with a straightforward fusion protein, termed 'CasTF' in Figure 5.9, consisting of an endonuclease-deficient Cas9 domain fused to a TF domain capable of binding core, relatively invariant elements of a promoter. When gRNA is used to guide a CasTF to a given promoter, the RNA polymerase component of the TF domain induces transcription of the downstream ORF. By binding to a host cell promoter site, CasTF usurps the natural TFs, decoupling the downstream ORFs from host cell regulatory pathways that influenced the natural TF and replacing them with factors of choice dictated by the CasTF designer. Transcription activator-like effectors (TALEs) are another system that can be used to direct transcription activation functions to a sequence of choice (Sanjana et al. 2012) and are discussed further in Section 5.5 of this chapter.

5.4.3 Computational Analysis of Promoter Sequences

Computational approaches to promoter analysis are valuable as they can identify novel genes, or novel promoters for known genes. The identification of novel promoter sequences results in a greater understanding of how natural gene regulatory networks (GRNs) function, which ultimately enables their manipulation via SGNs. Well-established computational approaches (Wang and Hannenhalli 2006) can calculate the probabilistic location of eukaryotic promoters and their TSS within a genome. They operate on the basis that several TFs and RNA polymerases must interact to initiate eukaryotic transcription, and this fact imposes location constraints on the positioning of genetic elements within a genome. Core promoter elements are often positioned at a specific distance from the TSS and from each other. Mathematical algorithms such as position-specific propensity analysis (PSPA), developed by Wang and Hannenhalli (2006), and PromoSer, developed by Halees (2003), calculate the probability of core promoter elements being present within a genome, along with their location and associated TSS.

5.4.4 Controlling Eukaryotic Translation

An increasingly popular route to control initiation of translation and subsequent translation rate in eukaryotes is to use artificial regulatory RNAs to co-opt mRNA to regulate either translation initiation or mRNA stability (Saito and Inoue 2007, 2009; Wieland and Fussenegger 2012). Pothoulakis et al. (2014) also use designed RNA molecules to help measure gene expression. They designed an mRNA molecule to incorporate both an ORF for red fluorescent protein (RFP) and, downstream of the ORF, an aptamer RNA sequence that fluoresces when bound to a fluorophore. This elegant combination of a fluorescent mRNA and fluorescent protein encoded by the mRNA enabled simultaneous measurement of transcription and translation.

5.4.5 Transcriptional Termination Control in Eukaryotes

The mechanism underlying transcription termination in eukaryotes is understood to a lesser degree (Proudfoot 2011) than the equivalent process in prokaryotes. In eukaryotes, a polyadenylation signal (PAS) encoded in DNA directs both transcription termination and addition of a poly-adenine tail to the nascent mRNA (Levitt et al. 1989; Huang et al. 2010). Despite the lack of mechanistic understanding of the PAS system, Choi et al. (2014) were able to reduce, amongst other elements, the size of a PAS in order to reduce the length of DNA needed to encode an adenoviral vectors gene therapy payload, without compromising payload delivery or function. There is an upper limit to the amount of genetic material that can be packaged into an adenoviral particle. By reducing the size of the PAS sites present in the engineered adenoviral genome, Choi et al. (2014) reduced the DNA sequence requirement by 399 bp. This, therefore, increased by 399 bp the spare capacity of the viral genome available for encoding further therapeutic payload genes.

5.5 Networks of Synthetic Genes in Mammalian Cells

Eukaryotic gene networks have been a fundamental topic of life science research since the early days of genetics, beginning with the mapping of gene linkage by reciprocal crossing and observations of oscillating catabolite utilisation in yeast (Schamhart et al. 1975). Epistatic effects on gene heritability within populations also hint at the importance of networks of genes for encoding complex and dynamic phenotypes. In principle, eukaryotic SGNs can control a greater degree of phenotypic complexity than those housed by prokaryotic chassis due to greater ultrastructural complexity of the host cells (Agapakis and Silver 2009) and divergent mechanisms of gene expression control (Haynes and Silver 2009; Lienert et al. 2014). Multiple organelles, some of which possess their own genomes (mitochondria, chloroplasts and other plastids) represent topologically distinct compartments in which SGNs and their encoded proteins can perform spatially localised cellular functions.

Fully exploiting this design space is an area of increasing research activity in synthetic biology (Greber and Fussenegger 2007; Tigges and Fussenegger 2009). Much of the elementary work on SGNs in eukaryotes, typically in mammalian and yeast chassis, builds on the formats established in *E. coli* and discussed in Chapter 4. As such, a preliminary aim of eukaryotic synthetic biology has been to recapitulate in eukaryotes the SGN-based construction of classical circuits achieved in prokaryotes (Greber and Fussenegger 2007; Tigges and Fussenegger 2009).

5.5.1 An Oscillatory SGN in Mammalian Cells

Tigges et al. (2009) constructed an oscillatory SGN in CHO cells using modified Cytomegalovirus (CMV) promoters to control expression of a tetracycline-dependent transcription transactivator (tTA), a GFP protein engineered for a shortened half-life and a pristinamycin-dependent transactivator (PIT), developed by Fussenegger et al. (2000), with each expression cassette distributed onto a separate plasmid (Figure 5.10). The PIT induces transcription from a pristinamycin-responsive promoter (P_{PIR}) oriented to control transcription of the reverse complement (antisense) strand of the tTA ORF, therefore downregulating tTA expression after a given amount of PIT has accumulated. Downregulation of tTA eventually leads to decreased tTA, GFP and PIT levels to the extent that the P_{PIR} is

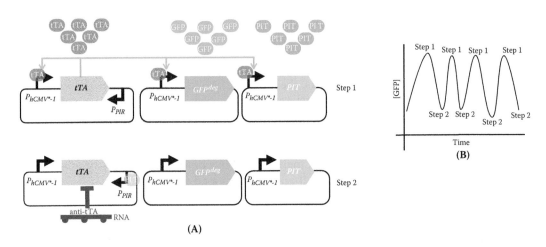

FIGURE 5.10

Synthetic mammalian oscillator circuit. (A) A tetracycline-dependent transactivator (tTA), a short half-life green fluorescent protein (GFP^{deg}) and a pristinamycin-dependent transactivator (PIT) are encoded across three separate plasmids. Once basal activity of $P_{hCMV^{*}-1}$ initiates tTA expression, tTA then induces all copies of the $P_{hCMV^{*}-1}$ promoter, causing a positive feedback loop of its own expression. GFP and PIT expression are also consequently expressed. GFP expression serves as a reporter of $P_{hCMV^{*}-1}$ promoter activity. The highest level of GFP expression can be referred to as 'Step 1'. Over time, sufficient PIT is present within the cell to induce expression from a PIT-sensitive promoter, P_{PIR}. Positioned in the antisense direction relative to the tTA open reading frame, induced P_{PIR} drives transcription of an RNA molecule that is antisense to the tTA-coding mRNA. This antisense RNA molecule will bind to its complimentary partner strand, the tTA mRNA, and abolish translation of tTA protein. This antisense downregulation gradually reduces expression of tTA, and consequently PIT and GFP^{deg}, down to a basal level that can be referred to as 'Step 2'. In the absence of PIT expression, PIT bound to the P_{PIR} promoter will degrade over time. Over time the antisense RNA will also degrade enabling expression of tTA to revive, restarting an oscillatory cycle. The activities of tTA and PIT can be 'tuned' with tetracycline and pristinamycin respectively to manipulate the dynamics of circuit behaviour (not depicted). (B) Cartoon of data that could be expected from such an oscillatory circuit.

no longer active, the anti-tTA antisense RNA eventually degrades and translation of tTA mRNA can start up again, beginning a new cycle of oscillation.

The use of tTA allows expression of ORFs controlled by the modified CMV promoter to be 'tuned', increased or decreased, by the level of tetracycline present in the media. A deterministic mathematical model of the system predicted that oscillation would be sensitive to the relative abundances of the different oscillator components. 'Tuning' of componentry, in order to modulate the oscillation behaviour, was therefore achievable by transfecting cells with different ratios of each of the three plasmids to effect different dosages of each expression cassette. Data showing GFP oscillation occurring over four or more cycles was captured for four individual adherent cells by time-lapse fluorescence analysis.

A key design consideration for SGNs, such as the Tigges et al. (2009) oscillator, is the requirement for 'orthogonal' switches: promoter and repressor devices. For instance, transactivators ideally will not activate unintended genes present in the cell whose activity may abolish or confound the intended oscillatory phenotype. Similarly, transcriptional effectors found naturally in the host chassis will ideally not activate promoters used in the SGN, causing them to be activated or repressed in manner that abolishes or disturbs the intended phenotype. For this reason, genetic elements are typically assembled from pathogens or divergent organisms such as herpes simplex virus (tTa), CMV and the bacterium *Streptomyces coelicolor* (PIT) to minimise the 'cross talk' between native host chassis elements and the inserted SGN.

5.5.2 Eukaryotic SGNs for Boolean Logic Functions

Computers are a ubiquitous feature of the modern world and are composed of circuits in which voltages are switched within a pattern of conducting lines 'printed' onto a small wafer or 'chip' using a semiconducting material, usually silicon. Voltage switching within these integrated circuits corresponds to the '1' and '0' of binary notation and also the 'true' and 'false' terms of a system of logical propositions or 'gates'. Logic gates, statements that are either true or false, were first codified by George Boole in 1849 and can have multiple inputs but typically only one output, 'true' or 'false'. Definitions of the seven key 'Boolean' logic gates, AND, OR, NOT, NAND, NOR, XOR and XNOR, are ubiquitous on the Internet, an unsurprising fact given the central role of computers in society today. Standardised graphical symbols that represent Boolean logic gates are widely adopted (Institute of Electrical and Electronics Engineers 1994). Early integrated circuits, developed in the 1960s, such as the 'small scale integration' (SSI) integrated circuit, comprised in the order of ten logic gates. Modern integrated circuits assemble logic gates at scales and within architectures that enable extreme miniaturisation and comprise millions of logic gates.

Although this miniaturisation of integrated circuits has now reached the nanoscale (Bianchi et al. 2015), the computational achievements of biology in terms of DNA- and cellular-based data storage, migration and execution of instructions, is still far superior to what humans have achieved with silicon and other non-biological materials. For example, one gram of DNA is capable of storing in the order of 700 terabytes of data. It is also the case that much of the gene regulation that occurs within cells remains incompletely understood with respect to how it controls complex phenotypes. As such, synthetic biologists seek to build the classic Boolean logic gates using SGNs in order to construct robust and accurate mechanisms of phenotype control and, ultimately, to help unlock the potential of biology as an option for materialising a new generation of computing.

The Boolean NOR and AND gates are key examples of logic gates that have been constructed with SGNs. NOR gates provide a binary output: 'on' or 'off', '0' or '1', based on the presence of two inputs (A and B). If either, or both, input is present, the output will be '0' and only if neither input is present will the output be '1' (Figure 5.11). Combining three NOR gates in a circuit creates an AND gate, where the presence of both inputs (A and B) is required for an output of 1 (Figure 5.11). Construction of analogous NOR and AND gates using biology is hindered by free diffusion of solutes in three dimensions throughout the cytosolic fluid of cells. Diffusion of TFs in this way allows them to interact simultaneously with all cognate promoters present within a given SGN or indeed an entire genome, preventing sequential activation or repression of gates. Therefore, rather than using the same regulator multiple times, a wide array of orthogonal regulators must be sourced or developed in order to allow correct gate activation and circuit operation.

The process of identifying naturally orthogonal gene regulatory elements currently tends not to involve standardised components selected in an automated, machine readable and executable manner. More often, it is the case that researchers familiar with both the basic and applied molecular biology research literature uncover knowledge of potentially orthogonal elements from research articles. Even if this component-sourcing process were to become automated, naturally orthogonal genetic elements can be heterogeneous in nature, exhibiting variation in their rates of activation, repression and de-repression. Such heterogeneity would likely add an unnecessary layer of complication to biological circuit design. To address these challenges, systems have been developed in which the same promoter can be modified with user-defined, sequence-specific regulatory elements. This approach enables a single-promoter type to be deployed within multiple expression cassettes, each copy of the promoter controlled by unique and orthogonal regulatory elements.

Rather than scouring research literature for naturally occurring orthogonal elements, Gaber et al. (2014) used multiple copies of the same CMV promoter, each rendered unique and orthogonal by incorporating unique binding sites for repressor proteins based on transcription activator-like effectors (TALEs), as indicated in Figure 5.11. TALEs were first described in the plant pathogen *Xanthomonas* (Boch et al. 2009). They were characterised as TFs that bind DNA sequences, typically 18 bp in length, according to a set of rules governing complementarity between the amino acid sequence of the DNA-binding domain and the base pair sequence of the bound region of DNA.

Elucidation of these rules enabled the construction of novel TFs specific for only a user-defined, therefore orthogonal, DNA sequence (Cong et al. 2012; Garg et al. 2012). Witzgall et al. (1994) previously showed the Kruppel-associated box (KRAB) protein domain to be an efficient suppressor of transcription. Gaber et al. (2014) showed this repression function is preserved in TALE-KRAB fusion proteins. Placing TALE-binding DNA sites next to a CMV promoter allows TALE-KRAB fusion proteins to be targeted to that promoter via their TALE domain. Once in position, the TALE-KRAB fusion protein can repress transcription from the nearby CMV promoter via the activity of the KRAB domain. The unique nature of each TALE-binding DNA site means that a unique TALE-DNA binding pair can be used to orthogonally repress each of a set of identical CMVs.

In the biological NOR gate in Figure 5.11, two TALE-binding DNA sites, A and B, are positioned upstream of the CMV promoter and reporter gene. The presence of TALE-KRAB repressor proteins A and/or B would repress transcription of the reporter gene by steric hindrance of further TF binding to CMV. Figure 5.11 shows how biological AND gate function can be achieved by linking three NOR gates. Repressors A and B repress expression of a third TALE-KRAB repressor C. Expression of repressor C would repress transcription of the 'output' gene. The construction of further, more complex logic gates based on inputs by six or more orthogonal, user-defined TALE-KRAB repressor/repressor site pairs could theoretically be used to achieve 256 three-input Boolean logic functions.

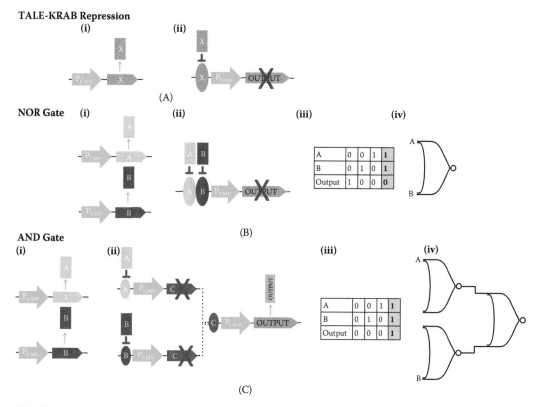

TALE-KRAB Repression

FIGURE 5.11

Constructing Boolean circuitry with genes. (A) Example of how a TALE-KRAB fusion protein-based can repress a P_{CMV} promoter by binding to a TALE-binding DNA sequence placed next to the promoter. (A) (i) Constitutive expression from a P_{CMV} promoter (P_{CMV}, green arrow) drives transcription of an open reading frame (ORF) (grey boxed arrow labelled 'x') encoding TALE-KRAB fusion protein 'X' (grey rectangle labelled 'x'). (A) (ii) Once expressed, TALE-KRAB fusion protein 'X' binds to its cognate TALE-binding DNA sequence (grey oval labelled 'x') in a sequence-specific, therefore orthogonal, manner. Once bound, the KRAB domain of the TALE-KRAB fusion protein acts to repress transcription from the nearby P_{CMV} promoter, preventing transcription (red letter 'X') of the ORF encoding a given output reported gene (grey-boxed arrow labelled 'OUTPUT'). (B) (i) Constitutive expression from P_{CMV} promoters (P_{CMV}, green arrow) drives transcription of an ORF (blue boxed arrow labelled 'A') encoding TALE-KRAB fusion protein 'A' (blue rectangle labelled 'A') and an ORF (red boxed arrow labelled 'B') encoding TALE-KRAB fusion protein 'B' (red rectangle labelled 'B'). (B) (ii) Once expressed, TALE-KRAB fusion proteins 'A' and 'B' bind their cognate TALE-binding DNA sequences (blue oval labelled 'A', red oval labelled 'B') in a sequence-specific, therefore orthogonal, manner. Once bound, the KRAB domains of the TALE-KRAB fusion proteins act to repress transcription from the nearby P_{CMV} promoter, preventing transcription (red letter 'X') of the ORF encoding a given output reported gene (grey boxed arrow labelled 'OUTPUT'). (B) (iii) Tabular representation of how the gene network functions as a NOR gate. The case in bold with grey background reflects the status of the network as illustrated in (B) (ii). (B) (iv) Standardised graphical representation of a NOR gate. (C) (i) Same scheme for expression of TALE-KRAB fusion proteins 'A' and 'B' as in (B) (i). (C) (ii) Once expressed, TALE-KRAB fusion proteins 'A' and 'B' bind their cognate TALE-binding DNA sequences (blue oval labelled 'A', red oval labelled 'B') in a sequence-specific, therefore orthogonal, manner. Once bound, the KRAB domains of the TALE-KRAB fusion proteins act to repress transcription from their nearby P_{CMV} promoters, preventing transcription (red letter 'X') of separate copies of an ORF (purple boxed arrow labelled 'C') encoding TALE-KRAB fusion protein 'C'. In the absence of TALE-KRAB fusion protein 'C', its cognate TALE-binding DNA sequence (purple oval labelled 'C') remains unbound, permitting expression of an ORF encoding a given output reported gene (grey boxed arrow labelled 'OUTPUT') into protein (grey rectangle labelled 'OUTPUT'). (C) (iii) Tabular representation of how the gene network functions as an AND gate. The case in bold with grey background reflects the status of the network as illustrated in (C) (ii). (C) (iv) Standardised graphical representation of an AND gate.

5.5.3 Future Prospects for Eukaryotic SGNs

The construction of biocircuits is still in its infancy. However, the capacity to generate large numbers of repressible promoters that are user-defined, unique and orthogonal is potentially a very significant milestone. Although there are still natural processes that may act to counteract circuit function, such as gene repression through chromatin silencing, real world applications such as controlled stem cell differentiation and maturation could benefit from this technology.

Exploration of eukaryotic SGN architecture *in silico* can provide useful insight (Longo et al. 2010) and a well-characterised set of tools is now available to materialise SGN designs (Arpino et al. 2013). One approach to incorporating larger numbers of biological logic gates is to distribute SGNs across multiple cell types (Weber et al. 2007) and Purnick and Weiss (2009) discuss how such systems may enable sophisticated new functions. Biomedical application could clearly benefit from the ability of designed biological circuits that can respond dynamically to changing inputs and this area is explored in Section 5.7.

5.6 Genome-Scale Engineering in Eukaryotes

Since the human genome was first sequenced in 2000, data sets derived from whole genome analysis have provided new perspectives that were previously unavailable from incomplete snapshots of genetic information and cell behaviour (Kanehisa 2001). Genome-scale data sets, encompassing entire proteomes, transcriptomes, oncomes and glycomes have accelerated our understanding of biological phenomena, not least by revealing roles for collections of genes and proteins that would not otherwise have been evident from studies limited to subgroups of genes or observations.

Genome synthesis by non-biological means and novel genome design are areas of synthetic biology that can draw scepticism, often borne from the fact that our knowledge of how natural genomes function is so incomplete. For example, how a given genome encodes subcellular phenomena, such as intracellular trafficking and metabolic homeostasis, or macro-scale, multicellular phenomena, such as the immune system and limb development, remain wide open questions. Nonetheless synthetic biologists have made real progress in developing the fundamental 'picks and shovels' of genome building as a robust and relatively predictable technology. Presently, these fundamental tools are used to assemble synthetic genomes composed of naturally occurring DNA sequences. However, the design of *de novo* genomes comprised of *de novo* genes and gene networks that encode predictable cellular phenotypes is a hugely challenging but exciting prospect.

5.6.1 Assembling Synthetic Regions of the Genome to Replace the Natural

The 14 Mb genome of *S. cerevisiae*, comprising 16 chromosomes, is fully sequenced and the organism is extremely effective as both a model for understanding eukaryote cell biology and as an industrial host cell for the biomanufacturing of fine chemicals and recombinant proteins. However, most efforts to bring new capabilities or optimised performance to yeasts involve relatively small-scale genetic modification, with truly genome-scale activities limited to mutagen treatment (Spencer and Spencer 1996) and *in silico* modelling of metabolism (Heavner et al. 2013). A greater level of control and understanding of the structure and function of the *S. cerevisiae* genome *in vivo* promises to deliver enhanced capabilities with respect to basic research and industrial exploitation. Practical goals for *in vivo* yeast genomics include better understanding of chromosomal properties, gene organisation and the functions of DNA that is untranscribed or encodes RNA of unknown function. To answer some of these questions 'synthetic genomics' has grown as a branch of synthetic biology that attempts to gain better understanding of genomes through their artificial synthesis and design. The hugely significant milestone, discussed in Section 4.4, of synthesising a biologically active genome based on the small prokaryotic organism, *Mycoplasma genitalium* (Gibson et al. 2010), has been built upon with the 'Sc 2.0' collaborative project to assemble an entirely synthetic chromosome in *S. cerevisiae* and eventually a synthetic yeast genome.

Dymond et al. (2011) constructed synthetic yeast chromosomes arms using *in vitro* assembly of 80 bp chemically synthesised oligonucleotides to form 750 bp fragments for ligation into a plasmid backbone. Remarkably, this work was carried out in part as a component of an undergraduate student synthetic biology teaching programme (Dymond et al. 2008). Subsequent rounds of assembly were used to combine 750 bp fragments into 10 kb then 30 kb fragments using homologous recombination-based assembly in *E. coli* and then yeast. Dymond et al. (2011) targeted the natural left arm of chromosome VI and the right arm of chromosome IX (Figure 5.12A) for replacement by synthetic, designed versions, generating SynVIL and SynIXR, respectively.

Assembled fragments in the order of 30 kb were designed to feature homology to the wild type region of the incumbent chromosome plus an auxotrophic marker such as LEU2, which enables only cells stably carrying the marker to grow on leucine-deficient media,

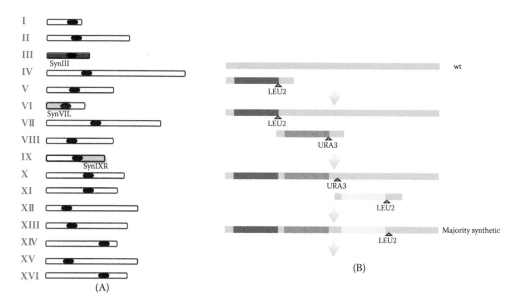

FIGURE 5.12

Synthetic chromosome assembly in yeast. (A) To date two separate yeast chromosome arms, SynVIL (middle green region) and SynIXR (lower green region), and all of chromosome III (pink region) have been refactored by assembly of synthetic DNA fragments followed by stepwise recombinative replacement of wild type genome segments. (B) Illustration of how large synthetic genome segments (purple, orange and yellow) are integrated into the wild type genome in a stepwise fashion. A given auxotrophic marker, such as LEU2, is used to confirm the presence of an initial recombined segment. Subsequent recombination steps (blue arrows) are configured to simultaneously remove LEU2 and insert a new marker, URA3. This enables repeated use of LEU2 and URA3 for multiple subsequent rounds of recombination until the entire wild type genome has been replaced with a synthetic genome.

when used in a LEU2 genotype background strain (Figure 5.12B). Once replacement of a natural region with a synthetic region is confirmed, a further round of 30 kb replacement can take place. Regions of homology again target the incoming synthetic fragment for recombination into the incumbent chromosome, such that the LEU2 gene is lost, but a further auxotrophic marker, such as URA3, is introduced. This time only cells stably carrying URA3 can grow on uracil-deficient media, in the context of a URA3 genotype background strain. Further rounds (Figure 5.12B) will remove URA3 and be selected for by LEU2, followed by the converse, until eventually an entire chromosome or chromosome arm has had its natural sequence replaced with synthetic sequence (Dymond and Boeke 2012). An entirely synthetic version of *S. cerevisiae* chromosome III, SynIII (Figure 5.12A), was generated by Annaluru et al. (2014) using this iterative approach.

The distributed nature of the DNA assembly steps up to the 30 kb level make this approach to synthetic genome construction well-suited to collaboration across multiple groups of researchers, each group responsible for a manageable 'chunk' of genome assembly. Having established robust physical means by which to replace large amounts of natural with synthetic DNA, the actual aims and intentions for the synthetic DNA sequence come into focus.

5.6.2 Writing Genomes: What to Write?

The ability to assemble synthetic DNA and use it to replace natural genomes is hugely significant. However, the next challenge is to how best to write DNA code that will affect new and useful capabilities. The Sc 2.0 designers of the synthetic *S. cerevisiae* chromosome elements set three design rules for their novel genetic material: (1) genetic changes should result in a near wild-type phenotype, (2) synthetic regions of the genome should be stable and therefore lack elements such as transposons and tRNA genes and (3) synthetic regions should be flexible with respect to future DNA recombination steps.

One strategy that supports all three of these aims is 'synthetic chromosome rearrangement and modification by loxPsym-mediated evolution' (SCRaMbLE). SCRaMbLE is a novel system of combinatorial mutagenesis that can rearrange and remove genes (Figure 5.13). The system operates through the action of the recombinase enzyme, Cre. Cre binds to and acts on LoxPsym recombination target sites to cause, with equal likelihood, inversion or deletions. The Sc 2.0 designers wrote multiple LoxPsym sites into the synthetic chromosomes and chromosome arms they had successfully incorporated into cells. Cre expression was then introduced in a transient pulse of expression from a plasmid and predicted to produce inversion or deletion of those regions of the synthetic genome flanked by LoxPsym sites. Cells surviving this pulse of Cre expression, and subsequent gene rearrangement, can then be screened by phenotype (Figure 5.13) and also sequenced on a region-by-region basis due the presence of convenient sites, PCRTags, introduced at the synthetic DNA assembly stage. SCRaMbLE was used in this way to help successfully preserve near wild type phenotype while significantly reducing the size of the synthetic region of the genome compared with its wild type counterpart due to the loss of nonessential genes and other genetic material.

5.6.3 Genome Editing

The ability to edit, by insertion or deletion, any desired region of a genome with ease and efficiency has been a holy grail for biotechnologists for many years. Several routes have been developed since the 1970s for base-pair resolution editing of genome targets *in situ*, including the use of DNA/RNA hybrid oligonucleotides developed in the late 1990s (Blaese 1999). Although DNA/RNA hybrid oligonucleotide-based approaches to

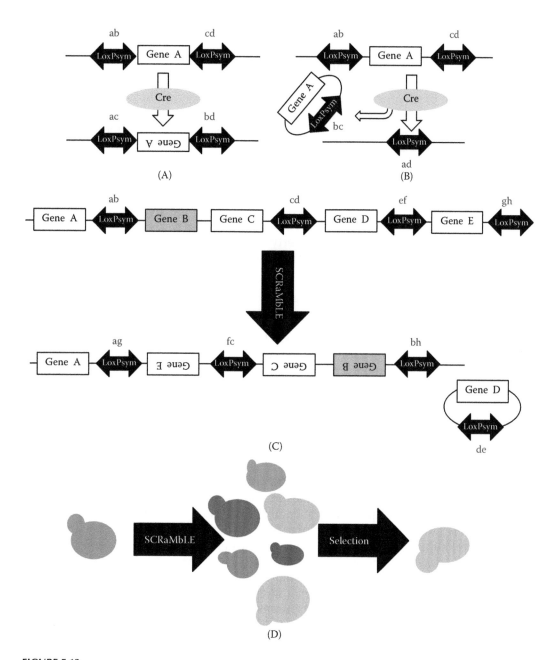

FIGURE 5.13
Recombinative genome self-optimisation. (A) Example of locus recombination within the SCRaMbLE system. Transient expression of the Cre recombinase can result in inversion of loci flanked by inserted LoxPsym sequences that are specifically targeted by Cre. The Cre recombines two halves of a LoxPsym site (letters above sites) in every reaction. (B) Cre can also 'loop out' (delete) regions flanked by LoxPsym sites. 'Looping out' or inversions occur with a 50:50 probability. (C) The locus illustrated here undergoes two recombination events: looping out of Gene D and inversion of the resultant Gene B-Gene C-Gene E locus to give the final configuration under the large arrow. (D) Synthetic genomes with strategically distributed LoxPsym sites can effectively undergo accelerated evolution by the action of multiple pulses of Cre expression. This generates a diverse population of genotypes (indicated by cell icons of different grey shade and size) that can be subjected to selective pressure to identify a genotype encoding useful cell properties.

in situ gene editing are still being developed, for example, in combination with the TALE technology (Wang et al. 2015) discussed in Section 5.5, they have not been widely adopted. Potentially the most powerful genome editing approach developed to date is based on the capabilities of the bacterial clustered regularly interspaced short palindromic repeats (CRISPR) system of adaptive immunity (Karginov and Hannon 2010).

The presence of unexplained regularly interspaced short palindromic repeats in the *E. coli* genome was first announced by Ishino et al. (1987). In the 1990s, the phenomenon was established as being common to all bacteria and archaea and was explained as a mechanism of storing DNA sequence information regarding invading predatory bacteriophage and also effecting the site-specific degradation of bacteriophage DNA as a defence strategy (Karginov and Hannon 2010). Since 2010, repurposing of CRISPR as a system for site-specific modification of user-defined sequences has become a major technological breakthrough. As the discovery and development of CRISPR-based technologies has involved many research groups, establishing the intellectual property rights to different applications has so far proved a complex process (Granahan and Loughran 2014; Sherkow 2015).

5.6.4 CRISPR/Cas9 as a Genome-Editing Tool

Jinek at al. (2012) discuss the natural function of the CRISPR system in excellent detail and also link the basic science with the convenient tools and strategies that have subsequently been developed. Figure 5.14 summarises the common approach to 'off the shelf' CRISPR/ Cas9 tools that can today be purchased from many commercial molecular biology companies. The Cas9 nuclease binds to an invariant region of a guide RNA (gRNA) transcript and is then guided to a specific DNA sequence by the remaining 17–20 bp of the gRNA, which are designed by the investigator for specificity to a gene or locus of choice. The only limitation on the target sequence is that it must be preceded by the triplet base pair sequence; G, G and N, where N is any base. This triplet is known as the protospacer adjacent motif (PAM). On being guided to the target sequence, the Cas9/gRNA complex will affect a double strand break (DSB). After this break has been achieved, the host non-homologous end join (NHEJ) DNA repair mechanisms tend to repair the break in a way that results in loss or insertion of bases so that the original sequence is lost (Figure 5.14). Within an ORF, this process is likely to lead to a missense mutation and loss of expression of the correct protein.

Commercial kits or services typically provide plasmids featuring an expression cassette for the Cas9 nuclease for transient expression in the target organism. A separate plasmid will encode expression cassettes for one or more gRNA molecules (Bao et al. 2015) for which the user defines the target specific region of 17–20 bases plus the PAM. In some organisms, Cas9/gRNA activity can be affected by promoter and/or transcription termination signal choice. To address this, Gao and Zhao (2014) proposed upstream- and downstream-flanking ribozyme sequences capable of efficient post-transcriptional self-processing whereby a stable initial RNA molecule is edited down to an active, final gRNA.

In theory, multiple related genes can be targeted simultaneously with a single gRNA design. Figure 5.15A details possible scenarios when sites are targeted for Cas9 breakage in consecutive ORFs. The promoters used for Cas9 and gRNA transcription and the codon usage bias applied for Cas9 translation can vary widely depending on the organism in which *in situ* genome editing is being attempted. As such, there is a significant amount of research effort being applied to developing robust CRISPR tools for a wide range of host organisms.

As well as gene knockdown, the CRISPR/Cas9 system can also be used for gene insertion. Co-transfecting cells with a plasmid containing a second copy of the Cas9/gRNA genomic target region (Auer and Del Bene 2014) will result in Cas9/gRNA making DSBs

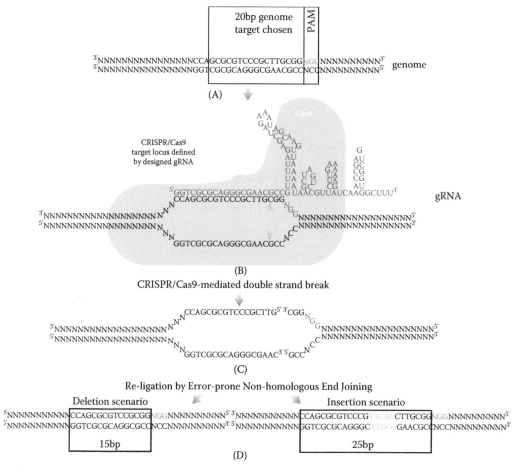

FIGURE 5.14
Targeted DNA strand breakage by CRISPR/Cas9. Overview of the major steps in a typical CRISPR/Cas9-mediated genome editing event. (A) A 20 bp target site must be selected including a protospacer adjacent motif (PAM) triplet sequence at its 5′ end. (B) An invariant region of secondary structure downstream of the 20 bp target and PAM completes the guide RNA (gRNA) molecule (red sequence). The gRNA binds to the Cas9 nuclease (light green shape) and the gRNA:Cas9 complex migrates to the locus defined by the target sequence within the gRNA. (C) Once in position, the gRNA:Cas9 complex will catalyse breakage of both strands of the substrate DNA molecule at a site within the target sequence. (D) Repair of the double-strand breaks is typically error-prone, resulting in loss (deletion) of base pairs or insertion of additional base pairs.

in both the intended genomic location and the target present in the plasmid. Homologous recombination of these DSBs can result in insertion of regions (Figure 5.15C) or the entire (Figure 5.15B) linearised plasmid into the genome.

Short donor oligonucleotides can also be used to direct the NHEJ process toward insertion of a short desired sequence (Figure 5.15D and E) or to effect complete gene deletion in yeasts for which DSBs are frequently lethal (Mans et al. 2015). In addition to DSBs, an aspartate-to-alanine mutant of Cas9 can be used to make only a single-strand break, or 'nick', within the gRNA-targeted genome location (Ran et al. 2013). When adjacent nicks are made using two separate Cas9/gRNAs, homology-directed repair mechanisms will then tend to incorporate sequence from a short donor oligonucleotide (Figure 5.15E).

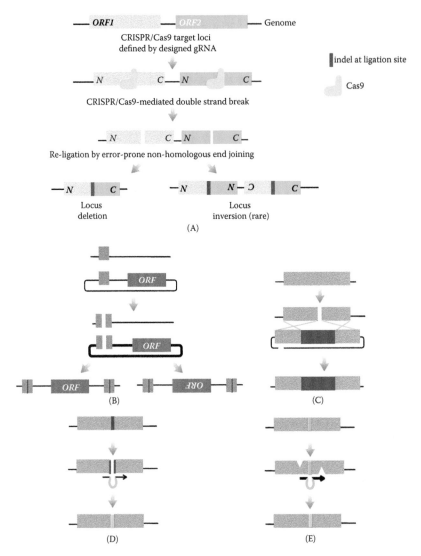

FIGURE 5.15

Genome editing with CRISPR/Cas9. A selection of CRISPR/Cas9 approaches for genome editing. (A) When two adjacent genome target sites undergo CRISPR/Cas9 double-strand breaks (DSBs) subsequent DNA repair mechanisms introduce insertions or deletions ('indels') at the join points and will also effect either complete deletion of the region that was flanked by DSBs or inversion. (B) A plasmid encoding a given open reading frame (orange box) and a second copy of the genomic sequence targeted by the gRNA. Cas9 will make DSBs in both the genome and the exogenous plasmid. Subsequent DNA repair events will in some cases result in integration of the entire plasmid into the genome target in one of two orientations, both with 'indels' at the join points (purple lines). (C) A plasmid-encoded transgene (purple box), intended for genome insertion, can be flanked by sequences with homology to those sections of genomic DNA immediately upstream and downstream (grey boxes) of a CRISPR/Cas9 DSB. Parallel CRISPR/Cas9-targeted DSB formation and transfection with linearised plasmid will efficiently direct homologous recombination resulting in transgene insertion. (D) Single-stranded oligonucleotides can also be incorporated into CRISPR/Cas9 strategies. Typically, the oligonucleotide will feature regions of homology that flank the DSB (red line) plus additional sequence (blue loop) intended for insertion into the resultant, rejoined locus (blue line). Such oligos are often essential to perform a 'repair' function when applying CRISPR for yeast genome editing. (E) Cas9 variants have been developed that catalyse only single-strand breaks (SSBs). Oligonucleotides can be used in a similar fashion to strategy (D) but spanning the region between the SSBs and encoding point mutations (blue loop and line) that replace the wild type sequence (green line) by action of host cell DNA repair functions.

5.6.5 Future of Genome Editing in Eukaryotes

Cas9/CRISPR tools are advancing at a rapid rate to included applications in targeted RNA degradation (O'Connell et al. 2014) and in forward genetics approaches to identify genes associated with a given phenotype (Shalem et al. 2015). Cas9/CRISPR has been used to genome-edit mice (Wang et al. 2013), human cells (Mali et al. 2013) and, controversially, human zygotes (Liang et al. 2015). The pace of this progress has been unsettling for many observers and has lead in some nations to a re-examination and reassertion of the legal limits on human genetic modification (Cyranoski 2015). The rate of development in CRISPR/Cas9 genome editing, combined with progress in control and understanding of genome-organising features such as chromatin (Pick et al. 2014), suggest that the capabilities of genome control could begin to approach the limits of societal consensus regarding acceptable practices.

5.7 Synthetic Biotherapeutics

The use of recombinant DNA technology in the mainstream pharmaceutical industry has made several major leaps since insulin derived from transgene expression in *E. coli* was first taken to market in the 1980s. Around half of the top ten global blockbuster drugs are now large, complex recombinant proteins. The use of viable organisms as therapeutics began in the 1950s with attenuated virus vaccines and this has ultimately lead to development of live vaccines consisting of viable human cells today. These cells typically express one or more transgenes encoding antigens or proteins that modulate the immune system. Human cell therapies can now be used to direct the immune system to arrest the growth of previously resistant cancers. The level of sophistication involved in harnessing the immune system in this way has moved beyond conventional concepts of vaccine development into the emerging field of 'immunotherapy'. Recombinant T cells, discussed earlier in this chapter, are in the vanguard of this approach, and the potential of the field is evidenced by the takeover attempt of AstraZeneca PLC by Pfizer Inc., reportedly with the intention to acquire enhanced product development capability in immunotherapy.

While the use of engineered human cells to reprogram the immune system is still at an early stage as a therapeutic approach, replenishing the body with unmodified cells that have been cultivated *ex vivo* has led to a significant body of clinical data regarding efficacy and reproducibility (Rao et al. 2015). This has enabled the application of stem cell technologies in fields such as regenerative medicine and neurological disease, with steady progress being made in recent years. In Section 5.7 we consider cases in which synthetic biology approaches can help establish routes to therapy that complement the ongoing efforts of the biotechnology, cancer therapy and biomedical sectors, both industrial and academic. The interested reader can also find excellent surveys of biomedical application of synthetic biology in articles by Ruder et al. (2011), Weber and Fussenegger (2011) and Gubeli et al. (2012).

5.7.1 Gain-of-Function Therapies

Gene therapy can involve delivery of cells that are already in a modified state or transduction *in vivo* of a targeted population of cells, typically using engineered viruses. Regardless of modification route, controlling gene expression in these cells can be challenging, particularly if a small molecule inducer or other factor must also be physically delivered to those cells in a way that does not compromise patient health or have 'off target' effects on other cells. Ye et al. (2011) used an elegant method to potentially enable control of therapeutic cells *in situ*, within the patient's body, by the non-invasive approach of merely shining blue light on the region where gene expression is required.

Their approach, illustrated in Figure 5.16, relied on coupling light-stimulated influx of calcium into the cytosol to gene expression driven by the calcium-sensitive 'nuclear factor of activated T cells' (NFAT) promoter. Crabtree and Schreiber (2009) showed that NFAT promoter-driven expression is induced by a calcium influx into the cytoplasm triggering a cascade that leads to dephosphorylation of the cytosolic NFAT protein. This dephosphorylation leads to conformational changes in NFAT that result in its migration to the nucleus where it initiates transcription at cognate NFAT promoter sites. Melyan et al. (2005) previously showed that the melanopsin receptor could be expressed as a transgene in mammalian cells and induce gene expression in response to light by way of a calcium-triggered signalling cascade.

Ye et al. (2011) also constructed a melanopsin receptor DNA expression cassette so that blue light would trigger the type of cytosolic calcium influx already shown to stimulate NFAT promoter-driven expression. Separate plasmids encoding the melanopsin receptor and NFAT-controlled reporters were used to co-transfect five mammalian cell lines, only one of which, human embryonic kidney line 293 (HEK-293), showed a sufficient degree of

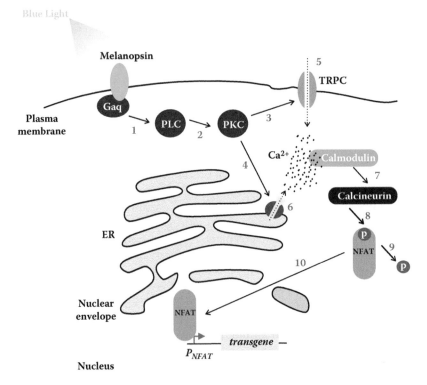

FIGURE 5.16

Optogenetic control cascade. Starting from the top left, blue light causes a conformational change in the cell surface protein Melanopsin, which then causes conformational changes in Gaq-type G protein (Gaq), that stimulates (1) phospholipase C (PLC) to interact with phosphokinase C (PKC) (2) that triggers (3, 4) an influx of calcium ions into the cytosol from the external milieu (5), via transient receptor potential channels (TRPCs), and the lumen of the endoplasmic reticulum (6). The increase in cytosolic calcium concentration causes conformational changes to calmodulin resulting in an interaction (7) with calcineurin which then binds (8) and dephosphorylates (9) nuclear factor of activated T cells (NFAT). Upon dephosphorylation, NFAT migrates to the nucleus (10) where it induces transcription of genes under control of the P$_{NFAT}$ promoter.

increase in reporter gene expression in response to blue light. They confirmed the function of the recombinant NFAT promoter and melanopsin receptor in HEK-293 cells using GFP, secreted alkaline phosphatase (SEAP) and luciferase as reporter proteins. Although a positive result in terms of the utility of the approach, the fact that, for unknown reasons, the synthetic circuit failed to function as intended in 80% of the recipient cell types illustrates the extent to which host chassis and circuit interface effects still need to be comprehensively mapped and understood by synthetic biologists.

The light-inducible HEK-293 cells were shown to have a desirable dynamic response range and control experiments indicated that light-actuated gene induction was indeed occurring as a result of the calcium-influx mediated mechanism as expected. Ye et al. (2011) then took the system further by substituting the reporter protein for a glucagon-like peptide-1 variant (shGLP-1), a protein whose function is deficient in people with Type II diabetes. When introduced, in an encapsulated format, to a mouse animal model of diabetes, the engineered HEK-293 cells secreted shGLP-1 in response only to blue light. The secreted shGLP-1 was able to bring about a beneficial effect, reducing blood sugar to levels comparable to those observed in healthy animals. Wend et al. (2013) developed alternative routes to light-induced gene expression, or 'optogenetics', in which a protein kinase, C-RAF, is activated by light as opposed to a surface receptor.

5.7.2 Controlled Cell Deletion

Biotechnological approaches to cancer therapy often involve the use of therapeutic agents, such as nucleic acids, cells, viruses and proteins (most often antibodies), to effect cancer cell death. A major technical challenge for this approach is to target only cancer cells while leaving non-cancer cells unharmed. A common tactic is to exploit the presence of cancer-specific patterns of surface protein expression, such as receptors, or overexpression of cancer-specific intracellular proteins or biomolecules. Options for achieving this are to either redesign the therapeutic agent or the delivery vehicle used to deliver the agent to cancer cells.

For therapies or vaccines involving naked DNA, the therapeutic agent is also the delivery vehicle. It is more often the case that therapeutic DNA must be coated or complexed with lipids or ionic polymers in order to be efficiently taken up by target cells. In such a scenario, the DNA is the therapeutic agent and the surrounding coating or complex can be thought of as the delivery vehicle. Viruses are a naturally occurring, highly evolved vehicle for delivery of specific nucleic acid sequences as 'payloads' to target cells and for this reason they have been widely exploited for gene therapy approaches to cancer treatment. Vaccinia virus has been modified for targeted destruction of cancer cells, or 'oncolysis', by restricting its replication to only those cells with a high a cytosolic concentration of nucleotides, a phenotype characteristic of many cancer cells. However, targeting only the presence or absence of a single cancer-associated factor is not always sufficient to discern cancer cells. Certain cancer cells may be defined by the presence of multiple factors or a characteristic pattern of abundance and scarcity across several factors.

Xie et al. (2011) sought to target a characteristic expression pattern of five micro-RNA (miR) molecules associated with HeLa cells, an immortal cancer line derived from of a glandular tumour sample taken from US citizen Henrietta Lacks in 1951 (Kloot 2010). They designed synthetic gene circuits that could, when successfully delivered into and expressed in human cells, react only to high levels of miR-21 and miR-17–30a plus the presence of miR-141, miR-142(3p) and miR-146a (Figure 5.17). Only when all five conditions, or inputs, were encountered would an output be triggered. In this case, the output was expression of a recombinant protein that triggers cell death by apoptosis. Exciting future goals for this 'classification/deletion' network include tuning the robustness and activity of the network and packaging the five expression cassettes into the relevant vehicles or formats for efficient delivery into animal models or *ex vivo* tissues to effect cancer cell deletion (Chen and Smolke 2011).

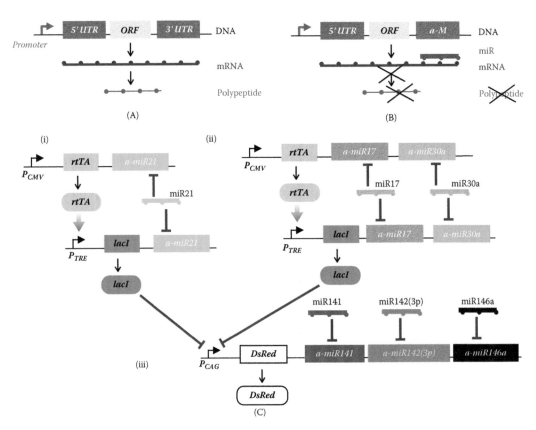

FIGURE 5.17

Biotherapeutic classifier gene circuit. A gene circuit designed to classify its cellular environment and direct pre-defined outputs based on its classification. The design utilises sequence-specific repression of translation by host cell micro-RNA (miR) molecules and recognition of a pattern of miR expression characteristic of a particular cell type. (A) Outline of typical eukaryotic gene expression in which messenger RNA comprises untranslated sequences upstream (5′ untranslated region [UTR]) and downstream (3′ UTR) of the coding region. (B) The 3′ UTR can be replaced with a sequence, designated 'α-M', that is antisense to a host cell miR. Translation of the mRNA will proceed normally when that miR is absent. However, when the miR is present it will bind to its complementary sequence within the α-M region and suppress translation from the mRNA. (C.i) The classifier consists, first, of an expression cassette for rtTA, driven by the constitutive P_{CMV} promoter. rtTA induces the P_{TRE} promoter in a second cassette that controls lacI expression. The mRNA 3′ UTRs of both cassettes are designed to be antisense to micro RNA 21 (a-miR21), which is highly abundant in cancer cells. As such, the presence of miR21 represses (red, flat head arrows) expression of rtTA and lacI. (C.ii) In the second element of the circuit, rtTA and lacI expression are repressed by miR17 and miR30a by the same mechanism used in (C.i). Parallel routes to suppression of both lacI and its rtTA inducer decrease sensitivity of the network so that only high concentrations of the targeted miR molecules will be sufficient to effect lacI downregulation. Only if abundance of miR21, miR17 and miR30a falls below a very high level will lacI be expressed. (C.iii) In the final element of the classifier, any lacI expressed from elements (C.i) and (C.ii) will suppress the P_{CAG} promoter controlling expression of the DsRed reporter protein. However, antisense sequences in the 3′ UTR of the DsRed mRNA make its expression sensitive to repression by even low levels of miR-141, miR142(3p) and miR146a. The cumulative effect of the overall circuit design is that only a signature of high miR21, miR17 and miR30a and low miR-141, miR142(3p) and miR146a will result in circuit output. A pro-apoptotic protein replaces DsRed in the functional circuit.

5.7.3 Future Synthetic Biotherapies

Many of the processes of oncogenesis involve dysregulation of signal transduction pathways and several approaches have been explored to address this from a synthetic biology perspective, including switching on and off certain steps in signal transduction pathways, as discussed in excellent detail by Gurevich and Gurevich (2012) and Heng et al. (2014). Direct reprogramming of cells using exogenous RNA is also a route being tested to control stem cell differentiation (Flynn and Chan 2014). Delivery methods for targeting subpopulations of cells must often be as sophisticated as the payloads themself, and their development is an active area of synthetic biology research (Zheng et al. 2011; Fussenegger et al. 2012). Drug discovery has long since been an activity for which high throughput automation and screening approaches is key. The standardisation and biosynthetic pathway assembly approaches characteristic of synthetic biology are suited to making a meaningful contribution to drug discovery technologies (Weber and Fussenegger 2009; Neumann and Neumann-Staubitz 2010).

Taken together, the technological approaches discussed in this chapter provide a complete pathway by which human cells can, either *ex vivo* or *in situ*, be re-programmed with synthetic gene circuits able to dynamically respond to multiple inputs (Horner et al. 2012; Folcher and Fussenegger 2012; Bacchus et al. 2014) to enable these cells to fulfil a therapeutic role within the body (Burrill et al. 2011; Keret 2013). While such progress allows ever more novel therapeutic approaches to be designed and tested (Shankar and Pillai 2011), oversight is also necessary to examine which methods and therapeutic purposes meet the approval of wider society. Meeting the financial, regulatory, ethical, legal and logistical challenges of bringing such technologies to the clinic and, ultimately, to market, already ensures a high level of scrutiny from a broad range of stakeholders.

References

Agapakis, C. M., H. Niederholtmeyer, R. R. Noche, T. D. Lieberman, S. G. Megason, J. C. Way, and P. A. Silver. 2011. Towards a synthetic chloroplast. *PLoS ONE* 6 (4): e18877. doi:10.1371/journal.pone.0018877.

Agapakis, C. M., and P. A. Silver. 2009. Synthetic biology: Exploring and exploiting genetic modularity through the design of novel biological networks. *Molecular BioSystems* 5 (7): 704. doi:10.1039/b901484e.

Alford, S. C., Y. Ding, T. Simmen, and R. E. Campbell. 2012. Dimerization-dependent green and yellow fluorescent proteins. *ACS Synthetic Biology* 1 (12): 569–75. doi:10.1021/sb300050j.

Annaluru, N., H. Muller, L. A. Mitchell, S. Ramalingam, G. Stracquadanio, S. M. Richardson, et al. 2014. Total synthesis of a functional designer eukaryotic chromosome. *Science* 344 (6179): 55–58. doi:10.1126/science.1249252.

Arpino, J. A., E. J. Hancock, J. Anderson, M. Barahona, G. B. Stan, A. Papachristodoulou, and K. Polizzi. 2013. Tuning the dials of synthetic biology. *Microbiology* 159 (Pt 7): 1236–53. doi:10.1099/mic.0.067975-0.

Aubel, D., and M. Fussenegger. 2010. Mammalian synthetic biology – From tools to therapies. *BioEssays* 32 (4): 332–45. doi:10.1002/bies.200900149.

Auer, T. O., and F. D. Bene. 2014. CRISPR/Cas9 and TALEN-mediated knock-in approaches in zebrafish. *Methods* 69 (2): 142–50. doi:10.1016/j.ymeth.2014.03.027.

Bacchus, W., D. Aubel, and M. Fussenegger. 2014. Biomedically relevant circuit-design strategies in mammalian synthetic biology. *Molecular Systems Biology* 9 (1): 691. doi:10.1038/msb.2013.48.

Bao, Z., H. Xiao, J. Liang, L. Zhang, X. Xiong, N. Sun, T. Si, and H. Zhao. 2015. Homology-integrated CRISPR–Cas (HI-CRISPR) system for one-step multigene disruption in *Saccharomyces cerevisiae*. *ACS Synthetic Biology* 4 (5): 585–94. doi:10.1021/sb500255k.

Behjousiar, A., C. Kontoravdi, and K. M. Polizzi. 2012. *In situ* monitoring of intracellular glucose and glutamine in CHO cell culture. *PLoS ONE* 7 (4): e34512. doi:10.1371/journal.pone.0034512.

Bianchi, M., E. Guerriero, M. Fiocco, R. Alberti, L. Polloni, A. Behnam, E. A. Carrion, E. Pop, and R. Sordan. 2015. Scaling of graphene integrated circuits. *Nanoscale* 7 (17): 8076–83. doi:10.1039/C5NR01126D.

Blaese, R. M. 1999. Optimism regarding the use of RNA/DNA hybrids to repair genes at high efficiency. *Journal of Gene Medicine* 1 (2): 144.

Boch, J., H. Scholze, S. Schornack, A. Landgraf, S. Hahn, S. Kay, T. Lahaye, A. Nickstadt, and U. Bonas. 2009. Breaking the code of DNA binding specificity of TAL-Type III effectors. *Science* 326 (5959): 1509–12. doi:10.1126/science.1178811.

Bogner, M., and U. Ludewig. 2007. Visualization of arginine influx into plant cells using a specific FRET-sensor. *Journal of Fluorescence* 17 (4): 350–60. doi:10.1007/s10895-007-0192-2.

Braun-Galleani, S., F. Baganz, and S. Purton. 2015. Improving recombinant protein production in the chlamydomonas reinhardtii chloroplast using vivid verde fluorescent protein as a reporter. *Biotechnology Journal* 10(8): 1289–97. doi:10.1002/biot.201400566.

Chen, Y. Y., M. C. Jensen, and C. D. Smolke. 2010. Genetic control of mammalian T-cell proliferation with synthetic RNA regulatory systems. *Proceedings of the National Academy of Sciences* 107 (19): 8531–36. doi:10.1073/pnas.1001721107.

Choi, J. H., N. K. Yu, G. C. Baek, J. Bakes, D. Seo, H. J. Nam, S. H. Baek, C. S. Lim, Y. S. Lee, and B. K. Kaang. 2014. Optimization of AAV expression cassettes to improve packaging capacity and transgene expression in neurons. *Molecular Brain* 7 (1): 1–10.

Cong, L., R. Zhou, Y. C. Kuo, M. Cunniff, and F. Zhang. 2012. Comprehensive interrogation of natural TALE DNA-binding modules and transcriptional repressor domains. *Nature Communications* 3: 968. doi:10.1038/ncomms1962.

Cyranoski, D. 2015. Ethics of embryo editing divides scientists. *Nature* 519 (7543): 272–272. doi:10.1038/519272a.

Daringer, N. M., R. M. Dudek, K. A. Schwarz, and J. N. Leonard. 2014. Modular extracellular sensor architecture for engineering mammalian cell-based devices. *ACS Synthetic Biology* 3 (12): 892–902. doi:10.1021/sb400128g.

David, F., and V. Siewers. 2014. Advances in yeast genome engineering. *FEMS Yeast Research*. doi:10.1111/1567-1364.12200.

Dymond, J. and Boeke, J. 2012. The Saccharomyces cerevisiae SCRaMbLE system and genome minimization. Bioengineered Bugs 3(3): 168–171.

Dymond, J. S., S. M. Richardson, C. E. Coombes, T. Babatz, H. Muller, N. Annaluru, et al. 2011. Synthetic chromosome arms function in yeast and generate phenotypic diversity by design. *Nature* 477 (7365): 471–76. doi:10.1038/nature10403.

Dymond, J. S., L. Z. Scheifele, S. Richardson, P. Lee, S. Chandrasegaran, J. S. Bader, and J. D. Boeke. 2008. Teaching synthetic biology, bioinformatics and engineering to undergraduates: The interdisciplinary build-a-genome course. *Genetics* 181 (1): 13–21. doi:10.1534/genetics.108.096784.

Edwards, P. A. 1985. Heterogeneous expression of cell-surface antigens in normal epithelia and their tumours, revealed by monoclonal antibodies. *British Journal of Cancer* 51 (2): 149–60.

Farzadfard, F., S. D. Perli, and T. K. Lu. 2013. Tunable and multifunctional eukaryotic transcription factors based on CRISPR/Cas. *ACS Synthetic Biology* 2 (10): 604–13. doi:10.1021/sb400081r.

Fukuda, N., and S. Honda. 2015. Rapid evaluation of tyrosine kinase activity of membrane-integrated human epidermal growth factor receptor using the yeast Gγ recruitment system. *ACS Synthetic Biology* 4 (4): 421–29. doi:10.1021/sb500083t.

Fussenegger, M., R. P. Morris, C. Fux, M. Rimann, B. von Stockar, C. J. Thompson, and J. E. Bailey. 2000. Streptogramin-based gene regulation systems for mammalian cells. *Nature Biotechnology* 18 (11): 1203–8. doi:10.1038/81208.

Gaber, R., T. Lebar, A. Majerle, B. Šter, A. Dobnikar, M. Benčina, and R. Jerala. 2014. Designable DNA-binding domains enable construction of logic circuits in mammalian cells. *Nature Chemical Biology* 10 (3): 203–8. doi:10.1038/nchembio.1433.

Gao, Y., and Y. Zhao. 2014. Self-processing of ribozyme-flanked RNAs into guide RNAs *in vitro* and *in vivo* for CRISPR-mediated genome editing: Self-processing of ribozyme-flanked RNAs into guide RNAs. *Journal of Integrative Plant Biology* 56 (4): 343–49. doi:10.1111/jipb.12152.

Garg, A., J. J. Lohmueller, P. A. Silver, and T. Z. Armel. 2012. Engineering synthetic TAL effectors with orthogonal target sites. *Nucleic Acids Research*, gks404. doi:10.1093/nar/gks404.

Gjetting, S. K., C. K. Ytting, A. Schulz, and A. T. Fuglsang. 2012. Live imaging of intra- and extracellular pH in plants using pHusion, a novel genetically encoded biosensor. *Journal of Experimental Botany* 63 (8): 3207–18. doi:10.1093/jxb/ers040.

Goodman, J. M., C. W. Scott, P. N. Donahue, and J. P. Atherton. 1984. Alcohol oxidase assembles post-translationally into the peroxisome of Candida boidinii. *The Journal of Biological Chemistry* 259 (13): 8485–93.

Granahan, P., and C. Loughran. 2014. CRISPR/Cas-9: An exciting addition to genomic editing. *Life Sciences Law & Industry Report*. Copyright © 2015 The Bureau of National Affairs, Inc. All Rights Reserved.

Greber, D., and M. Fussenegger. 2007. Mammalian synthetic biology: Engineering of sophisticated gene networks. *Journal of Biotechnology* 130 (4): 329–45. doi:10.1016/j.jbiotec.2007.05.014.

Ha, J.-S., J. J. Song, Y.-M. Lee, S.-J. Kim, J.-H. Sohn, C.-S. Shin, and S.-G. Lee. 2007. Design and application of highly responsive fluorescence resonance energy transfer biosensors for detection of sugar in living *Saccharomyces cerevisiae* cells. *Applied and Environmental Microbiology* 73 (22): 7408–14. doi:10.1128/AEM.01080-07.

Halees, A. S. 2003. PromoSer: A large-scale mammalian promoter and transcription start site identification service. *Nucleic Acids Research* 31 (13): 3554–59. doi:10.1093/nar/gkg549.

Hartenbach, S., and M. Fussenegger. 2006. A novel synthetic mammalian promoter derived from an internal ribosome entry site. *Biotechnology and Bioengineering* 95 (4): 547–59. doi:10.1002/bit.21174.

Hartner, F. S., C. Ruth, D. Langenegger, S. N. Johnson, P. Hyka, G. P. Lin-Cereghino, J. Lin-Cereghino, K. Kovar, J. M. Cregg, and A. Glieder. 2008. Promoter library designed for fine-tuned gene expression in *Pichia Pastoris*. *Nucleic Acids Research* 36 (12): e76. doi:10.1093/nar/gkn369.

Hasty, J., D. McMillen, F. Isaacs, and J. J. Collins. 2001. Computational studies of gene regulatory networks: In numero molecular biology. *Nature Reviews. Genetics* 2 (4): 268–79. doi:10.1038/35066056.

Haynes, K. A., and P. A. Silver. 2009. Eukaryotic systems broaden the scope of synthetic biology. *The Journal of Cell Biology* 187 (5): 589–96. doi:10.1083/jcb.200908138.

Heavner, B. D., K. Smallbone, N. D. Price, and L. P. Walker. 2013. Version 6 of the consensus yeast metabolic network refines biochemical coverage and improves model performance. *Database* 2013. bat059–bat059. doi:10.1093/database/bat059.

Hirokawa, Y., H. Kawano, K. Tanaka-Masuda, N. Nakamura, A. Nakagawa, M. Ito, H. Mori, T. Oshima, and N. Ogasawara. 2013. Genetic manipulations restored the growth fitness of reduced-genome *Escherichia Coli*. *Journal of Bioscience and Bioengineering* 116 (1): 52–58. doi:10.1016/j.jbiosc.2013.01.010.

Huang, Y., X. Weng, and I. M. Russu. 2010. Structural energetics of the adenine tract from an intrinsic transcription terminator. *Journal of Molecular Biology* 397 (3): 677–88. doi:10.1016/j.jmb.2010.01.068.

Ishino, Y., H. Shinagawa, K. Makino, M. Amemura, and A. Nakata. 1987. Nucleotide sequence of the IAP gene, responsible for alkaline phosphatase isozyme conversion in *Escherichia Coli*, and identification of the gene product. *Journal of Bacteriology* 169 (12): 5429–33.

Jinek, M., K. Chylinski, I. Fonfara, M. Hauer, J. A. Doudna, and E. Charpentier. 2012. A programmable dual-RNA-guided DNA endonuclease in adaptive bacterial immunity. *Science* 337 (6096): 816–21.

Jossé, L., C. M. Smales, and M. F. Tuite. 2010. Transient expression of human TorsinA enhances secretion of two functionally distinct proteins in cultured Chinese hamster ovary (CHO) cells. *Biotechnology and Bioengineering* 105 (3): 556–66. doi:10.1002/bit.22572.

Kalos, M., and C. H. June. 2013. Adoptive T cell transfer for cancer immunotherapy in the era of synthetic biology. *Immunity* 39 (1): 49–60. doi:10.1016/j.immuni.2013.07.002.

Kelwick, R., J. T. MacDonald, A. J. Webb, and P. Freemont. 2014. Developments in the tools and methodologies of synthetic biology. *Frontiers in Bioengineering and Biotechnology* 2. doi:10.3389/fbioe.2014.00060.

Kershaw, Michael H., Jennifer A. Westwood, and Phillip K. Darcy. 2013. Gene-engineered T cells for cancer therapy. *Nature Reviews Cancer* 13 (8): 525–41. doi:10.1038/nrc3565.

Kloss, C. C., M. Condomines, M. Cartellieri, M. Bachmann, and M. Sadelain. 2013. Combinatorial antigen recognition with balanced signaling promotes selective tumor eradication by engineered T cells. *Nature Biotechnology* 31 (1): 71–75. doi:10.1038/nbt.2459.

Lanitis, E., M. Poussin, A. W. Klattenhoff, D. Song, R. Sandaltzopoulos, C. H. June, and D. J. Powell. 2013. Chimeric antigen receptor T cells with dissociated signaling domains exhibit focused antitumor activity with reduced potential for toxicity *in vivo*. *Cancer Immunology Research* 1 (1): 43–53. doi:10.1158/2326-6066.CIR-13-0008.

Lee, J., S.-I. Choi, J. S. Jang, K. Jang, J. W. Moon, C. S. Bae, D. S. Yang, and B. L. Seong. 1999. Novel secretion system of recombinant *Saccharomyces cerevisiae* using an N-terminus residue of human IL-1β as secretion enhancer. *Biotechnology Progress* 15 (5): 884–90. doi:10.1021/bp9900918.

Levitt, N., D. Briggs, A. Gil, and N. J. Proudfoot. 1989. Definition of an efficient synthetic poly (A) site. *Genes and Development* 3 (7): 1019–25.

Liang, P., Y. Xu, X. Zhang, C. Ding, R. Huang, Z. Zhang, et al. 2015. CRISPR/Cas9-mediated gene editing in human tripronuclear zygotes. *Protein and Cell* 6 (5): 363–72. doi:10.1007/s13238-015-0153-5.

Lienert, F., J. J. Lohmueller, A. Garg, and P. A. Silver. 2014. Synthetic biology in mammalian cells: Next generation research tools and therapeutics. *Nature Reviews Molecular Cell Biology* 15 (2): 95–107. doi:10.1038/nrm3738.

Lim, K. H., I. Hwang, and S. Park. 2012. Biotin-assisted folding of streptavidin on the yeast surface. *Biotechnology Progress* 28 (1): 276–83. doi:10.1002/btpr.721.

Lin, C., V. L. Kolossov, G. Tsvid, L. Trump, J. J. Henry, J. L. Henderson, et al. 2011. Imaging in real-time with FRET the redox response of tumorigenic cells to glutathione perturbations in a microscale flow. *Integrative Biology: Quantitative Biosciences from Nano to Macro* 3 (3): 208–17. doi:10.1039/C0IB00071J.

Ling, J., R. Daoud, M. J. Lajoie, G. M. Church, D. Soll, and B. F. Lang. 2014. Natural reassignment of CUU and CUA sense codons to alanine in Ashbya mitochondria. *Nucleic Acids Research* 42 (1): 499–508. doi:10.1093/nar/gkt842.

Liu, L., Y. Liu, H. D. Shin, R. R. Chen, N. S. Wang, J. Li, G. Du, and J. Chen. 2013. Developing *Bacillus* spp. as a cell factory for production of microbial enzymes and industrially important biochemicals in the context of systems and synthetic biology. *Applied Microbiology and Biotechnology* 97 (14): 6113–27. doi:10.1007/s00253-013-4960-4.

Longo, D. M., A. Hoffmann, L. S. Tsimring, and J. Hasty. 2010. Coherent activation of a synthetic mammalian gene network. *Systems and Synthetic Biology* 4 (1): 15–23. doi:10.1007/s11693-009-9044-5.

Mali, P., L. Yang, K. M. Esvelt, J. Aach, M. Guell, J. E. DiCarlo, J. E. Norville, and G. M. Church. 2013. RNA-guided human genome engineering via Cas9. *Science* 339 (6121): 823–26. doi:10.1126/science.1232033.

Mans, R., H. M. van Rossum, M. Wijsman, A. Backx, N. G. A. Kuijpers, M. van den Broek, P. Daran-Lapujade, J. T. Pronk, A. J. A. van Maris, and J.-M. G. Daran. 2015. CRISPR/Cas9: A molecular Swiss army knife for simultaneous introduction of multiple genetic modifications in *Saccharomyces cerevisiae*. *FEMS Yeast Research* 15 (2): fov004–fov004. doi:10.1093/femsyr/fov004.

Markson, J. S., and M. B. Elowitz. 2014. Synthetic biology of multicellular systems: New platforms and applications for animal cells and organisms. *ACS Synthetic Biology* 3 (12): 875–76. doi:10.1021/sb500358y.

Maul, J. E. 2002. The chlamydomonas reinhardtii plastid chromosome: Islands of genes in a sea of repeats. *The Plant Cell Online* 14 (11): 2659–79. doi:10.1105/tpc.006155.

Melyan, Z., E. E. Tarttelin, J. Bellingham, R. J. Lucas, and M. W. Hankins. 2005. Addition of human melanopsin renders mammalian cells photoresponsive. *Nature* 433 (7027): 741–45. doi:10.1038/nature03344.

Nagaraj, S., E. Mills, S. S. Wong, and K. Truong. 2013. Programming membrane fusion and subsequent apoptosis into mammalian cells. *ACS Synthetic Biology* 2 (4): 173–79. doi:10.1021/sb3000468.

Nandagopal, N., and M. B. Elowitz. 2011. Synthetic biology: Integrated gene circuits. *Science* 333 (6047): 1244–48.

Nesbeth, D., S. L. Williams, L. Chan, T. Brain, N. K. Slater, F. Farzaneh, and D. Darling. 2006. Metabolic biotinylation of lentiviral pseudotypes for scalable paramagnetic microparticle-dependent manipulation. *Molecular Therapy* 13 (4): 814–22. doi:10.1016/j.ymthe.2005.09.016.

Nett, J. H., T. A. Stadheim, H. Li, P. Bobrowicz, S. R. Hamilton, R. C. Davidson, et al. 2011. A combinatorial genetic library approach to target heterologous glycosylation enzymes to the endoplasmic reticulum or the Golgi apparatus of *Pichia Pastoris*. *Yeast* 28 (3): 237–52. doi:10.1002/yea.1835.

O'Connell, M. R., B. L. Oakes, S. H. Sternberg, A. East-Seletsky, M. Kaplan, and J. A. Doudna. 2014. Programmable RNA recognition and cleavage by CRISPR/Cas9. *Nature* 516 (7530): 263–66. doi:10.1038/nature13769.

Pick, H., S. Kilic, and B. Fierz. 2014. Engineering chromatin states: Chemical and synthetic biology approaches to investigate histone modification function. *Biochimica et Biophysica Acta (BBA)–Gene Regulatory Mechanisms* 1839 (8): 644–56. doi:10.1016/j.bbagrm.2014.04.016.

Pothoulakis, G., F. Ceroni, B. Reeve, and T. Ellis. 2014. The spinach RNA aptamer as a characterization tool for synthetic biology. *ACS Synthetic Biology* 3 (3): 182–87. doi:10.1021/sb400089c.

Potzkei, J., M. Kunze, T. Drepper, T. Gensch, K. E. Jaeger, and J. Buechs. 2012. Real-time determination of intracellular oxygen in bacteria using a genetically encoded FRET-based biosensor. *BMC Biology* 10 (1): 28. doi:10.1186/1741-7007-10-28.

Proudfoot, N. J. 2011. Ending the message: Poly(A) signals then and now. *Genes and Development* 25 (17): 1770–82. doi:10.1101/gad.17268411.

Purnick, P. E., and R. Weiss. 2009. The second wave of synthetic biology: From modules to systems. *Nature Reviews Molecular Cell Biology* 10 (6): 410–22. doi:10.1038/nrm2698.

Ronchi, P., C. Tischer, D. Acehan, and R. Pepperkok. 2014. Positive feedback between Golgi membranes, microtubules and ER exit sites directs de Novo biogenesis of the Golgi. *Journal of Cell Science* 127 (21): 4620–33. doi:10.1242/jcs.150474.

Ruth, C., T. Zuellig, A. Mellitzer, R. Weis, V. Looser, K. Kovar, and A. Glieder. 2010. Variable production windows for porcine trypsinogen employing synthetic inducible promoter variants in *Pichia Pastoris*. *Systems and Synthetic Biology* 4 (3): 181–91. doi:10.1007/s11693-010-9057-0.

Saito, H., and T. Inoue. 2007. RNA and RNP as new molecular parts in synthetic biology. *Journal of Biotechnology* 132 (1): 1–7. doi:10.1016/j.jbiotec.2007.07.952.

Saito, H., and T. Inoue. 2009. Synthetic biology with RNA motifs. *The International Journal of Biochemistry and Cell Biology* 41 (2): 398–404. doi:10.1016/j.biocel.2008.08.017.

Sanjana, N. E., L. Cong, Y. Zhou, M. M. Cunniff, G. Feng, and F. Zhang. 2012. A transcription activator-like effector toolbox for genome engineering. *Nature Protocols* 7 (1): 171–92. doi:10.1038/nprot.2011.431.

Schamhart, D. H., A. M. Ten Berge, and K. W. Van De Poll. 1975. Isolation of a catabolite repression mutant of yeast as a revertant of a strain that is maltose negative in the respiratory-deficient state. *Journal of Bacteriology* 121 (3): 747–52.

Shalem, O., N. E. Sanjana, and F. Zhang. 2015. High-throughput functional genomics using CRISPR–Cas9. *Nature Reviews Genetics* 16 (5): 299–311. doi:10.1038/nrg3899.

Sherkow, J. S. 2015. Law, history and lessons in the CRISPR patent conflict. *Nature Biotechnology* 33 (3): 256–57.

Tang, B. L. 2015. Synthetic mitochondria as therapeutics against systemic aging: A hypothesis: synthetic mitochondria. *Cell Biology International* 39 (2): 131–35. doi:10.1002/cbin.10362.

Tannous, B. A., J. Grimm, K. F. Perry, J. W. Chen, R. Weissleder, and X. O. Breakefield. 2006. Metabolic biotinylation of cell surface receptors for *in vivo* imaging. *Nature Methods* 3 (5): 391–96. doi:10.1038/nmeth875.

Tey, S. K. 2014. Adoptive T-cell therapy: Adverse events and safety switches. *Clinical and Translational Immunology* 3 (6): e17. doi:10.1038/cti.2014.11.

Tigges, M., and M. Fussenegger. 2009. Recent advances in mammalian synthetic biology—Design of synthetic transgene control networks. *Current Opinion in Biotechnology* 20 (4): 449–60. doi:10.1016/j.copbio.2009.07.009.

Tornøe, J., P. Kusk, T. E. Johansen, and P. R. Jensen. 2002. Generation of a synthetic mammalian promoter library by modification of sequences spacing transcription factor binding sites. *Gene* 297 (1): 21–32.

Vogl, T., F. S Hartner, and A. Glieder. 2013. New opportunities by synthetic biology for biopharmaceutical production in *Pichia Pastoris*. *Current Opinion in Biotechnology* 24 (6): 1094–1101. doi:10.1016/j.copbio.2013.02.024.

Vogl, T., C. Ruth, J. Pitzer, T. Kickenweiz, and A. Glieder. 2014. Synthetic core promoters for *Pichia Pastoris*. *ACS Synthetic Biology* 3 (3): 188–91. doi:10.1021/sb400091p.

Wang, H., H. Yang, C. S. Shivalila, M. M. Dawlaty, A. W. Cheng, F. Zhang, and R. Jaenisch. 2013. One-step generation of mice carrying mutations in multiple genes by CRISPR/Cas-mediated genome engineering. *Cell* 153 (4): 910–18. doi:10.1016/j.cell.2013.04.025.

Wang, J., and S. Hannenhalli. 2006. A mammalian promoter model links cis elements to genetic networks. *Biochemical and Biophysical Research Communications* 347 (1): 166–77. doi:10.1016/j.bbrc.2006.06.062.

Wang, M., Y. Liu, C. Zhang, J. Liu, X. Liu, L. Wang, et al. 2015. Gene editing by co-transformation of TALEN and chimeric RNA/DNA oligonucleotides on the rice OsEPSPS gene and the inheritance of mutations. *PLoS One* 10 (4): e0122755. doi:10.1371/journal.pone.0122755.

Weber, W., W. Bacchus, F. Gruber, M. Hamberger, and M. Fussenegger. 2007. A novel vector platform for vitamin H-inducible transgene expression in mammalian cells. *Journal of Biotechnology* 131 (2): 150–58. doi:10.1016/j.jbiotec.2007.06.008.

Wei, P., W. W. Wong, J. S. Park, E. E. Corcoran, S. G. Peisajovich, J. J. Onuffer, A. Weiss, and W. A. Lim. 2012. Bacterial virulence proteins as tools to rewire kinase pathways in yeast and immune cells. *Nature* 488 (7411): 384–88. doi:10.1038/nature11259.

Wieland, M., and M. Fussenegger. 2012. Engineering molecular circuits using synthetic biology in mammalian cells. *Annual Review of Chemical and Biomolecular Engineering* 3 (1): 209–34. doi:10.1146/annurev-chembioeng-061010-114145.

Witzgall, R., E. O'Leary, A. Leaf, D. Onaldi, and J. V. Bonventre. 1994. The Krüppel-associated Box-A (KRAB-A) domain of zinc finger proteins mediates transcriptional repression. *Proceedings of the National Academy of Sciences* 91 (10): 4514–18.

Wuest, D. M., S. W. Harcum, and K. H. Lee. 2012. Genomics in mammalian cell culture bioprocessing. *Biotechnology Advances* 30 (3): 629–38. doi:10.1016/j.biotechadv.2011.10.010.

Xie, Z., L. Wroblewska L, L. Prochazka L, R. Weiss, and Y. Benenson. 2011. Multi-input RNAi-based logic circuit for identification of specific cancer cells. *Science* 2; 333 (6047): 1307–11.

Yang, Y.-W., and M. D. Koob. 2012. Transferring isolated mitochondria into tissue culture cells. *Nucleic Acids Research* 40 (19): e148–e148. doi:10.1093/nar/gks639.

Ye, H., M. D.-E. Baba, R.-W. Peng, and M. Fussenegger. 2011. A synthetic optogenetic transcription device enhances blood-glucose homeostasis in mice. *Science* 332 (6037): 1565–68. doi:10.1126/science.1203535.

Yoon, S., S. Kim, and J. Kim. 2010. Secretory production of recombinant proteins in *Escherichia Coli*. *Recent Patents on Biotechnology* 4 (1): 23–29. doi:10.2174/187220810790069550.

6

Synthetic Plants

Nicola J. Patron

CONTENTS

Technologies for inserting cloned and modified DNA sequences into the genomes of plants have been in use for over three decades but, until recently, the assembly of complex multi-gene constructs was prohibitively difficult and limited the scale of plant genome engineering projects. In the two decades following the production of the first transgenic plants,[1] the majority of engineered plants were made by inserting a selectable marker gene along with a single gene-of-interest, usually under the control of plant viral promoter to ensure constitutive over-expression, into a random position in the genome. For plants destined for the biotechnology market, it is preferable both to remove the selectable marker and, for the purpose of simultaneous deregulation of multiple traits, to insert more than one transgene on the same piece of DNA. Until recently, however, it was uncommon to insert more than one transgene for modification of the same trait. Even so, transgenic plants have had an enormous impact on agriculture. In 2012, 1,700,000 km² (395 million acres) of land in 28 countries were used to grow 11 different species of genetically modified plants.[2] In addition, vaccines and therapeutics with clinical and veterinary applications have been produced in recombinant plant systems.[3]

To date, most synthetic biology has been done in single-celled organisms, but progression of the technology in multicellular organisms is the inevitable next step. Compared with animals, plants are easy to work with being sessile and having low requirements for healthy growth. In the past two decades, almost 100 genome-sequencing projects have been initiated for plant and algal species and there is no longer a paucity of plant genetic resources. These data are being mined for coding and regulatory sequences for use in numerous ambitious genetic engineering projects with the potential to revolutionise plant biotechnology.

Plants are an intrinsic part of human life and society, being a source of building materials, medical products and paper, as well providing oxygen and food. The application of synthetic biology promises to accelerate plant science and to provide solutions to global challenges such as food security for a growing global population, malnutrition, the safe production of therapeutic compounds, reducing reliance on fossil fuels, and the contamination of ecosystems with agrochemicals and macronutrients. Additionally, new roles for plants in our environment are being explored, such as sentinels for chemical threats and cleansers of contaminated land. A crowd-funded start-up venture in San Francisco, 'Glowing Plant', is exploring the possibility of engineering plants to be sustainable sources of light. Genetic manipulation of plants is generally more complex than that of bacteria and yeast, with some species being particularly recalcitrant to transgenesis. In the plant research and biotechnology community, a small number of plant species that are most easily modified have been selected as models. Synthetic biologists are taking advantage of the genetic and molecular resources available for these model chassis as well as engineering new characteristics into species of particular social and economic importance (Figure 6.1). In this chapter, scientific endeavours employing synthetic biology techniques to engineer plants are discussed along with the tools and resources currently available.

Arabidopsis thaliana
thale cress

A dicotyledenous plant in the brassica family popular with plant scientists for several decades due to its diminutive stature, short life-cycle and small genome.

Nicotiana benthamina

A relative of tobacco and a popular model due to its susceptibility to a wide range of plant pathogens and the ease at which it can be genetically transformed and regenerated. It is also used as a chassis for pharming.

Camelina sativa
false flax

A brassica species cultivated as an oilseed, it has recently been engineered to produce the omega-3 long chain polyunsaturated fatty acids usually produced by marine algae.

Oryza sativa
rice

Providing one-fith of the calories consumed worldwide, this grain crop is the subject of engineering projects that aim to improve its nutritious value, carbon fixation and yield.

Tritcum aestivum
bread wheat

One of the first cereals to be domesticated, wheat is the world's most traded crop. Many ongoing projects aim to engineer improved disease resistance, yield and other agronomic traits.

FIGURE 6.1
Several plant species are being engineering by plant synthetic biologists. Species have been chosen either for their ease of engineering or for their current importance in food production.

6.1 Tools and Resources

Direct delivery of genes to plant genomes is achieved by the application of purified, cloned DNA to cells. Plant cell walls and membranes can be penetrated with the aid of a mechanical DNA-carrier such as gold nanoparticles delivered with a gene gun. This is known as 'bombardment' or 'biolistic particle delivery'. DNA can also be delivered by mixing recipient cells with inorganic fibres such as silicon carbide whiskers. Alternatively, macerozymes and cellulases can be used to digest the cell wall, releasing protoplasts.[4–6] Prepared protoplasts are incubated with DNA, a high molecular weight polycation such as polyethylene glycol thought to compact DNA and aid association with membranes, and divalent ions to destabilise the membrane.[7] For all of these methods, the DNA to be inserted can be delivered as a circular plasmid molecule or the desired sequences can be separated from plasmid backbone with restriction endonucleases or amplified by polymerase chain reaction and delivered as linear molecules, free of the plasmid origin of replication and bacterial selection genes. Other than a high-copy origin of replication to provide sufficient quantities of DNA, there are no specific requirements for the backbones of plasmid vectors in which synthetic genes or transgenes are cloned for these direct delivery methods.

Plastid genomes are also transformed by direct delivery, typically using a biolistic particle delivery system to deliver heavy metal particles coated with DNA directly to plant cells, but also by direct DNA delivery to protoplasts.[8] Integration into plastid genomes is achieved by recA-mediated homologous recombination[9] therefore plastid transformation vectors are designed with homologous flanking sequences on either side of the DNA to be inserted to facilitate double recombination (Figure 6.2A). Targeting sequences have no special properties other than that they are homologous to the chosen target site and are generally about 1 kb in size. Plant plastids are widely accepted to have arisen from the endosymbiosis of cyanobacteria, probably only once, in the common ancestor of the Archaeaplastida.[10] As such, plastid genomes are bacterial in origin and it is therefore

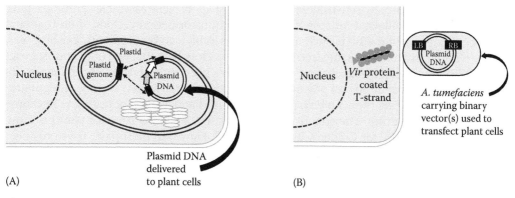

(A) (B)

FIGURE 6.2

Methods for inserting DNA into plant genomes. (A) Vectors for the transformation of plant plastids include cognate sequences from the plastid to enable homologous recombination (black boxes), enabled by the plastidial recA-dependent DNA repair mechanism. The genes of interest are inserted into the plasmid genome between these sequences. (B) Binary vectors are assembled in *Escherichia coli* with the DNA to be inserted flanked by left and right border sequences (LB and RB). The plasmids are transferred to *Agrobacterium tumefaciens*, cultures of which are used to transfect plants. A T-strand is produced from the region between the borders and coated with *Vir* proteins produced by the bacterium that enables it to enter the plant cell and traffic to the nucleus where it interacts with plant proteins and is inserted into the plant genome.

possible to engineer plastidial origins of replication into the backbones of cloning vectors to produce shuttle vectors capable of replicating both in laboratory bacteria for ease of construct but also within plastids after delivery. This approach increases the amount of DNA template available in the cell for homologous recombination.[11]

6.1.1 Agrobacterium as a Gene Delivery Tool

The most widely used method to transfer DNA into plant cells, due to the relatively low number of insertion events obtained, is *Agrobacterium*-mediated delivery. *Agrobacterium tumefaciens* is a pathogenic soil bacterium of the family Rhizobiaceae that infects plants by inserting a section of a tumour-inducing (Ti) plasmid, known as the T-region, into the genome of the plant host. T-regions are defined by 25 base pair T-DNA border sequences and it is possible to clone almost any sequence between these borders. In the 1980s, cloning into T-regions was simplified by the removal of the *vir* genes that encode the proteins required to transfer the T-region to the plant cell from the Ti plasmid to a separate plasmid known as the virulence plasmid, maintained in the same *Agrobacterium* cell.[12] A number of strains with non-oncogenic virulence plasmids were developed in the 1980s and 1990s and are still in common use.[12]

Plasmids containing T-regions, but not *vir* genes, are known as binary vectors and, as well as an origin of replication for *A. tumefaciens*, they often contain a high-copy origin of replication for *E. coli*. Constructs are assembled in *E.coli* as it is comparatively much easier to work with in the laboratory. Completed assemblies are transferred to *A. tumefaciens* and used to transfect plant cells enabling the transfer of the engineered T-strand to the nuclear genome (Figure 6.2B). Several series of binary vectors have been developed, differing mainly in their origins of replication, plant regulatory elements, selectable markers and sites for different cloning methods.[13] Until recently, the laboriousness of cloning meant that the T-regions of binary vectors usually contained one of a handful of selectable marker cassettes, encoding products to confer resistance to an antibiotic or herbicide. It was also common for them to contain promoter and terminator sequences flanking a multiple cloning site or Gateway™ cassette into which new genes of interest could be cloned.[13] Construction of large, multi-gene vectors was therefore limited. New construction methods that allow facile assemble of multiple parts in a single step have allowed for more flexible plant transformation vectors. Multi-fragment, sequence-independent methods such as Gibson Assembly can be used with existing plant vectors with no modifications and, recently, binary vectors that enable BioBrick, uracil excision (USER) and Golden Gate cloning methods have been published.[14–18] Additionally, the availability and reduction in cost of gene synthesis means that, as new coding or regulatory sequences are identified, they can be rapidly synthesised and cloned for testing with characterised, standard modules. The quantity of characterised, standard parts for plants is growing rapidly, particularly for Golden Gate cloning.[17,19–21] However, most parts have only been tested for function in a few plant species. The majority of plant engineering experiments aim to integrate genes into the nuclear genome where the position and context of the insertion affect the function of regulatory elements. There has therefore been particular interest in the detection of operon-like clusters of plant genes that code for the synthesis of secondary metabolites. The regulatory sequences from these clusters seem to be 'self-contained', and have been shown to function remarkably similarly in several different species.[22,23]

6.1.2 Virus-Based Vector Systems

With synthesis and manipulation of DNA becoming cheap and efficient, the bottleneck of engineering plants is transferred downstream to the delivery of constructs to plant cells and regeneration of transgenic plants. In some plant species, it is not uncommon for the generation of primary transformants to take months. Non-integrative methods are therefore very useful for testing constructs. Non-integrative methods are also used when the primary aim is the rapid production of very large quantities of foreign proteins such as antibodies. Several vectors have been developed, many based on plant viruses.[24-26] To avoid the use of insects and other specialised methods of viral transmission, all or part of the viral genome can be cloned into the T-region of a binary vector. Once transferred to plant cells, expression will result in infectious, autonomously replicating nucleic acids. Viruses, such as potato virus X (PVX), will tolerate the insertion of an extra open reading frame cloned within their genomes.[27] To facilitate cloning, however, it has become more common to use only a part of the viral genome. For example, the TMV genome has been engineered into two modules, one containing the viral RNA-dependent RNA polymerase and movement proteins, and a second to carry the gene of interest and viral 3′ untranslated region (UTR) required for efficient replication and amplification of the vector. The two modules are co-delivered to plant cells along with a vector encoding a site-specific recombinase that assembles the two viral modules together *in vivo*.[28] Geminiviruses, single-stranded, circular DNA viruses that replicate through a rolling circle process in plant nuclei, have also been used as platforms for transient expression of heterologous proteins in plant cells. The geminiviral genome, for example bean yellow dwarf virus, is depleted of movement and coat-protein genes and foreign sequences are cloned between the two intragenic repeat regions necessary for the sustenance of rolling circle replication.[29] The production of high levels of heteromultimeric proteins has been achieved using the cowpea mosaic virus-based HyperTrans (CPMV-HT) system.[30] CPMV is a comovirus with a genome consisting of two positive sense RNAs separately encapsidated. The system involves cloning the desired genes of interest between 5′ leader and 3′ untranslated regions of RNA 2.[31,32]

6.2 Food Security and Sustainable Agriculture

In the middle of the 20th century fears of global population growth outstripping food supply were widespread. As late as 1968, worldwide starvation events in the 1970s and 1980s were believed to be likely.[33,34] Instead, high-yielding varieties of maize, wheat and rice, along with the transfer of technologies such as irrigation, pesticides and synthetic nitrogen fertiliser to developing countries enabled what has become known as the Green Revolution. With advances in molecular genetics, the genes responsible for the traits that made it possible to double the production of cereal crops in developing nations between 1961 and 1985 have been identified and isolated. These include genes responsible for reduction in stem growth, redirection of photosynthetic assimilates to developing grains and disease resistance.[35] Global population growth in the 21st century is arguably even more challenging. World population is currently growing by approximately 74 million people per year and is projected to reach 9 billion by 2050. The population of developing nations is expected to rise from 5.3 billion to 7.8 billion while the population of developed nations, apart from the United States, is expected to stabilise.[36] Global development, particularly agricultural expansion, has led to mass deforestation. Along with industrialisation, this has contributed to increased concentrations of atmospheric carbon dioxide.[37] Norman Borlaug's hypothesis that 'increasing the productivity of agriculture on the best farmland can help to control deforestation by reducing the demand for new farmland' is as pertinent today as it was in the middle of the previous century.[38] Critics of intensive monocultural agriculture claim that the use of high-yielding varieties has enforced reliance on irrigation, pesticides and fertilisers and suggest that, in their absence, traditional varieties may out-perform them.[39] However, while labour-intensive multi-species cropping and harvesting of mixed varietals that vary in size and maturation times are suited to smallholdings and subsistence agriculture, as populations grow and become increasingly urbanised, a pref-erence for cropping systems that allow large harvests with minimal labour is inevitable.

Synthetic biology offers the hope that a solution can be found in crops that yield more food but have reduced requirements for water, fertilisers and other agrochemicals. The main tar-gets are the mass-calorie grain crops of wheat, rice and maize, which together provide more than half the global caloric consumption. The goals are to improve yield by the sophisticated engineering of major food crops to fix carbon, use nutrients and evade diseases.

6.2.1 Challenge of Re-Engineering Photosynthesis

The maximal yield of grain crops is dictated by the photosynthetic capacity of the leaves to supply carbohydrates produced from photosynthesis to the developing seed. Ninety per cent of plants, including rice and wheat, utilise C_3 fixation in which ribulose-1, 5-bisphosphate carboxylase/oxygenase (RuBisCO) catalyses the net fixation of atmospheric CO_2 into organic molecules during photosynthesis by the carboxylation of ribulose-1,5-bisphosphate into the three-carbon molecule 3-phosphoglycerate (Figure 6.3). When CO_2 levels are high RuBisCO operates relatively efficiently but when temperatures reach 30°C the partial pressure of atmospheric O_2 is over 100 times greater than CO_2. At these concentrations RuBisCO can also oxygenate ribulose-1,5-bisphosphate producing phos-phoglycolate, which cannot be progressed through the Calvin–Benson cycle. Conversion of phosphoglycolate results in the loss of 25% of carbon atoms.[40–44] To adapt to the high rates of photorespiration and carbon deficiency experienced in warm, dry climates, some plant families have made biochemical and anatomical adjustments to concentrate CO_2 to a level that nearly saturates the RuBisCO active site (Figure 6.3). This is called C_4 photo-synthesis because inorganic carbon is fixed by phosphoenolpyruvate (PEP) carboxylase and the resulting four-carbon acids are moved to a compartment in which RuBisCO is localised to release CO_2 by decarboxylation of the four-carbon acid.[45,46] It is hypothesised that introducing the C_4 photosynthetic cycle into a C_3 crop such as rice might increase crop yield by as much as 50%.[47,48]

The scale of genome engineering required to engineer C_4 photosynthesis into rice sur-passes anything achieved in plants to date, but since it might enable gains as great as those from the Green Revolution of the 20th century, a consortium of plant scientists, bioinfor-maticians, computer modellers and molecular biologists are attempting to do exactly this. The C_4 Rice Project (http://c4rice.irri.org) aims to modify the C_3 leaf structure to form the concentration compartment in which RuBisCO is situated by creating a version of the two-celled system where mesophyll and bundle sheath cells are arranged in the so-called kranz anatomy (kranz being the German word for 'wreath')[49] (Figure 6.3). This will require reducing the interveinal distance to decrease the diffusion distance between the meso-phyll and bundle sheath cells, engineering transport between the two cell types, as well as modifications and an increase in the density of chloroplasts in the bundle-sheath cells (BS).[47,48,50,51] To recreate C4 fixation across these engineered cell types, carbonic anhydrase and PEP carboxylase must be expressed in mesophyll cells to convert CO_2 to bicarbonate and fix it to PEP forming oxaloacetate (OAA). OAA must then be converted to malate by malate-dehydrogenase before being transported to the bundle-sheath chloroplasts where

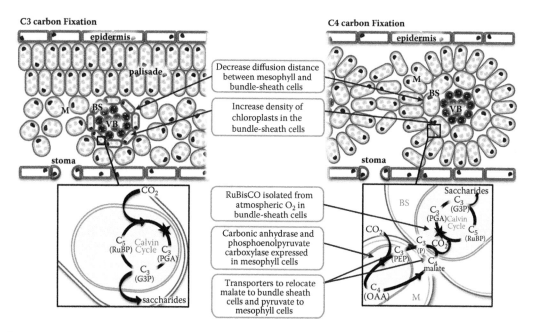

FIGURE 6.3

Targets for re-engineering the carbon-fixing mechanism found in leaves of C3 plants (left) to acquire the capabilities of those found in C4 plants (right). In C3 plants, CO_2 diffuses from the stoma to mesophyll (M) and bundle-sheath cells (BS) where it competes with O_2 for ribulose-1,5-bisphosphate carboxylase/oxygenase (RuBisCO, indicated by a black star). CO_2 enters the Calvin cycle by fixation into 3-phosphoglycerate (PGA) by the carboxylation of ribulose-1-bisphosphate (RuBP). PGA is converted to glyceraldehyde 3-phosphate (G3P), which is used to make saccharides that are stored as starch or transported as sugars in the vascular bundle (VB). G3P is also converted to RuBP to complete the cycle. In C4 plants, CO_2 is fixed to phosphoenolpyruvate (PEP) to form oxaloacetate (OAA) in M cells. This is converted to malate and transported to the chloroplasts of BS cells to be decarboxylated, concentrating CO_2 around RuBisCO.

it can be decarboxylated to concentrate CO_2 around RuBisCO. The resulting pyruvate must be moved back to the mesophyll cells to be converted to PEP. As well as engineering the coding sequences of all the required enzymes and transporters, the identification of cell-specific regulatory elements is also critical for success (Figure 6.3). In the opening stages of this project, it has been shown that promoters that express mesophyll cells in maize can also direct mesophyll-specific expression in rice and that the cell-specific expression of some genes in C_4 leaves is mediated by elements not present in the promoter.[52–57]

6.2.2 Engineering Synthetic Symbioses

The application of synthetic fertiliser leads to environmental contamination by nitrogen. Most of the nitrogen applied, however, is lost to the environment. Nitrogen-containing fertilisers produce 9% of the UK's total ammonia emissions and contribute to the release of nitrous oxide into the atmosphere, which is responsible for almost 5% of the UK's greenhouse gas emissions.[58] At the same time, developing countries, particularly those in Africa, lack the resources to purchase or apply fertilisers.[59] Data collected by the World Bank show that cereal crops yield 5–7 tonnes per hectare in Europe, North America and China, but less than 2 tonnes per hectare in Africa. A solution to both of these problems might be found in an effort to engineer the cereal crops for improved nitrogen use efficiency.

The roots of legume species associate with certain species of rhizobia, soil bacteria that are able to fix atmospheric, molecular nitrogen into ammonia. Fixed nitrogen is delivered to the plant though this symbiosis, which is dependent on the intracellular uptake of bacteria and the formation of root nodules.[60,61] Efforts are underway both to engineer plants with the bacterial capability to fix nitrogen in their own cells, and also to transfer the legume-rhizobial symbiosis to cereal crops.[62] The former requires engineering nitrogenase, an enzyme complex comprised of a heterotetrameric protein, MoFe and a homodimer known as the Fe-protein. The catalytic components of nitrogenase are impaired by the presence of oxygen, and require an energy-hungry electron transfer process to reduce its substrate. As many as 18 genes are required for the assembly and function of the nitrogenase complex.[63] Engineering a synthetic symbiosis into cereals requires conferring the ability to recognise the molecular signals (nodulation factors) produced by rhizobia, the ability to establish root nodules and the establishment of the bacteria in the nodules. This strategy, however, will take advantage of a related pathway for forming symbiotic relationships with mycorrhizal fungi.[62,64] Mycorrhizal fungi associate with the roots of most plant species, including cereals, providing the fungus with carbohydrates and the plant with an enhanced surface area from which to absorb water and nutrients.[65] Mycorrhizal symbiosis evolutionarily predates nodulation symbiosis and several genes have been shown to have dual functions in both mycorrhizal and nodulation signalling.[66–69] Several of these genes are conserved in cereals and homologues cloned from the rice genome can be used to restore mycorrhization and nodulation in reciprocal mutants of model legumes.[70] A major aim for plant synthetic biologists is to augment the wheat mycorrhizal pathway with genes for recognition of bacterial nodulation factors that are then capable of triggering sufficient nodulation to impact nitrogen use in wheat.[62,64]

6.2.3 Towards Durable Resistance

In countries with developed agricultural infrastructure, many disease outbreaks may be controlled by the application of pesticides or fungicides; however, it is often uneconomic to do so in developing countries. The costs of chemicals, fuel, vehicle maintenance and machinery frequently exceed agricultural profits. Additionally, fewer chemicals are available for disease control as stricter environmental legislation is enforced. Plant cells recognise conserved molecules of pathogens via receptors on their surface. Recognition triggers a defensive response that, to complete infection, a pathogen must suppress, usually by delivering molecules known as effectors to disrupt the plant's signalling mechanisms. To counteract specific effectors, plants encode disease resistance (R) genes. Pathogen effectors and host R genes co-evolve such that in a naturally evolving plant population, most individual plants will encode a suite of R genes that enable them to resist most of the pathogens.[71] However, in monocultural agriculture systems, pathogens are able to invade entire populations of genetically identical plants if they lack a suitable R gene. After devastating outbreaks of stem rust decimated wheat production in North America in the mid-20th century, the varieties bred in the Green Revolution were selected for their resistance to these biotrophic fungi.[72] This resistance has endured for almost half a century, but new races of rust fungi are evolving for which current commercial varietals have no resistance.[73] Additionally, in the past 30 years, outbreaks of stripe rust have caused losses of 30%–50% of the expected grain yield in almost every continent.[74] Rust epidemics cause economic devastation and, in developing countries, humanitarian disasters.

Wild relatives of crop plants are a source of genes for disease resistance. However, because of complex genetics, not least the polyploidy common in many crop species, it is rare to find *R* genes in species that can be directly bred with commercial cultivars. Breeding also separates combinations of alleles that have been previously selected to confer desirable traits such as yield or early-season vigour (Figure 6.4). Desirable genes may also be closely linked to deleterious genes that are difficult to eliminate. The process can take years and single *R* genes are sometimes defeated in less time than it took to breed them into the population. New sequencing technologies have led to easier identification of *R* genes.[75,76] But, although it may now be possible to identify multiple *R* genes in a short period of

Classical plant breeding is the deliberate crossing of plants to produce populations with desired characteristics such as increased yield, disease resistance or tolerance to environmental stress. Closely or distantly related individuals with specific desired traits are chosen to breed with an existing cultivar, typically one rich in agronomically desirable traits. Progeny from this cross are backcrossed with individuals from the parent population until a progeny population is obtained in which the new trait does not segregate, the desirable characteristics of the original population are retained and any undesirable traits from the new individual have been bred-out.

Molecular marker assisted breeding techniques identify unique gene sequences in close proximity to genes or genetic loci that confer desirable traits. Large populations of plants can be screened to identify individuals that possess the trait of interest to be employed in breeding programs. Progeny can easily be screened for the presence of the marker reducing the need for continual field conditions for the selection of desirable traits.

Mutagenesis is used to generate cultivars with novel traits, not known in natural populations. Mutagenised populations can be screened for specific desirable characteristics are produced using chemicals such as EMS or with radiation or transposons. Thousands of new varieties containing traits identified from from mutagenic populations have been marketed over the last century. Recently, targeted genome-editing techniques that employ synthetic, programmable molecular tools such as zinc finger nucleases (ZFNs), transcription activator-like effector nucleases (TALENs) and clustered regularly interspaced short palindromic repeats (CRISPR/cas) have allowed the mutation of specific genes known to be responsible for particular traits.

Transgenesis employs recombinant DNA technology to insert specific genes into the genome. Genes are delivered to plant cells directly or by using the bacterium Agrobacterium tumefaciens to shuttle DNA from a plasmid to the nucleus where it is inserted into the genome. A selectable marker gene is usually introduced with the gene of interest to allow the selection of transgenic plants. Synthetic biology techniques allow the facile assembly of large genetic constructs for delivery to plant genomes.

Cultivars are populations of plants that have been selected for particular characters, are distinct, uniform and stable in those characters, and retain those characters through propagation

Linkage Drag is the inheritance of unwanted donor alleles in the same genomic region as the desired locus.

Pathovars are pathogens with proved differences in host range. Differences in symptomatology allows divisions of pathovars into races, which are a collection of strains differing from others in their host specialisation.

Polyploids are cells and organisms that contain more than more than the two paired, homologous sets of chromosomes found in the nucleus of a eukaryotic diploid cell. Polyploidy is common in plants. For example banana and many citrus are triploid; cotton, potato and rapeseed are tetraploid and bread wheat, oats and kiwi-fruit and hexaploid

Variety is a legal term that provides the breeder of a cultivar with exclusive control over the propagation of that cultivar for a number of years.

FIGURE 6.4

Terminology used in plant breeding. The terms in the white box describe technologies currently used in production of new crops.

time, breeding multiple unlinked genes into a population is extremely difficult. By comparison, inserting *R* genes into plant genomes as transgenes is relatively straightforward. Field resistance with transgenic *R* genes has been shown for several pathogens including bacterial spot in tomato,[77] powdery mildew in wheat[78] and late blight in potatoes.[71] As parallel, high-throughput DNA assembly techniques have been adopted for plant science, building multi-gene stacks for transfer into plant genomes has become far simpler.[17-20] Stacked *R* genes reduce the selection pressure on each individual gene. If each *R* gene has a unique effector target or mode of action, the pathogen cannot easily defeat the locus, and long-term, durable resistance is more likely.[79]

Many *R* genes encode nucleotide-binding leucine-rich repeat (NB-LRR) proteins. The LRRs of NB-LRR proteins are highly irregular and poorly conserved as a consequence of the rapid evolution required to continually combat new pathogen races. However, composite models of LRR domains can be constructed from multiple known structures.[80,81] The specific interactions between NB-LRRs and pathogen effectors are only understood in a small number of pathogen-host systems, making rational design of synthetic NB-LRRs a difficult prospect. The emergence of DNA foundry companies offering low-cost custom gene synthesis means that site-saturated mutagenesis libraries of synthetic NB-LRR genes can be built and rapidly screened for new candidates that confer resistance to effectors.

6.3 Improving Nutritive Value

6.3.1 Engineering the Bioavailability of Micronutrients in Food Crops

As well as energy malnutrition caused by caloric deficit, poor diets can also result in micronutrient deficiencies. Grain crops are often the sole food source for the world's poorest people and, though they may prevent hunger, they are poor sources for many essential nutrients. Iron deficiency is the most common nutritional disorder in the world; over 30% of the world's population is anaemic, due in part to low iron intake.[82] An estimated 250 million preschool children are deficient in vitamin A, causing up to 500,000 children to lose their sight each year.[82] Traditional strategies to alleviate micronutrient deficiencies include the distribution of micronutrient supplements and the fortification of food products such as flour. Such strategies require ongoing investment in production as well as transportation. An alternative is to engineer staple cereal crops such as rice for increased bioavailability of micronutrients in the edible parts. Several transgenic strategies that aimed to do this have already been successful. The most notable of these is Golden rice, which employs a two-gene cassette consisting of phytoene synthase (*psy*) and a bacterial phytoene desaturase (*crt I*) that enable rice plants to synthesise β-carotene in the grain as well as in the leaves.[83-89] Various methods have been employed to improve the storage capacity of iron in rice grains, including expression of the iron storage protein, ferritin, under the control of endosperm-specific promoters[90-92]; enhancing translocation of iron by expression of nicotianamin[93,94]; increasing the flux of iron into the endosperm by expression of the Fe(II)-nicotianamine transporter gene *OsYSL2*[95]; and the introduction of genes to improve the uptake of iron from the soil.[96,97] Combining the multiple sets of genes from more than one of these approaches within a single construct has enabled the production of rice plants with iron concentrations up to four times that of wild type rice, with no reduction in yield.[98]

6.3.2 Engineering to Optimise the Fat and Fibre Content of Food Crops

Concurrent with poor and developing nations experiencing starvation, malnutrition and food insecurity, the populations of more affluent countries are over-consuming. In 2010, the International Obesity Taskforce estimated that 475 million adults were obese.[99] Excess adiposity is a strong risk factor for type II diabetes and recent studies have shown that diet and physical activity are effective in delaying the onset of the disease.[100–102] Lowering the intake of saturated fat and increasing consumption of dietary fibre have both been shown to increase sensitivity to insulin. Additionally, starch that is easily digested has a high glycaemic index but resistant starches have a limited effect on glucose absorption from the gastrointestinal tract and have been shown to increase insulin sensitivity.[103] Further, human colonic bacteria ferment digestion-resistant starch, a source of insoluble fibre, to short-chain fatty acids including butyrate, acetate and propionate. These acids stimulate colonic blood flow and electrolyte uptake, and butyrate inhibits cell growth in human colonic cancer cells.[104,105] Plant starch is comprised of two types of polysaccharides: highly branched amylopectin and the relatively linear amylose. There have been several successful endeavours to produce starch with different ratios of amylose and amylopectin and modified chain lengths by manipulating the levels of enzymes involved in metabolism. A starch engineered for the economical conversion of starch to ethanol for biofuel production has been commercialised.[106,107] There is now also growing interest in crops enriched with starch polymers that are resistant to digestion following a study that found high-amylose wheat starch produced by engineering the levels of starch branching enzymes increased the large-bowel health of rats.[108]

Obesity also increases the risk of cardiac failure. Studies have shown that several plant compounds, including anthocyanins, flavonoid pigments that give red and purple fruit their colour, can affect lipid metabolism, reduce oxidative stress and increase levels of high-density-lipoprotein cholesterol (sometimes known as 'good' cholesterol), all of which may be cardio-protective.[109–111] For example, it was found that juice from blood oranges is more effective than both water and juice from regular, 'blond' oranges at limiting the development of fat cells and weight gain in mice.[112] Production of blood oranges is geographically limited to a small portion of citrus-producing environments, as very specific growth requirements are required to produce the bloody fruit phenotype. Anthocyanin production in blood orange fruit is temperature-dependent; a cold-stress induced retroelement is responsible for expression of a transcriptional activator of anthocyanin production.[113] It is therefore not possible to produce a variety that is free from cold-dependency by conventional breeding. Re-engineering the gene with a different promoter that frees it from the requirement for cold may result in blood oranges that can be grown in all citrus-growing regions. Several attempts to engineer fruits and grains with the ability to produce anthocyanins have been successful. Soybean grains and purple tomatoes have both been engineered to produce anthocyanins and were found to have anti-inflammatory effects and to slow the progression of soft-tissue carcinoma in cancer-prone mice.[114,115]

Consumption of omega-3 long chain polyunsaturated fatty acids (omega-3 LC-PUFAs) has also been linked to improved cardiovascular health.[116,117] Although plants produce polyunsaturated fatty acids, it is mainly the consumption of oily fish that provide humans with omega-3 LC-PUFAs. Fish themselves do not make omega-3 LC-PUFAs: they are biosynthesised by the algae and phytoplankton at the bottom of aquatic food chains. As global fish-stocks decline and wild fishing becomes both economically and environmentally unsustainable, the aquaculture industry has grown at 9% per year since the 1970s and now produces 40% of all fish consumed.[118] Providing farmed fish with an alternative source of omega-3 LC-PUFAs to replace the practice of harvesting wild fish to feed farmed stock is critical for the sustainability of this industry.

Camelina sativa or false flax, a brassica species cultivated as an oil crop, has been engineered to produce omega-3 LC-PUFAs by the addition of either a five-gene pathway to produce eicosapentaenoic acid (20:5 n-3, EPA) or a seven-gene pathway to produce both EPA and docosahexaenoic acid (22:6 n-3, DHA).[119] Synthetic genes were made by placing coding sequences of the required enzymes sourced from green algae, moss and several heterokont species under the control of seed-specific promoters of plant origin. Engineered *C. sativa* chassis have been developed that can successfully catalyse the conversion of α-linolenic acid to omega-3 LC-PUFAs, resulting in seeds with up to 31% EPA or 12% EPA and 14% DHA. These engineered chassis are also able to prevent accumulation of unwanted omega-6 acids, which are intermediates of the pathway. Omega-6 acids are not present in fish oils but were observed in previous attempts to engineer omega-3 LC-PUFA production into plants.[119]

6.4 Synthetic Biology and Green Biopharms

Throughout human history plants have been an important source of medicines. Many commonly used therapeutic compounds are found in plants including digoxin and quinidine used to treat cardiac arrhythmia; tubocurarine for anaesthetics; the anti-psychotic reserpine; chemotherapies such as irinotecan and topotecan and analgesics such as codeine, aspirin and morphine. The ability to express foreign genes in plant cells has enabled the production of recombinant human and animal proteins and antibodies in plants. The advantages of plant systems over animals for the production of human therapeutics are both the absence of diseases that can infect animals and the relatively simple requirements for growth and nutrition. Additionally, oral vaccines and therapeutics need not be extensively purified: if produced in edible tissues they can merely be minimally processed to preserve the antigen. Plants also have advantages over bacterial systems for the production of biotherapeutics as plants make many of the post-translational modifications observed in human proteins such as disulfide bond formation and lipidations. These advantages have led to multiple successes with the production of therapeutic agents in plants. Plant chloroplasts have been utilised for the production of more than 20 antigens against 16 different diseases as well as a number of biopharmaceutical proteins.[3] Transgenic plants expressing recombinant antigens for cholera, entertoxigenic *Escherichia coli*, hepatitis B and respiratory syncytial virus and others have induced the production of specific antibodies when fed to mice and, in some cases, humans.[120–125]

There is significant interest in both increasing the availability and lowering the cost of production of type I interferons (IFNs), which have several clinical applications including the inhibition of viral replication and suppression of tumour cell growth. In 2007, tobacco plants in which the chloroplast genome had been engineered for the production of inteferon-alpha2b (IFN-alpha2b) were grown in the field.[126] Up to 20% of the soluble protein purified from these plants was found to be IFN-alpha2b which has similar biological activity to commercially produced PEGintron™, produced by chemically attaching human IFNs to molecules of polyethylene glycol.[126] Human IFN-gamma, or type II IFN, is essential for immunity and tumour control. IFN-gamma protein expressed in plant chloroplasts was shown to be as biologically active as that produced in bacteria, and to protect a carcinoma cell line against infection with the encephalomyocarditis virus.[127]

Stable transformation of plants, however, is currently too time-consuming to be compatible with the rapid response times required for the mass production of vaccines for emerging diseases. Following the outbreak of the swine H1N1 influenza, interest in systems that can produce vaccines faster than the standard egg-based vaccine manufacturing process has grown. The ability to produce virus-like particles (VLPs) in leaves within weeks of disease sequence information becoming available is attractive as an alternative production system. Transient expression in leaves of the model species *Nicotiana benthamiana* of VLPs containing the hemagglutinin protein of HN1 influenza stimulated production of cross-reactive antibodies in both ferrets and humans and was able to reduce pathology following infection.[128] Expression in *N. benthamiana* of a surface protein from cells of *Plasmodium falciparum*, the causative agent of malaria, fused to the coat protein of cowpea mosaic virus resulted in VLPs that, when used to immunise mice, induced the production of serum antibodies with complete transmission-blocking properties.[129]

Plants, or their crude extracts, have been used to treat human diseases since pre-history. While the concept of plant-products in medicine is old, synthetic biology is allowing plants to become factories for the production of natural products and therapeutic compounds. When supported by parallel DNA assembly methods, the scalability and low-production costs make plant-based pharming an attractive alternative to production in animals and microorganisms.

6.5 Fibre and Fuel

Although there is still concern about the use of transgenic and synthetic biology technologies for improving food plants, there is less unease about methods used to manufacture structural and industrial materials.[130] As well as re-engineering economically important fibres naturally produced by plants such as cotton, hemp (cannabis) and linen (flax), there is also interest in exploiting the ease with which plants can be employed for the large-scale production of materials not of plant origin. For example, there is considerable interest in the remarkable properties of spider silk. Spiders produce a range of silks for different functions including those that, even across a wide range of temperatures, retain a tensile strength equivalent to that of high-grade steel but with only one-sixth of the density and which can stretch up to five times in length before breaking. Spiders produce small quantities of these remarkable materials and, as territorial animals, are unsuited for domestication at the scale used for silk production by the larvae of silk moths.

Spider silk polymers consist of non-repetitive N-terminal and C-terminal domains that flank a repetitive middle core domain that consists of alternating blocks of crystalline and non-crystalline polypeptides.[131] There have been multiple endeavours to produce spider silks in bacteria and yeast, but yields have generally been low, due either to the instability of the repetitive gene structure or the inability of the cell to supply sufficient building blocks for translation of the long repetitive domains.[132,133] Silk production in plant cells has been successful, with high-molecular weight multimers being produced in model systems such as arabidopsis and tobacco as well as in potato and soybean.[134–136] Most successfully, spider silk coding sequences used for expression in tobacco have been engineered to circumvent the problems of RNA instability inherent with highly repetitive sequences by the introduction of inteins.[137] Inteins are intervening protein elements that are able to autocatalytically splice themselves from the protein sequences that flank them, known as exteins. This enabled the production of fibre-forming, high molecular weight spider silk multimers.[137] Tobacco has also been used for the production of bioplastics in the form of biodegradable polyhydroxybutyrate (PHB). Plants that accumulated up to 18% dry weight PHB in their leaves were produced by transferring bacterial genes encoding the PHB pathway enzymes, selected for similar codon usage and GC content to tobacco chloroplast genomes. The genes were inserted into the host's psbA operon and thus did not require an additional promoter to drive transgene expression.[138,139]

There has been considerable effort applied to improving the ability of microorganisms to biotransform specific crops or waste from food crops into useful biofuels. Many attempts have also been made to engineer plants that are easier for microbes to digest.[140] Lignin, a complex heteropolymer found in the cell walls of vascular plants, acts as a physical barrier to the microbial enzymes that digest polysaccharides, limiting the efficiency of microbial conversion of cellulose and hemicellulose to biofuels.[141] While several efforts have been made to modify the lignin biosynthetic pathway,[142–147] severe disruption of this pathway has frequently led to phenotypic abnormalities such as reduced stature or developmental retardation.[148–155] A synthetic approach has recently enabled the disconnection of lignification from the regulatory network that directs the formation of cell fibres in the model plant *Arabidopsis thaliana*. This was done by replacing the promoter of a lignin-biosynthetic gene, cinnamic acid 4-hydroxylase, with a promoter from a vascular-vessel-specific gene regulated by the transcription factor VND6. An artificial feedback loop was created using the promoter from a glycosyltransferase gene essential for secondary wall synthesis to express a new copy of the fibre transcription factor NST1.[156] The result was increased levels of digestible, polysaccharide-rich tissues without the associated high levels of lignin (Figure 6.5).

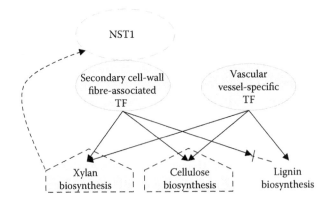

FIGURE 6.5

Digestible, polysaccharide-rich plant tissues engineered in the model plant, *Arabidopsis thaliana*. Solid arrows indicate activation by transcription factors (TF) present in the wild-type chassis. By replacing the promoter of a lignin biosynthetic gene, most lignin accumulation was limited to vascular vessels (indicated by a dashed T). Polysaccharide synthesis was then increased by introducing another copy of the cell-wall TF, under the control of a promoter activated by NST1 itself, closing the feedback loop (dashed arrow). (From Yang F., Mitra P., Zhang L., et al. *Plant Biotechnol. J.* 11:325–335, 2013).

6.6 Sensing and Phytoremediation

Environmental pollution from industrial activities, mining, agriculture, war and military exercises is a global problem, but particularly widespread in India, China and Eastern Europe. Plants are adept at extracting chemicals from the soil and translocating them to their aerial organs. A plant capable of such phytoextraction may either degrade it to harmless compounds or concentrate it in the leaves and stems, which can then be removed from the contaminated site. Some plants are naturally tolerant of toxic elements and compounds: *Astragalus bisulcatus*, milk vetch, accumulates selenium by incorporating it into amino acids in place of sulphur[157]; *Thlaspi caerulescens* and *Arabidopsis halleri* are able to accumulate zinc and cadmium[158] and *Pteris vittata*, brake fern, can hyperaccumulate arsenic to up to 1% of its dry weight.[159] For effective phytoremediation of large areas of land contaminated with toxins that easily penetrate soils, characteristics such as a deep and extensive root system and unpalatability to insects, birds and mammals are desirable. Since such traits are complex and introducing non-native species may be environmentally damaging, it is more straightforward to confer the ability to accumulate or transform toxic chemicals into plants with desirable growth habits for specific environments. Several plant species have been transformed with bacterial genes encoding organo-mercularial lyase and mercuric ion reductase to enable them to remove the organic ligand from the highly toxic methylmercury, producing the less reactive elemental mercury.[160–162] Similarly, arabidopsis plants engineered to express arsenate reductase in their leaves were able to convert arsenate to arsenite. Constitutive co-expression of arsenate reductase and γ-glutamylcysteine enabled an up-regulation of the phytochelatin pathway, absorbing the toxic arsenite into thiol-peptide compounds.[163,164]

Explosives are particularly difficult to remove from natural environments, but the manufacture, testing and deployment of weapons has led to vast areas of environmental pollution by compounds such as 2,4,6-trinitrotoluene (TNT), hexahydro-1,3,5-trinitro-1,3,5-triazine (RDX) and octahydro-1,3,5,7-tetranitro-1,3,5,7-tetrazocine (HMX). Many plant species are able to take up small amounts of TNT and several species, including several water plants, have been trialled as phytoremediators. Bacterial enzymes have been identified that are up-regulated by exposure to TNT or that are unique to strains that persist in TNT-contaminated soils, such as pentaerythritol tetranitrate reductase from *Enterobacter cloacae* and nitroreductase from *E. coli*. Plants transformed to overexpress these enzymes exhibited both enhanced tolerance and growth, indicating the incorporation of nitrogen liberated from TNT into biomass production.[165–168]

RDX is also taken up by several plant species, including major crop species such as rice and maize, albeit with less efficiency than TNT.[165] Although RDX is translocated from roots to aerial parts of the plant, it remains unchanged and negative effects such as chlorosis and necrosis of leaves are observed.[169] Arabidopsis plants engineered to express a flavodoxin-cytochrome P_{450}-like enzyme from a strain of *Rhodococcus rhodochrous* found in hexogen-contaminated soil were able to decrease RDX levels five- to tenfold more than wild-type plants.[170,171]

The ability to detect or 'biosense' toxins is also critical, both to recognise when phytoremediation pathways should be activated and also for the production of synthetic sentinels that might be deployed to inform humans of threats to the environment. Bacterial periplasmic binding proteins are cell receptors that usually mediate chemotaxis. These receptor-ligand complexes have affinity to chemotactic receptors that mediate the signal to the cytoplasm. Computational redesign of periplasmic binding proteins and coupling to histidine-kinase signalling pathways enabled the creation of bespoke biosensors,

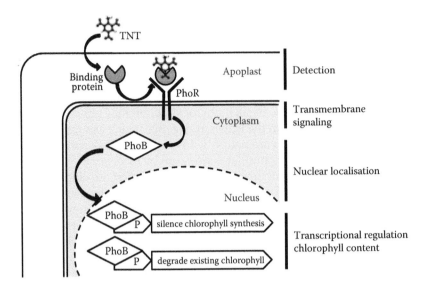

FIGURE 6.6
A synthetic signal transduction pathway detects the presence of 2,4,6-trinitrotoluene (TNT) and responds by 'de-greening'. A periplasmic binding protein from bacteria was computationally redesigned to detect TNT and localised to the apoplast of *Arabidopsis thaliana* plant cells. Interaction of TNT with the receptor causes binding to a modified histidine kinase (PhoR). PhoB, which accepts the signal from PhoR, was engineered to target the nucleus where it activates genes to silence the chlorophyll synthesis pathway and prevent chlorophyll production. (From Antunes M.S., Morey K.J., Smith J.J., et al. *PLoS One*. 6:e16292, 2011.)

including a sensor of TNT.[172] Bacterial biosensors, however, while suitable for test-kits, are difficult to incorporate into natural human environments. The re-designed TNT receptor has been coupled to response circuits in plants to make plants that respond rapidly with a detectable visual signal to the presence of this toxin in the environment[173] (Figure 6.6). This was done by localising the receptor to the apoplast of plant cells by fusing it to a plant secretory signal and using a bacterial chemotactic receptor fused to a modified histidine kinase protein (PhoR) localised across the inner membrane such that interaction of the TNT with the receptor activates the histidine kinase domain. The bacterial response regulator, PhoB, is activated by PhoR and was modified for use in eukaryotic cells by adding a nuclear targeting signal (NLS) in the engineered variant, PhoB[NLS]. When genes that block chlorophyll synthesis and initiate chlorophyll breakdown were placed downstream of a PhoB binding sequence the presence of TNT resulted in plants that 'de-greened' within 48 hours.[174] In future, biosensing capacities might be linked with phytoremediation enhancement to produce plants that can both alert the presence of and remove environmental threats.

6.7 Future of Plant Synthetic Biology

Although there is reasonable public support for the use of biotechnology for the production of bioactive molecules for pharmaceutical and industrial application, there is less public faith in the benefits of genetically engineered plants, particularly of food crops. This may hamper the ability of synthetic biology to solve problems such as sustainable agricultural practices and food security. A recent report into plant synthetic biology in the United Kingdom noted that commercial impact from plant synthetic biology will be limited if EU regulations on genetically modified organisms do not change.[175] Several other challenges also need to be addressed to realise the potential of emerging plant technologies. Tools and standard parts are not yet available for plants on the same scale as for bacterial systems. At time of publication, the largest part repositories, such as the Registry of Standard Biological Parts (http://parts.igem.org/Main_Page), have yet to add many of the sequences most commonly used in plant science. Attracting scientists skilled in the range of molecular, computational and biochemical skills required to execute synthetic biology projects is critical for the future of plant synthetic biology. As noted earlier, efficient transformation of plants is the new logjam for engineering plant genomes. Improvements in the speed and efficiency of transferral of DNA to plant cells and in the production of synthetic plants would be of great benefit.

References

1. Fraley R.T., Rogers S.G., Horsch R.B., et al. 1983. Expression of bacterial genes in plant cells. *Proc. Natl. Acad. Sci. U. S. A.* 80:4803–4807.
2. James C. 2012. ISAAA Briefs BRIEF 44. Global Status of Commercialized Biotech/GM Crops.
3. Daniell H., Singh N., Mason H., et al. 2009. Plant-made vaccine antigens and biopharmaceuticals. *Trends Plant Sci.* 14:669–679.
4. Altpeter F., Baisakh N., Beachy R., et al. 2005. Particle bombardment and the genetic enhancement of crops: Myths and realities. *Mol. Breed.* 15:305–327.
5. Asad S. and Arshad M. 2011. Silicon carbide whisker-mediated plant transformation. In Gerhardt, R, ed., *Properties and Application of Silicon Carbide*. Rijeka: InTech Europe, pp 345–359.
6. Davey M.R., Anthony P., Power J.B., et al. 2005. Plant protoplasts: Status and biotechnological perspectives. *Biotechnol. Adv.* 23:131–171.
7. Shillito R. 1999. Methods of genetic transformation: electroporation and polyethylene glycol treatment. In Vasil, IK, ed., *Molecular Improvement of Cereal Crops*. Springer, p. 402.
8. Verma D. and Daniell H. 2007. Chloroplast vector systems for biotechnology applications. *Plant Physiol.* 145:1129–1143.
9. Cerutti H. and Johnson A. 1995. Inhibition of chloroplast DNA recombination and repair by dominant negative mutants of *Escherichia coli* RecA. *Mol. Cell. Biol.* 15:3003–3011.
10. Burki F., Shalchian-Tabrizi K., Minge M., et al. 2007. Phylogenomics reshuffles the eukaryotic supergroups. *PLoS One.* 2:e790.
11. Lugo S.K., Kunnimalaiyaan M., Singh N.K., et al. 2004. Required sequence elements for chloroplast DNA replication activity *in vitro* and in electroporated chloroplasts. *Plant Sci.* 166:151–161.
12. Gelvin S. 2003. Agrobacterium-mediated plant transformation: The biology behind the "gene-jockeying" tool. *Microbiol. Mol. Biol. Rev.* 67:16–37.
13. Lee L.-Y. and Gelvin S.B. 2008. T-DNA binary vectors and systems. *Plant Physiol.* 146:325–332.
14. Nour-Eldin H.H., Geu-Flores F. and Halkier B.A. 2010. USER cloning and USER fusion: The ideal cloning techniques for small and big laboratories. In Fett-Neto, AG, ed., *Plant Secondary Metabolism Engineering* 643, Totowa, NJ: Humana Press, pp 185–200.
15. Boyle P.M., Burrill D.R., Inniss M.C., et al. 2012. A BioBrick compatible strategy for genetic modification of plants. *J. Biol. Eng.* 6:8.
16. Werner S., Engler C., Weber E., et al. 2012. Fast track assembly of multigene constructs using Golden Gate cloning and the MoClo system. *Bioeng. Bugs.* 3:38–43.
17. Engler C., Youles M. and Grützner R. 2014. A Golden Gate modular cloning toolbox for plants. *ACS Synth. Biol.*

18. Patron N.J. 2014. DNA assembly for plant biology: Techniques and tools. *Curr. Opin. Plant Biol.* 19C:14–19.

19. Lampropoulos A., Sutikovic Z., Wenzl C., et al. 2013. GreenGate – A novel, versatile, and efficient cloning system for plant transgenesis. *PLoS One.* 8:e83043.

20. Emami S., Yee M.-C. and Dinneny J.R. 2013. A robust family of Golden Gate Agrobacterium vectors for plant synthetic biology. *Front. Plant Sci.* 4:339.

21. Weber E., Gruetzner R., Werner S., et al. 2011. Assembly of designer TAL effectors by Golden Gate cloning. *PLoS One.* 6:e19722.

22. Mugford S.T., Louveau T., Melton R., et al. 2013. Modularity of plant metabolic gene clusters: A trio of linked genes that are collectively required for acylation of triterpenes in oat. *Plant Cell.* 25:1078–1092.

23. Qi X., Bakht S., Leggett M., et al. 2004. A gene cluster for secondary metabolism in oat: Implications for the evolution of metabolic diversity in plants. *Proc. Natl. Acad. Sci. U. S. A.* 101:8233–8238.

24. Gleba Y., Klimyuk V. and Marillonnet S. 2007. Viral vectors for the expression of proteins in plants. *Curr. Opin. Biotechnol.* 18:134–141.

25. Desai P.N., Shrivastava N. and Padh H. 2010. Production of heterologous proteins in plants: Strategies for optimal expression. *Biotechnol. Adv.* 28:427–435.

26. Lico C., Chen Q. and Santi L. 2008. Viral vectors for production of recombinant proteins in plants. *J. Cell. Physiol.* 216:366–377.

27. Chapman S., Kavanagh T. and Baulcombe D. 1992. Potato virus X as a vector for gene expression in plants. *Plant J.* 2:549–557.

28. Marillonnet S., Giritch A., Gils M., et al. 2004. In planta engineering of viral RNA replicons: Efficient assembly by recombination of DNA modules delivered by Agrobacterium. *Proc. Natl. Acad. Sci. U. S. A.* 101:6852–6857.

29. Chen Q., He J., Phoolcharoen W., et al. 2011. Geminiviral vectors based on bean yellow dwarf virus for production of vaccine antigens and monoclonal antibodies in plants. *Hum. Vaccin.* 7:331–338.

30. Thuenemann E.C., Meyers A.E., Verwey J., et al. 2013. A method for rapid production of heteromultimeric protein complexes in plants: Assembly of protective bluetongue virus-like particles. *Plant Biotechnol. J.* 11:839–846.

31. Sainsbury F. and Lomonossoff G.P. 2008. Extremely high-level and rapid transient protein production in plants without the use of viral replication. *Plant Physiol.* 148:1212–1218.

32. Sainsbury F., Thuenemann E.C. and Lomonossoff G.P. 2009. pEAQ: Versatile expression vectors for easy and quick transient expression of heterologous proteins in plants. *Plant Biotechnol. J.* 7:682–693.

33. Ehrlich P.R. 1968. *The Population Bomb.* New York, NY: Sierra Club/Ballantine Books.

34. Ehrlich P.R. and Ehrlich A.H. 2009. The Population Bomb Revisited. *Electron. J. Sustain. Dev.* 1:63–71.

35. Hedden P. 2003. The genes of the Green Revolution. *Trends Genet.* 19:5–9.

36. UNFPA. State of world population 2011: People and possibilities in a world of 7 billion.

37. Stocker T., Qin D., Plattner G.-K., et al. 2013. *Climate Change 2013: The Physical Science Basis.* Cambridge, UK: Cambridge University Press.

38. Angelson A. and Kaimowitz D., eds. 2001. *Agricultural Technologies and Tropical Deforestation.* Wallingford: Cabi.

39. Igbozurike U.M. 1978. Polyculture and monoculture: Contrast and analysis. *GeoJournal.* 2:443–449.

40. Von Caemmerer S. and Quick W. 2000. Rubisco: Physiology *in vivo.* In Leegood, R, Sharkey, T and von Cammerer, S, eds, *Photosynthesis: Physiology and Metabolism.* pp 86–113. Dordrecht, Netherlands: Springer.

41. Andrews T.J. and Lorimer G.H. 1987. Rubisco: Structure, mechanisms, and prospects for improvement. In MD, H and Boardman, N, eds, *The Biochemistry of Plants: A Comprehensive Treatise, Vol. 10 Photosynthesis.* San Diego, CA: Academic Press.

42. Jordan D. and Ogren W. 1984. The CO2/O2 specificity of ribulose 1,5-bisphosphate carboxylase/oxygenase. *Planta.* 161:308–313.

43. Sharkey T. 1988. Estimating the rate of photorespiration in leaves. *Physiol. Plant.* 73:147–152.

44. Ehleringer J.R., Cerling T.E. and Helliker B.R. 1997. C4 photosynthesis, atmospheric CO2, and climate. *Oecologia.* 112:285–299.

45. Hatch M. 1988. C4 photosynthesis: A unique blend of modified biochemistry, anatomy and ultrastructure. *Biochim. Biophys. Acta.* 895:81–106.

46. Furbank R.T. 2011. Evolution of the C4 photosynthetic mechanism: Are there really three C4 acid decarboxylation types? *J. Exp. Bot.* 62:3103–3108.

47. Mitchell P. and Sheehy J. 2006. Supercharging rice photosynthesis to increase yield. *New Phytol.* 171:688–693.

48. Hibberd J.M., Sheehy J.E. and Langdale J.A. 2008. Using C4 photosynthesis to increase the yield of rice-rationale and feasibility. *Curr. Opin. Plant Biol.* 11:228–231.

49. Sage R.F., Christin P.-A. and Edwards E.J. 2011. The C(4) plant lineages of planet Earth. *J. Exp. Bot.* 62:3155–3169.

50. Fouracre J.P., Ando S. and Langdale J.A. 2014. Cracking the Kranz enigma with systems biology. *J. Exp. Bot.*1–13.

51. Muhaidat R. 2007. Diversity of Kranz anatomy and biochemistry in eudicots. *Am. J. Bot.* 94:362–381.

52. Kajala K., Covshoff S., Karki S., et al. 2011. Strategies for engineering a two-celled C(4) photosynthetic pathway into rice. *J. Exp. Bot.* 62:3001–3010.

53. Ku M., Agarie S. and Nomura M. 1999. High-level expression of maize phosphoenolpyruvate carboxylase in transgenic rice plants. *Nat. Biotechnol.* 17.

54. Miyao M. 2003. Molecular evolution and genetic engineering of C4 photosynthetic enzymes. *J. Exp. Bot.* 54:179–189.

55. Nomura M., Higuchi T., Ishida Y., et al. 2005. Differential expression pattern of C4 bundle sheath expression genes in rice, a C3 plant. *Plant Cell Physiol.* 46:754–761.

56. Hibberd J.M. and Covshoff S. 2010. The regulation of gene expression required for C4 photosynthesis. *Annu. Rev. Plant Biol.* 61:181–207.

57. Matsuoka M., Furbank R.T., Fukayama H., et al. 2001. Molecular engineering of C4 photosynthesis. *Annu. Rev. Plant Physiol. Plant Mol. Biol.* 52:297–314.

58. DEFRA (Department for Environment Food and Rural Affairs). 2011. Greenhouse Gas Emission Projections for UK Agriculture to 2030.

59. Mueller N.D., Gerber J.S., Johnston M., et al. 2012. Closing yield gaps through nutrient and water management. *Nature.* 490:254–257.

60. Sprent J.I. 2007. Evolving ideas of legume evolution and diversity: A taxonomic perspective on the occurrence of nodulation. *New Phytol.* 174:11–25.

61. Wall L. and Berry A. 2008. Early interactions, infection and nodulation in actinorhizal symbiosis. In Pawlowski, K and Newton, W, eds, *Nitrogen Fixation: Origins, Applications, and Research Progress, Vol. 6.* Springer Netherlands, pp 147–166.

62. Oldroyd G.E. and Dixon R. 2014. Biotechnological solutions to the nitrogen problem. *Curr. Opin. Biotechnol.* 26C:19–24.

63. Temme K., Zhao D. and Voigt C.A. 2012. Refactoring the nitrogen fixation gene cluster from Klebsiella oxytoca. *Proc. Natl. Acad. Sci. U. S. A.* 109:7085–7090.

64. Rogers C. and Oldroyd G.E.D. 2014. Synthetic biology approaches to engineering the nitrogen symbiosis in cereals. *J. Exp. Bot.* 65:1939–46.

65. Bonfante P. and Anca I.-A. 2009. Plants, mycorrhizal fungi, and bacteria: A network of interactions. *Annu. Rev. Microbiol.* 63:363–383.

66. Oldroyd G.E.D. 2013. Speak, friend, and enter: Signalling systems that promote beneficial symbiotic associations in plants. *Nat. Rev. Microbiol.* 11:252–263.

67. Gobbato E., Marsh J.F., Vernié T., et al. 2012. A GRAS-type transcription factor with a specific function in mycorrhizal signaling. *Curr. Biol.* 22:2236–2241.

68. Maillet F., Poinsot V., André O., et al. 2011. Fungal lipochitooligosaccharide symbiotic signals in arbuscular mycorrhiza. *Nature.* 469:58–63.

69. Chabaud M., Genre A., Sieberer B.J., et al. 2011. Arbuscular mycorrhizal hyphopodia and germinated spore exudates trigger Ca2+ spiking in the legume and nonlegume root epidermis. *New Phytol.* 189:347–355.

70. Banba M., Gutjahr C., Miyao A., et al. 2008. Divergence of evolutionary ways among common sym genes: CASTOR and CCaMK show functional conservation between two symbiosis systems and constitute the root of a common signaling pathway. *Plant Cell Physiol.* 49:1659–1671.

71. Jones J.D.G., Witek K., Verweij W., et al. 2014. Elevating crop disease resistance with cloned genes. *Philos. Trans. R. Soc. B.* 365:20130087.

72. Ortiz R. and Mowbray D. 2007. Norman E. Borlaug: The humanitarian plant scientist who changed the world. *Plant Breed. Rev.* 28:1–38.

73. Singh R.P., Hodson D.P., Huerta-Espino J., et al. 2011. The emergence of Ug99 races of the stem rust fungus is a threat to world wheat production. *Annu. Rev. Phytopathol.* 49:465–481.

74. (ICARDA) International Center for Agricultural Research in the Dry Areas. 2011. Strategies to reduce the emerging wheat stripe rust disease.

75. Jupe F., Witek K., Verweij W., et al. 2013. Resistance gene enrichment sequencing (RenSeq) enables reannotation of the NB-LRR gene family from sequenced plant genomes and rapid mapping of resistance loci in segregating populations. *Plant J.* 76:530–544.

76. Tollenaere R., Hayward A., Dalton-Morgan J., et al. 2012. Identification and characterization of candidate Rlm4 blackleg resistance genes in *Brassica napus* using next-generation sequencing. *Plant Biotechnol. J.* 10:709–715.

77. Horvath D.M., Stall R.E., Jones J.B., et al. 2012. Transgenic resistance confers effective field level control of bacterial spot disease in tomato. *PLoS One.* 7:e42036.

78. Brunner S., Hurni S., Herren G., et al. 2011. Transgenic Pm3b wheat lines show resistance to powdery mildew in the field. *Plant Biotechnol. J.* 9:897–910.

79. Zhu S., Li Y., Vossen J., et al. 2012. Functional stacking of three resistance genes against *Phytophthora infestans* in potato. *Transgenic Res.* 21:89–99.

80. Ellis J., Dodds P. and Pryor T. 2000. Structure, function and evolution of plant disease resistance genes. *Curr. Opin. Plant Biol.* 278–284.

81. Marone D., Russo M.A., Laidò G., et al. 2013. Plant nucleotide binding site-leucine-rich repeat (NBS-LRR) genes: Active guardians in host defense responses. *Int. J. Mol. Sci.* 14:7302–7326.

82. Micronutrient Initiative. 2009. Investing in the future GLOBAL REPORT 2009. Micronutrient Initiative.

83. Stein A.J., Sachdev H.P.S. and Qaim M. 2008. Genetic engineering for the poor: Golden Rice and public health in India. *World Dev.* 36:144–158.

84. Tang G., Qin J., Dolnikowski G.G., et al. 2009. Golden Rice is an effective source of vitamin A. *Am. J. Clin. Nutr.* 89:1776–1783.

85. Al-Babili S. and Beyer P. 2005. Golden Rice—five years on the road—five years to go? *Trends Plant Sci.* 10:565–573.

86. Paine J.A., Shipton C.A., Chaggar S., et al. 2005. Improving the nutritional value of Golden Rice through increased pro-vitamin A content. *Nat. Biotechnol.* 23:482–487.

87. Beyer P. 2010. Golden Rice and "Golden" crops for human nutrition. *N. Biotechnol.* 27:478–481.

88. Beyer P., Al-Babili S., Ye X., et al. 2002. Golden Rice: Introducing the beta-carotene biosynthesis pathway into rice endosperm by genetic engineering to defeat vitamin A deficiency. *J. Nutr.* 132:506S–510S.

89. Dove A. 2000. Golden rice. *Nat. Biotechnol.* 18:135.

90. Goto F., Yoshihara T., Shigemoto N., et al. 1999. Iron fortification of rice seed by the soybean ferritin gene. *Nat. Biotechnol.* 17:282–286.

91. Qu L.Q., Yoshihara T., Ooyama A., et al. 2005. Iron accumulation does not parallel the high expression level of ferritin in transgenic rice seeds. *Planta.* 222:225–233.

92. Vasconcelos M., Datta K., Oliva N., et al. 2003. Enhanced iron and zinc accumulation in transgenic rice with the ferritin gene. *Plant Sci.* 164:371–378.

93. Lee S., Jeon U.S., Lee S.J., et al. 2009. Iron fortification of rice seeds through activation of the nicotianamine synthase gene. *Proc. Natl. Acad. Sci. U. S. A.* 106:22014–22019.

94. Johnson A.A.T., Kyriacou B., Callahan D.L., et al. 2011. Constitutive overexpression of the OsNAS gene family reveals single-gene strategies for effective iron- and zinc-biofortification of rice endosperm. *PLoS One.* 6:e24476.

95. Ishimaru Y., Masuda H., Bashir K., et al. 2010. Rice metal-nicotianamine transporter, OsYSL2, is required for the long-distance transport of iron and manganese. *Plant J.* 62:379–390.

96. Masuda H., Suzuki M., Morikawa K.C., et al. 2008. Increase in iron and zinc concentrations in rice grains via the introduction of barley genes involved in phytosiderophore synthesis. *Rice.* 1:100–108.

97. Xiong H., Guo X., Kobayashi T., et al. 2014. Expression of peanut Iron Regulated Transporter 1 in tobacco and rice plants confers improved iron nutrition. *Plant Physiol. Biochem.* 80C:83–89.

98. Masuda H., Aung M.S. and Nishizawa N.K. 2013. Iron biofortification of rice using different transgenic approaches. *Rice (N. Y.).* 6:40.

99. World Health Organisation. 2013. Fact Sheet N°311: Obesity and Overweight.

100. Lindström J., Peltonen M., Eriksson J.G., et al. 2006. High-fibre, low-fat diet predicts long-term weight loss and decreased type 2 diabetes risk: The Finnish Diabetes Prevention Study. *Diabetologia.* 49:912–920.

101. Knowler W. and Barrett-Connor E. 2002. Reduction in the incidence of type 2 diabetes with lifestyle intervention or metformin. *N. Engl. J. Med.* 346:393–403.

102. Tuomilehto J. and Lindström J. 2001. Prevention of type 2 diabetes mellitus by changes in lifestyle among subjects with impaired glucose tolerance. *N. Engl. J. Med.* 344:1343–1350.

103. Robertson M.D., Bickerton A.S., Dennis A. L., et al. 2005. Insulin-sensitizing effects of dietary resistant starch and effects on skeletal muscle and adipose tissue metabolism. *Am. J. Clin. Nutr.* 82:559–567.

104. Kobayashi H., Tan E. and Fleming S. 2003. Sodium butyrate inhibits cell growth and stimulates p21WAF1/CIP1 protein in human colonic adenocarcinoma cells independently of p53 Status. *Nutr. Cancer.* 46:202–211.

105. Blouin J.-M., Penot G., Collinet M., et al. 2011. Butyrate elicits a metabolic switch in human colon cancer cells by targeting the pyruvate dehydrogenase complex. *Int. J. Cancer.* 128:2591–2601.

106. Lanahan M.B., Basu S.S., Batie C.J., et al. 2006. Self-processing plants and plant parts. US Patent application no. US 7102057 B2.

107. Sonnewald U. and Kossmann J. 2013. Starches-from current models to genetic engineering. *Plant Biotechnol. J.* 11:223–232.

108. Regina A., Bird A., Topping D., et al. 2006. High-amylose wheat generated by RNA interference improves indices of large-bowel health in rats. *Proc. Natl. Acad. Sci. U. S. A.* 103:3546–3551.

109. Hassellund S.S., Flaa A., Kjeldsen S.E., et al. 2012. Effects of anthocyanins on cardiovascular risk factors and inflammation in pre-hypertensive men: A double-blind randomized placebo-controlled crossover study. *J. Hum. Hypertens.*

110. Wallace T. 2011. Anthocyanins in cardiovascular disease. *Adv. Nutr. An Int. Rev. J.* 2:1–7.

111. Basu A., Rhone M. and Lyons T. 2010. Berries: Emerging impact on cardiovascular health. *Nutr. Rev.* 68:168–177.

112. Titta L., Trinei M., Stendardo M., et al. 2010. Blood orange juice inhibits fat accumulation in mice. *Int. J. Obes. (Lond).* 34:578–588.

113. Butelli E., Licciardello C., Zhang Y., et al. 2012. Retrotransposons control fruit-specific, cold-dependent accumulation of anthocyanins in blood oranges. *Plant Cell.* 24:1242–1255.

114. Kovinich N., Saleem A., Rintoul T.L., et al. 2012. Coloring genetically modified soybean grains with anthocyanins by suppression of the proanthocyanidin genes ANR1 and ANR2. *Transgenic Res.* 21:757–771.

115. Butelli E., Titta L., Giorgio M., et al. 2008. Enrichment of tomato fruit with health-promoting anthocyanins by expression of select transcription factors. *Nat. Biotechnol.* 26:1301–1308.

116. Kris-Etherton P.M. 2003. Fish consumption, fish oil, omega-3 fatty acids, and cardiovascular disease. *Arterioscler. Thromb. Vasc. Biol.* 23:e20–e30.

117. Siscovick D.S., Raghunathan T., Irena King I., et al. 1995. Dietary intake and cell membrane levels of long-chain n-3 polyunsaturated fatty acids and the risk of primary cardiac arrest. *J. Am. Medical Assoc.* 274:1363–1367.

118. Food and Agriculture Organisation of the United Nations. 2010. The state of world fisheries and aquaculture.

119. Ruiz-Lopez N., Haslam R.P., Napier J.A., et al. 2013. Successful high-level accumulation of fish oil omega-3 long chain polyunsaturated fatty acids in a transgenic oilseed crop. *Plant J.* 77:198–208.

120. Kong Q., Richter L., Yang Y.F., et al. 2001. Oral immunization with hepatitis B surface antigen expressed in transgenic plants. *Proc. Natl. Acad. Sci. U. S. A.* 98:11539–11544.

121. Thanavala Y., Mahoney M., Pal S., et al. 2005. Immunogenicity in humans of an edible vaccine for hepatitis B. *Proc. Natl. Acad. Sci. U. S. A.* 102:3378–3382.

122. Richter L.J., Thanavala Y., Arntzen C.J., et al. 2000. Production of hepatitis B surface antigen in transgenic plants for oral immunization. *Nat. Biotechnol.* 18:1167–1171.

123. Sandhu J.S., Krasnyanski S.F., Domier L.L., et al. 2000. Oral immunization of mice with transgenic tomato fruit expressing respiratory syncytial virus-F protein induces a systemic immune response. *Transgenic Res.* 9:127–135.

124. Tacket C.O., Mason H.S., Losonsky G., et al. 1998. Immunogenicity in humans of a recombinant bacterial antigen delivered in a transgenic potato. *Nat. Med.* 4:607–609.

125. Arakawa T., Chong D.K. and Langridge W.H. 1998. Efficacy of a food plant-based oral cholera toxin B subunit vaccine. *Nat. Biotechnol.* 16:292–297.

126. Arlen P.A., Falconer R., Cherukumilli S., et al. 2007. Field production and functional evaluation of chloroplast-derived interferon-α2b. *Plant Biotechnol. J.* 5:511–525.

127. Leelavathi S. and Reddy V.S. 2003. Chloroplast expression of His-tagged GUS-fusions: A general strategy to overproduce and purify foreign proteins using transplastomic plants as bioreactors. *Mol. Breed.* 11:49–58.

128. Shoji Y., Bi H., Musiychuk K., et al. 2009. Plant-derived hemagglutinin protects ferrets against challenge infection with the A/Indonesia/05/05 strain of avian influenza. *Vaccine.* 27:1087–1092.

129. Jones R.M., Chichester J.A., Mett V., et al. 2013. A plant-produced Pfs25 VLP malaria vaccine candidate induces persistent transmission blocking antibodies against Plasmodium falciparum in immunized mice. *PLoS One.* 8:e79538.

130. Fesenko E. and Edwards R. 2014. Plant synthetic biology: A new platform for industrial biotechnology. *J. Exp. Bot.* 65:1927–1937.

131. Römer L. and Scheibel T. 2008. The elaborate structure of spider silk: Structure and function of a natural high performance fiber. *Prion.* 2:154–161.

132. Tokareva O., Michalczechen-Lacerda V.A., Rech E.L., et al. 2013. Recombinant DNA production of spider silk proteins. *Microb. Biotechnol.* 6:651–663.

133. Xia X.-X., Qian Z.-G., Ki C.S., et al. 2010. Native-sized recombinant spider silk protein produced in metabolically engineered *Escherichia coli* results in a strong fiber. *Proc. Natl. Acad. Sci. U. S. A.* 107:14059–14063.

134. Scheller J., Gührs K.H., Grosse F., et al. 2001. Production of spider silk proteins in tobacco and potato. *Nat. Biotechnol.* 19:573–577.

135. Barr L.A., Fahnestock S.R. and Yang J. 2004. Production and purification of recombinant DP1B silk-like protein in plants. *Cell.* 13:345–356.

136. Yang J., Barr L.A., Fahnestock S.R., et al. 2005. High yield recombinant silk-like protein production in transgenic plants through protein targeting. *Transgenic Res.* 14:313–324.

137. Hauptmann V., Weichert N., Menzel M., et al. 2013. Native-sized spider silk proteins synthesized in planta via intein-based multimerization. *Transgenic Res.* 22:369–377.

138. Matsumoto K., Morimoto K., Gohda A., et al. 2011. Improved polyhydroxybutyrate (PHB) production in transgenic tobacco by enhancing translation efficiency of bacterial PHB biosynthetic genes. *J. Biosci. Bioeng.* 111:485–488.

139. Bohmert-Tatarev K., McAvoy S., Daughtry S., et al. 2011. High levels of bioplastic are produced in fertile transplastomic tobacco plants engineered with a synthetic operon for the production of polyhydroxybutyrate. *Plant Physiol.* 155:1690–1708.

140. Connor M.R. and Atsumi S. 2010. Synthetic biology guides biofuel production. *J. Biomed. Biotechnol.* 2010.

141. Jung H.G. and Allen M.S. 1995. Characteristics of plant cell walls affecting intake and digestibility of forages by ruminants. *J. Anim. Sci.* 73:2774–2790.

142. Hisano H., Nandakumar R. and Wang Z.-Y. 2009. Genetic modification of lignin biosynthesis for improved biofuel production. *Vitr. Cell. Dev. Biol. – Plant.* 45:306–313.

143. Halpin C., Knight M.E., Foxon G.A., et al. 1994. Manipulation of lignin quality by downregulation of cinnamyl alcohol dehydrogenase. *Plant J.* 6:339–350.

144. Guo D., Chen F., Inoue K., et al. 2001. Downregulation of caffeic acid 3-O-methyltransferase and caffeoyl CoA 3-O-methyltransferase in transgenic alfalfa. Impacts on lignin structure and implications for the biosynthesis of G and S lignin. *Plant Cell.* 13:73–88.

145. Chen L., Auh C., Chen F., et al. 2002. Lignin deposition and associated changes in anatomy, enzyme activity, gene expression, and ruminal degradability in stems of tall fescue at different developmental stages. *J. Agric. Food Chem.* 50:5558–5565.

146. Sewalt V.J.H., Ni W., Jung H.G., et al. 1997. Lignin impact on fiber degradation: Increased enzymatic digestibility of genetically engineered tobacco (*Nicotiana tabacum*) stems reduced in lignin content. *J. Agric. Food Chem.* 45:1977–1983.

147. Sattler S.E., Funnell-Harris D.L. and Pedersen J.F. 2010. Brown midrib mutations and their importance to the utilization of maize, sorghum, and pearl millet lignocellulosic tissues. *Plant Sci.* 178:229–238.

148. Jones L., Ennos A.R. and Turner S.R. 2001. Cloning and characterization of irregular xylem4 (irx4): A severely lignin-deficient mutant of Arabidopsis. *Plant J.* 26:205–216.

149. Mir Derikvand M., Sierra J.B., Ruel K., et al. 2008. Redirection of the phenylpropanoid pathway to feruloyl malate in Arabidopsis mutants deficient for cinnamoyl-CoA reductase 1. *Planta.* 227:943–956.

150. Van der Rest B., Danoun S., Boudet A.-M., et al. 2006. Down-regulation of cinnamoyl-CoA reductase in tomato (Solanum lycopersicum L.) induces dramatic changes in soluble phenolic pools. *J. Exp. Bot.* 57:1399–1411.

151. Prashant S., Srilakshmi Sunita M., Pramod S., et al. 2011. Down-regulation of Leucaena leucocephala cinnamoyl CoA reductase (LlCCR) gene induces significant changes in phenotype, soluble phenolic pools and lignin in transgenic tobacco. *Plant Cell Rep.* 30:2215–2231.

152. Piquemal J., Lapierre C., Myton K., et al. 1998. Down-regulation of Cinnamoyl-CoA Reductase induces significant changes of lignin profiles in transgenic tobacco. *Plant Biotechnol.* 13:71–83.

153. Goujon T., Ferret V., Mila I., et al. 2003. Down-regulation of the AtCCR1 gene in Arabidopsis thaliana: Effects on phenotype, lignins and cell wall degradability. *Planta.* 217:218–228.

154. Dauwe R., Morreel K., Goeminne G., et al. 2007. Molecular phenotyping of lignin-modified tobacco reveals associated changes in cell-wall metabolism, primary metabolism, stress metabolism and photorespiration. *Plant J.* 52:263–285.

155. Pedersen J.F., Vogel K.P. and Funnell D.L. 2005. Impact of reduced lignin on plant fitness. *Crop Sci.* 45:812.

156. Yang F., Mitra P., Zhang L., et al. 2013. Engineering secondary cell wall deposition in plants. *Plant Biotechnol. J.* 11:325–335.

157. Valdez Barillas J.R., Quinn C.F., Freeman J.L., et al. 2012. Selenium distribution and speciation in the hyperaccumulator *Astragalus bisulcatus* and associated ecological partners. *Plant Physiol.* 159:1834–1844.

158. Cosio C., Martinoia E. and Keller C. 2004. Hyperaccumulation of cadmium and zinc in *Thlaspi caerulescens* and *Arabidopsis halleri* at the leaf cellular level. *Plant Physiol.* 134:716–725.

159. Ellis D.R., Gumaelius L., Indriolo E., et al. 2006. A novel arsenate reductase from the arsenic hyperaccumulating fern Pteris vittata. *Plant Physiol.* 141:1544–1554.

160. Rugh C.L., Wilde H.D., Stack N.M., et al. 1996. Mercuric ion reduction and resistance in transgenic *Arabidopsis thaliana* plants expressing a modified bacterial merA gene. *Proc. Natl. Acad. Sci. U. S. A.* 93:3182–3187.

161. Ruiz O.N., Hussein H.S., Terry N., et al. 2003. Phytoremediation of organomercurial compounds via chloroplast genetic engineering. *Plant Physiol.* 132:1344–1352.

162. Bizily S.P., Kim T., Kandasamy M.K., et al. 2003. Subcellular targeting of methylmercury lyase enhances its specific activity for organic mercury detoxification in plants. *Plant Physiol.* 131:463–471.

163. Dhankher O.P., Li Y., Rosen B.P., et al. 2002. Engineering tolerance and hyperaccumulation of arsenic in plants by combining arsenate reductase and gamma-glutamylcysteine synthetase expression. *Nat. Biotechnol.* 20:1140–1145.

164. Dhankher O.P., Rosen B.P., McKinney E.C., et al. 2006. Hyperaccumulation of arsenic in the shoots of Arabidopsis silenced for arsenate reductase (ACR2). *Proc. Natl. Acad. Sci. U. S. A.* 103:5413–5418.

165. Panz K. and Miksch K. 2012. Phytoremediation of explosives (TNT, RDX, HMX) by wild-type and transgenic plants. *J. Environ. Manage.* 113:85–92.

166. Hannink N., Rosser S.J., French C.E., et al. 2001. Phytodetoxification of TNT by transgenic plants expressing a bacterial nitroreductase. *Nat. Biotechnol.* 19:1168–1172.

167. Zhu B., Peng R.H., Fu X.Y., et al. 2012. Enhanced transformation of TNT by arabidopsis plants expressing an old yellow enzyme. *PLoS One.* 7.

168. Khatisashvili G., Gordeziani M., Adamia G., et al. 2009. Higher plants ability to assimilate explosives. *World Acad. Sci. Eng. Technol.* 33:256–270.

169. Vila M., Lorber-Pascal S. and Laurent F. 2007. Fate of RDX and TNT in agronomic plants. *Environ. Pollut.* 148:148–154.

170. Seth-Smith H.M.B., Rosser S.J., Basran A., et al. 2002. Cloning, sequencing, and characterization of the hexahydro-1,3,5-Trinitro-1,3,5-triazine degradation gene cluster from *Rhodococcus rhodochrous. Appl. Environ. Microbiol.* 68:4764–4771.

171. Rylott E.L., Budarina M. V, Barker A., et al. 2011. Engineering plants for the phytoremediation of RDX in the presence of the co-contaminating explosive TNT. *New Phytol.* 192:405–413.

172. Looger L.L., Dwyer M.A., Smith J.J., et al. 2003. Computational design of receptor and sensor proteins with novel functions. *Nature.* 423:185–190.

173. Antunes M.S., Morey K.J., Smith J.J., et al. 2011. Programmable ligand detection system in plants through a synthetic signal transduction pathway. *PLoS One.* 6:e16292.

174. Antunes M.S., Ha S.-B., Tewari-Singh N., et al. 2006. A synthetic de-greening gene circuit provides a reporting system that is remotely detectable and has a re-set capacity. *Plant Biotechnol. J.* 4:605–622.

175. Cook C., Bastow R. and Martin L. 2014. *Developing Plant Synthetic Biology in the UK–Challenges and Opportunities.* 37.

Section III

Constructing New Biologies

7

Theory and Construction of Semi-Synthetic Minimal Cells

Pasquale Stano and Pier Luigi Luisi

CONTENTS

7.1 Introduction

Although the term synthetic biology was coined by the French chemist Stéphane Leduc (1912), only recently, it has become an umbrella term to describe one of the most attractive and emerging current research fields at the interface between molecular biology and bioengineering. A possible definition of synthetic biology refers to approaches aimed at engineering and studying biological systems that do not exist as such in nature, and using these approaches for (1) achieving better understanding of life processes, (2) generating and assembling functional modular components or (3) developing novel applications or processes.

In addition to classical bioengineer-oriented synthetic biology (Endy 2005), this new discipline embraces studies oriented to basic research that can be unified under the tag of *'chemical synthetic biology'* (Luisi 2007; Luisi and Chiarabelli 2011). In such a context, biological-like structures are synthesised thanks to typical 'synthetic' methodologies in order to understand why nature did them in that way, and not in another one. Several previous studies on this topic can now be recognised within the chemical synthetic biology approach. For example the classical work of Eschenmoser on nucleic acids containing pyranose instead of ribose (Bolli et al. 1997a,b), or Benner's research on modified nucleic acid bases (Benner and Sismour 2005), or the creation of proteins with a reduced amino acid 'alphabet' (Akanuma et al. 2002; Doi et al. 2005) can all be defined as *ante litteram* chemical synthetic biology approaches. A short, comparative discussion on the epistemological aspects of engineering-oriented versus chemical synthetic biology can be found in Luisi (2011).

Our group has been quite active in recent years exploring key biological questions by means of a chemical synthetic biology approach (Chiarabelli et al. 2009, 2012). In particular, we are interested in investigating two major issues, namely, (1) synthesizing novel proteins or nucleic acids whose sequences have no similarity with the known proteins, with the aim of understanding whether or not such novel structures are also capable of functioning like existing proteins (folding, binding, catalysis, etc.) and (2) the origin of cellular life by constructing simple cell-like entities, in order to verify under which physico-chemical conditions it is possible to observe the transition from non-living to living systems, using plausible primitive compartmentalised systems as experimental models. The first research line is often referred to as one of 'never born RNAs' and 'never born proteins', and it will be not treated in this chapter. Interested readers should refer to the original articles (Thomas et al. 2001; Chiarabelli et al. 2006a,b; De Lucrezia et al. 2006a,b; Anella et al. 2011). The second research line deals with the construction of cellular models that actually are 'synthetic' or 'semi-synthetic' cells. Because these cells are supposed to be built by using the minimal and sufficient number of molecules to reconstruct any function, they are also called 'minimal' – i.e. representing the simplest constructs capable of imitating the living cell functions – and for that reason they are often taken as models of primitive cells (Luisi 2002, 2006; Luisi et al. 2006; Luisi and Stano 2010). This chapter describes the use of lipid vesicles (i.e. liposomes) as cell models and the technology of *semi-synthetic minimal cells*. This will be done by illustrating the origin of the funding concepts – rooted in origin of life research – and the most interesting developments, as well as by sketching some possible future applications.

7.2 Autopoiesis and Minimal Life

7.2.1 Premise

Let us start our discussion on the origin and early developments of the 'minimal cell' concept. As we have anticipated, this concept stems from open questions regarding the origins of life.

Contemporary simpler microorganisms, including the smallest known living organisms, are based on thousands of genes and molecular components. This elicits the question whether such complexity is really necessary for life or whether a living organism can exist with a much smaller number of genes and other molecules, also considering that early cells were certainly simpler than contemporary cells.

The basic assumption is that life on Earth originated from inanimate matter by a very long and lengthy series of steps of prebiotic molecular evolution. Basing on the increase of molecular complexity, one can envisage the transition from small molecules (the building blocks) to large molecules, from the latter to coordinated and integrated macromolecular systems and from these to cells. The cell is the minimal living entity. By further increasing complexity, one then finds multicellular organisms and association of organisms (i.e. communities). The formation of self-reproducing protocells that are able to display biological autonomy marks the transition between non-life and life. That such transition occurred without any transcendent help is a commonly accepted concept in science but it has not been demonstrated yet by any scientific experiment. Therefore, an investigation of the construction of minimal *living* cells aimed at demonstrating that the transition from non-living to living systems (by using a minimal number of molecules) faces indeed a very important scientific question and can be realised in the laboratory.

We generally focus on the notion of the minimal living cell, defined as a system that has minimal and sufficient structural conditions for life (number of molecules, spatial arrangement and number of reactions). But what does 'minimal life' mean? The definition of life is intrinsically complex and it is still a question that divides experts and laypersons (Luisi 1998; Szostak 2012). Many scientists agree, however, that a living system (i.e. a cell) could be defined alive when it displays homeostatic self-maintenance, self-reproduction and the capability to evolve. The next question is on how would one conceive the organization scheme of a minimal living cell in order to display such features. Here a powerful general theory of living systems helps considerably: the theory of autopoiesis.

7.2.2 Autopoiesis

Autopoiesis (from the Greek *auto*: self; and *poiesis*: construction) is a word coined by Humberto Maturana and Francisco Varela – two Chilean biologists – in the 1970s, when they proposed their theory of living systems (Maturana and Varela 1972; Varela et al. 1974). Autopoiesis deals with the question 'what is life?' and attempts to define, beyond the diversity of all living organisms, a common denominator that allows for the discrimination of the living from the non-living. Autopoiesis is not concerned with the origin of life per se, rather, it is based on the direct observation of how extant cells work, and it is concerned with the various processes connected with life, such as the interaction with the environment, evolution and 'cognition'. The reader can find several Maturana–Varela original texts on autopoiesis (Maturana and Varela 1980, 1992), as well as a recently published commentary (Luisi 2003) – quite useful for chemists and biologists – and other interesting examples (Damiano and Luisi 2010; Bich and Damiano 2012). Here we present a brief overview of autopoietic theory (see also Luisi 2006).

The central feature that characterises all living entities (and in particular unicellular organisms) is their self-maintenance. By self-maintenance here, we mean the self-generation of all components by chemical reactions, occurring within a boundary (e.g. the cell membrane). The boundary is also produced by the internal metabolic system. This self-generation is due to the peculiar form of chemical organization, that is, a dynamical organization typical of autopoietic systems. It is defined as a network of processes of productions of the molecular components, which (1) participate recursively in the network of processes for their own production and (2) occur in a defined region (physical space) delimited by a physical boundary (Figure 7.1). The physical boundary also belongs to the autopoietic organization. The so obtained physical entity is an autopoietic unit. Note that the autopoietic organization is a collective, distributed property of the whole system, and does not reside in any particular molecule.

Maturana and Varela arrived at the generalization of living systems by observing that the living cell is the simplest autopoietic unit. In fact, a cell is a physical object separated by a semi-permeable membrane from its environment. The cell continuously produces (and destroys) its own components (including those that form the membrane boundary) thanks to a network of biochemical transformation occurring inside the cell. Such a set of transformations would lie out of equilibrium, at the expense of a continuous flow of chemical and/or physical energy from their environment (the cell is a dissipative, open system). Despite the continuous turnover of matter and the consumption of energy, the cell maintains its identity, persisting in a homeostatic, yet dynamical, state.

In autopoietic theory, the 'operational closure' is often included in the discussion, to mean that the autopoietic unit reacts to environmental changes in order to maintain its own inner autopoietic organization, and that the unit tolerates only changes that can be accommodated within the autopoietic organization.

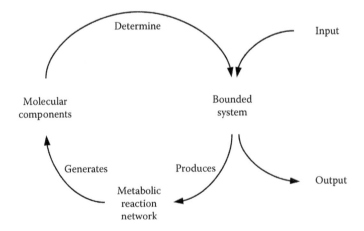

FIGURE 7.1

The cyclic logic of autopoiesis. Molecular components self-assemble and give rise to a bounded systems that enclose a reaction network which generates the molecular components that determine the bounded system and so on. The whole autopoietic system is kept out-of-equilibrium thanks to the continuous uptake of input building blocks from the environment. It also releases by-products (output) in the environment. (Reprinted from Luisi, P. L., *Die Naturwissenschaften*, 90, 49–59, 2003).

If we accept that the cell is an autopoietic unit, it is tempting to speculate as to whether the criteria for autopoietic organization are also the minimal criteria for defining life in the most general way. For a critical discussion on this issue, see Bitbol and Luisi (2004) and Bourgine and Stewart (2004). The necessary conditions for life can be simplified as follows: (1) the system should have a semi-permeable boundary, which (2) is produced from within the system and (3) that encompasses reactions that regenerate all components of the system (Varela 2000). Thanks to the elegance and simplicity of these conditions, it is indeed possible to use the autopoietic theory as a theoretical framework for constructing minimal living systems in the laboratory. Clearly, technical difficulties will hamper this enterprise, but the goal of autopoietic organization stands distinct from, and irrespective of the actual practical implementation of, the diverse approaches.

Perhaps surprisingly, many life scientists are unfamiliar with autopoietic theory and certainly the concept does not often fall within what could be termed mainstream biology. This may in part be due to the fact that autopoietic theory does not directly deal with concepts such as information flow or codification – two concepts that underlie the nucleic acid paradigm and which are often the starting point for theoretical biological enquiry (Luisi 2003).

Autopoiesis is a system theory and, before the work of Maturana and Varela in the 1970s, Robert Rosen proposed an even more general theory of living systems, 'relational theory' or the 'theory of (M,R) systems' (Rosen 1958a,b). Recent studies have revealed that several aspects of autopoiesis and relational theory that overlap (Lee et al. 2006; Letelier et al. 2006). Also in the 1970s, the Hungarian chemical engineer, Tibor Gánti, a contemporary of Maturana and Varela, proposed a theory of a fluid chemical automaton, also known as 'chemoton' theory, based on the integration, in a physically distinct space, of three sub-systems: replicable membrane, autocatalytic metabolism and replicable information-carrier molecule, described quantitatively by cyclic stoichiometry (Gánti 1975, 2003). Simplifying, it can be said that in principle, the chemoton theory allows the realization – in the chemical domain – of J. von Neumann self-replicating automata (1966) in the electro/mechanical or computer domains.

7.2.3 Chemical Autopoiesis I – Early Work with Micelles

Early attempts to develop chemical autopoietic systems were carried out in the laboratory of Maturana in Santiago, with a project called 'molecular protobion', based on a kind of coacervates, with at best modest results (Guiloff 1981).

In the late 1980s Pier Luigi Luisi, co-author of this chapter, embarked on a collaboration with Francisco Varela to experimentally investigate the minimal chemical componentry of autopoietic systems (Luisi and Varela 1989). This work, which took place at ETH Zurich, was very ambitious and pioneering for its time and initially focused on well-characterised chemical self-compartmentalizing 'reverse micelle' systems (Luisi 1985; Luisi et al. 1988).

Reverse micelles are tiny water droplets (typical diameter ~5–20 nm) surrounded by surfactants, and suspended in apolar solvents, like hydrocarbons. They are formed by self-organization of water and surfactant molecules in an apolar solvent, to give a stable, self-bounded compartment consisting of a surfactant monolayer around a small aqueous core. Fatty acid (caprylate) reverse micelles were fed with ethyl caprylate or with caprylic alcohol. Due to a catalyst (i.e. a reactant) contained within the reverse micelle core (sodium hydroxide or potassium permanganate, respectively), these substrates were transformed – by the reverse micelles themselves – in the boundary-forming surfactant (caprylate) (vA).

As the number of interfacial molecules increases, the reverse micelles become unstable, and spontaneously break in smaller reverse micelles.

The whole process corresponds to a quasi-autopoietic self-reproduction of reverse micelles. This system cannot be defined as fully autopoietic because the reactants and catalyst within the initial reverse micelles are not reproduced and are therefore consumed over time (Bachmann et al. 1990, 1991a,b, 1992).

In the following years, similar autopoietic dynamical systems were assembled by using normal micelles, which are very small surfactant assemblies (diameter ~4 nm) suspended in aqueous solutions. Fatty acids were used also in this case, and they were fed with the corresponding ester or anhydrides, following a strategy similar to the reverse micelle cases. Strikingly, in the case of normal micelles, it was observed that their self-reproduction proceeds auto-catalytically (Bachmann et al. 1992) (Figure 7.2B).

7.2.4 Chemical Autopoiesis II – Progressing from Micelles to Vesicles

Fatty acid vesicles are cell-like structures consisting of semi-permeable, phospholipid-like bilayers and are considered by many in the field to represent the most plausible configuration of the early cell-like entities that arose on Earth (Figure 7.2C). Furthermore, fatty acids can be chemically synthesised in purportedly abiotic conditions (Rushdi and Simoneit 2001; Simoneit 2004) and have been found in meteorites (Yuen and Kvenvolden 1973; Deamer 1985).

An accepted definition of autopoietic growth, established in micelle/reverse micelle studies, was that the following phenomena had to be observed: (i) uptake of precursor, (ii) conversion of precursor into a boundary-forming molecule, (iii) consequent vesicle growth and (iv) division (Figure 7.3). It has been shown that that fatty acid vesicles can grow in a manner that meets with these criteria when they are 'fed' with fatty acid anhydrides

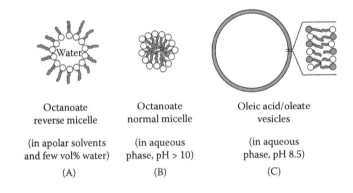

Octanoate reverse micelle	Octanoate normal micelle	Oleic acid/oleate vesicles
(in apolar solvents and few vol% water)	(in aqueous phase, pH > 10)	(in aqueous phase, pH 8.5)
(A)	(B)	(C)

FIGURE 7.2
Supramolecular structures formed by self-assembly of fatty acids, like octanoic acid or oleic acid, that have been used for creating chemical autopoietic systems (reviewed in Stano and Luisi 2010). Reverse micelles (A) are tiny surfactant-stabilised water droplets in apolar solvents. Normal micelles (B) are also very small and they form in aqueous solutions by surfactants, hiding their hydrophobic tails in the micelle core, and exposing their polar head groups; in the case of fatty acids, this structure is stable only when carboxylate groups are ionised, that is, at high pH values. Fatty acid vesicles (C) also form in water phase, but only when about 50% of the fatty acid molecules are protonated. Actually, the fatty acid bilayer is a mixture of dissociated and undissociated molecules; an extensive hydrogen bond network is established among the carboxylates and the carboxylic acid groups (Haines 1983). Note that the structures are drawn not to scale. Reverse micelles and micelles are a few nanometre large, whereas vesicle size can range from 30–40 nm to about 50–100 μm.

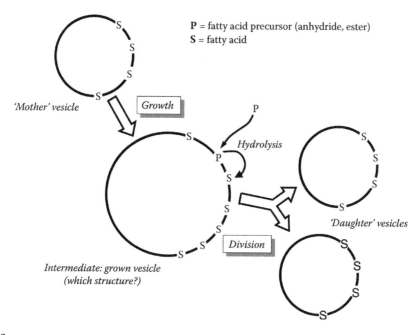

FIGURE 7.3
Mechanism of fatty acid vesicles self-reproduction (Walde et al. 1994b). Pre-formed vesicles were fed with a fatty acid precursor, like the insoluble fatty acid anhydride. Fatty acid vesicles adsorb the insoluble anhydride molecules in their hydrophobic membrane, where they are hydrolysed due to the high pH of the solution. In this way, new fatty acid molecules are created and the membrane grows accordingly. The grown vesicle becomes unstable and divides giving rise to two or more 'daughter' vesicles. Overall, the mother fatty acid vesicle self-reproduces and gives new fatty acid vesicles. The structure of the intermediate grown vesicle is not grown, although it is believed that departure from the spherical symmetry plays an important role.

(Walde et al. 1994a,b; Oberholzer et al. 1995b) or fatty acid micelles (Blochliger et al. 1998; Lonchin et al. 1999; Berclaz et al. 2001a,b; Rasi et al. 2003).

Such behaviour has been observed both for 'empty' fatty acid vesicles, as well as with enzyme-containing vesicles (in particular, polynucleotide phosphorylase [PNPase] and Qbeta replicase), as will be described in Section 7.3.1. The autopoietic self-reproduction of fatty acid vesicles occurs despite vesicle size remaining constant, a phenomenon (Figure 7.4) called the 'matrix effect' (Blochliger et al. 1998). The fact that fatty acid vesicles also exhibit autopoietic dynamics initially excited the 'origin-of-life' community, because this mechanism provides a 'free-ticket' to primitive cell proliferation without the need of the complex biochemical machineries typically used by modern cells to coordinate their reproduction.

Autopoiesis, however, does not only mean self-production of components. In order to have a homeostatic system, production must be accompanied by component removal, either by destruction or by utilisation to build other components. The only attempt to experimentally model such balance between 'catabolic' and 'anabolic' processes, to the best of our knowledge, was carried out by our group. In particular, the production of fatty acids derived from anhydride hydrolysis was coupled with their degradation by an oxidative reaction (osmium tetroxide was used to oxidise the unsaturated chain of ole-ate molecules, to produce a compound not able to form vesicles). The two concurrent reactions occurred simultaneously and their rate could be fine-tuned by regulating the

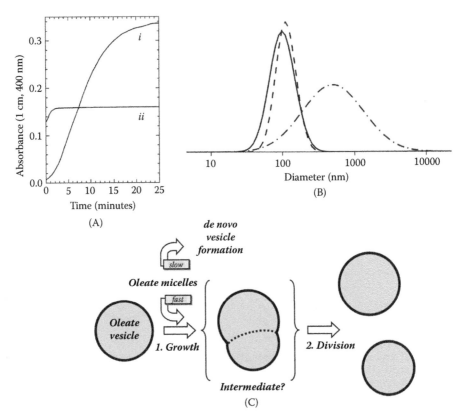

FIGURE 7.4
Fatty acid self-reproduction and the 'matrix' effect. Oleate vesicles can grow and self-reproduce by feeding them with fatty acid micelles. Fatty acid micelles are stable at high pH and convert spontaneously in vesicles when diluted in pH 8.5 buffer. The spontaneous micelle-to-vesicle transition, as evidenced by the turbidity versus time profile, follows a slow sigmoidal trend ($t_{1/2}$ ca. 10 minutes; panel A, curve *i*). On the other hand, when micelles are added to a suspension of pre-formed vesicles, the turbidity increase is small, and much faster (panel A, curve *ii*), ($t_{1/2}$ ca. 1 min). Dynamic light scattering analysis (panel B) shows that the spontaneous micelle-to-vesicle process gives rise to vesicles with a broad size distribution, centred at high vesicle size (panel B, dash-dot curve). In contrast, when the micelles are added to pre-formed vesicles with a narrow size distribution (panel B, continue curve), the resultant oleate vesicle population has a size distribution very similar to that of one of the pre-formed vesicles (panel B, dashed curve). The mechanism of matrix effect originates from the competition between the spontaneous micelle-to-vesicle path, which is slow, and the fast oleate micelle uptake by the pre-formed vesicles. The intermediate grown vesicle could consist of a peculiar two-lobes pseudo-symmetric structure that has been visualised by freeze-fracture electron microscopy (Stano et al. 2006), which splits into two daughter vesicles whose size is approximately similar to that of the mother vesicle. Panel A has been redrawn after Bloechliger et al. (1997), Rasi et al. (2003); panel B represents a generic dataset; panel C is a cartoon inspired from freeze-fracture images published by Stano et al. (2006).

concentration of reactants or their rate of addition (Zepik et al. 2001). Quite intriguingly, it was shown that the amount of fatty acid vesicles could increase, remain constant, or decrease, depending on the ratio between the fatty acid synthesis (V_s) and fatty acid degradation (V_d) rates. When $V_s > V_d$, a net growth was observed; the contrary was true if $V_s < V_d$. In the very special case of $V_s = V_d$, the fatty acid vesicles could sustain a kind of homeostatic state that is typical of autopoiesis. Despite the continuous concurrent production and degradation of the molecular components, fatty acid vesicles could maintain their structure – mimicking in a very simple way the cellular homeostasis.

A comprehensive discussion on experimental approaches to autopoietic and self-reproducing chemical systems is available in a recently published review article (Stano and Luisi 2010).

A common aspect of the chemical autopoiesis is the simplicity of such systems. This is their beauty, especially from the viewpoint of origin of life. Modern complex cells base their material growth on enzyme-catalysed reactions, and the enzymes themselves are produced by the metabolic network. Every molecule of the cellular autopoietic organization is continuously produced by the cell. This aspect was missing in the chemical autopoietic systems based on micelles, reverse micelles, and vesicles. In fact, the chemical transformation that converts the precursor (fatty acid ester or anhydride, a fatty alcohol, fatty acid micelles) into the boundary-forming compound (a fatty acid molecule) relies on reactant/catalysts (hydroxide ion, permanganate) that are themselves not produced by the autopoietic network. This is the general problem of all synthetic autopoietic systems here described: they are all 'burst' reactors. Reactions go until the concentration of the limiting component goes to zero. Until now none of them has been transformed into a continuous bioreactor. All reported attempts to construct continuous micro-reactors, as opposed to burst reactors, have to date failed to surmount significant technical difficulties and as such this remains a critically important goal for researchers seeking to construct synthetic cells both for applied research and in order to elucidate origin-of-life mechanisms.

However, efforts to date that have ended up with 'burst' reactor systems have still been both elegant and helpful in expanding the range of tools and approaches for achieving the ultimate goal of constructing vesicles that function as continuous micro-reactors and as such achieve *bona fide* autopoiesis.

7.2.5 Mechanisms of Vesicle Self-Reproduction

In this section, we discuss a selection of the mechanistic features of vesicle self-reproduction. Ideally, the self-reproduction of vesicles consists of two steps: (i) growth and (ii) division, although it might be difficult to strictly separate such steps as certain modes of vesicle division might derive from a specific way of growing.

First, we must return to the 'matrix effect' (Figure 7.4) which predicts the phenomenon where fatty acid vesicles can grow autopoietically by addition of a fatty acid precursor (Blochliger et al. 1998; Lonchin et al. 1999; Rasi et al. 2003). In early work by Walde et al. (1994a,b), it was shown that added fatty acid anhydrides can be converted to fatty acid molecules via anhydride hydrolysis occurring preferentially on the vesicle membrane.

In addition to the anhydride hydrolysis route, Walde et al. (1994a,b) showed that fatty acid vesicles can grow at the expenses of fatty acid micelles provided as feedstock to the vesicles. This can occur as a purely physical process with no requirement for a chemical reaction. The anhydride hydrolysis approach also requires that the system is biphasic, whereas when micelles are provided as a feedstock the system is homogeneous and therefore tractable to analysis by photometric techniques.

Typically feedstock micelles are made in solutions of pH up to ca. 10–11, in which they are fully stable, before being added to a solution of fatty acid vesicles at pH 8.5. The fatty acid vesicles then uptake fatty acids from the micelles, grow and then divide. In the absence of pre-formed vesicles, fatty acid micelles have also been observed to give rise to vesicles, following a spontaneous transition triggered by fatty acid protonation, due to the pH shift from about 10–11 to 8.5. The size distribution of fatty acid vesicles obtained by this spontaneous micelles-to-vesicle transformation is very broad and contains both small and very large vesicles. Formation of vesicles in this way is also slow, with a half-life of about 10–15 minutes, and follows a sigmoidal profile (Figure 7.4A). When, however, fatty acid micelles are added to pre-formed vesicles with a certain (narrow) size distribution, a very different behaviour is observed. First, the rate of vesicle formation increases notably (half-life of around one minute or less), the sigmoidal profile almost disappears and – most impressively – the size distribution of the resultant vesicles largely overlaps the (narrow) size distribution of pre-formed vesicles present in the solution (Figure 7.4B).

Cryogenic transmission electron microscopy (cryoTEM) studies carried out on ferritin-containing vesicles revealed that pre-formed vesicles uptake oleate molecules (from the micelles) and grow. During growth, the vesicles reach an unstable state and divide to form daughter vesicles, as evident by measuring the average number of entrapped ferritin per vesicle, and by comparing the size distribution between the parent and the resulting vesicles (Berclaz et al. 2001a,b). By means of a freeze-fracture electron microscopy study, we revealed the structure of a possible intermediate: a vesicle narrowed in the equatorial plane (Stano et al. 2006) (Figure 7.4C). Kinetic studies suggest that vesicle growth can be limited by the number of micelles that can cover the pre-formed vesicles (Chen and Szostak 2004). It has been suggested that vesicle division is favoured by the initial non-spherical growth (Markvoort et al. 2010), possibly deriving from differential surface-versus-volume growth; in this respect, the importance of the buffer permeability has been also pointed out (Zhu and Szostak 2009). Vesicle pearling has been observed in the case of giant fatty acid vesicles (Zhu and Szostak 2009, 2012).

Despite the large amount of work devoted to self-reproduction of fatty acid vesicles, mechanistic explanations are still open to debate.

7.2.6 Self-Reproduction of Non-Fatty Acid Vesicles

In addition to fatty acid vesicles, other surfactants can give rise to a self-reproduction mechanism. The group of Sugawara introduced an *ad hoc* designed surfactant that forms giant vesicles (GVs) and that can be synthesized *in situ* (within vesicle, thanks to a simple catalyst) so that it can be used to model primitive cell growth (Takakura et al. 2003; Toyota et al. 2008; Takahashi et al. 2010). The self-reproduction of these synthetic vesicles has been carried out more recently on DNA-containing vesicles (DNA was produced by intraliposome polymerase chain reaction [PCR]) in work that demonstrated that the process of DNA replication not only is compatible with this mode of chemical vesicle division, but can also affect it (Kurihara et al. 2011).

7.2.7 Vesicles Fusion

We have seen that vesicle division plays a major role for the proliferation of vesicles by a self-reproducing mechanism. The opposite of vesicle division is vesicle fusion, a process that could also have played a role in the origins of cellular life. In fact, vesicle fusion has been proposed as a primitive mechanism that might lead to an increase in the molecular complexity of protocells by combining different and maybe complementary molecular sets, initially present in two different vesicles, giving rise to a symbiogenesis-like pattern (Luisi 2006; Caschera et al. 2010).

The key and critical event in vesicle fusion is the physical approximation of the membranes of two different vesicles so that the lipid molecules can contact with each other. This is a difficult process because the innermost hydration shells prevent such close contact. Vesicle fusion is therefore promoted by enhancing the attractive forces between bilayers, thanks to bridging agents, dehydrating agents, specialised peptides/

proteins or electrostatics (Duzgunes et al. 1989; Cevc and Richardsen 1999). In the latter case, a viable strategy involves the use of oppositely charged vesicles (Pantazatos and MacDonald 1999; Pantazatos et al. 2003). In a recent study, we demonstrated that positively and negatively charged vesicles – performed with plausibly prebiotic molecules such as fatty acids and tetraalkyl ammonium surfactants – can indeed fuse together, as revealed by the classical terbium-dipicolinic acid assay (Caschera et al. 2010). This approach was further extended and brought to the successful programmable fusion (Sunami et al. 2010a) of vesicles, and it was used to reconstitute a cellular 'function' from separated parts. In particular, the fusion between DNA-containing vesicles and vesicles containing the transcription-translation machinery gave vesicles that, as expected, were functional and able to produce a protein (Tsumoto et al. 2006; Caschera et al. 2011).

Association and fusion of oleate-containing GVs, typically defined as being 1–100 μm in diameter, was recently reported, introducing the concept of primitive cell communities (Carrara et al. 2012). In particular, isolated GVs composed by a mixture of phosphatidylcholine and anionic fatty acids can form 'GVs colonies' by the simple addition of a bridging agent, namely poly(arginine). The GVs in the colony mixed their content by fusion, and the permeability of the membrane was enhanced, allowing the capture and uptake of negatively charged biomolecules present in the environment.

Far from being fully understood, division and fusion mechanisms still require extensive investigation. In particular, it is interesting to understand what conditions (if any) lead a spherical vesicle to spontaneously deform to favour division, and if primitive peptides could somehow mediate or facilitate vesicle division and/or vesicle fusion.

7.3 Semi-Synthetic Approach to the Construction of Minimal Cells

Is it possible to study the emergence of autopoietic cells from the association and organization of a minimal number of chemical components in the laboratory?

We have seen that simple chemical autopoietic systems can be indeed constructed, but in order to assemble cell-like systems a novel degree of sophistication is needed. Here synthetic biology enters into the game, with a methodology that resembles the typical thinking way of chemists, which build their molecules by assembling building blocks. Can biological systems be built similarly, that is, from their building blocks?

It would be indeed very interesting to construct 'minimal' cells from a defined selection of components. Minimal cells might be extraordinarily simple compared with modern biological cells, but presumably would still 'imitate' some of their core functions. The process of constructing minimal cells is likely to provide valuable insights into the constraints of biological organization. A number of approaches are being used to construct such minimal cells, including so-called 'top-down' and 'bottom-up' routes. Typically these methods involve using combinations of small, simple molecules and large biomolecules, both the naturally occurring and the non-natural/synthetic (DeClue et al. 2009). The resultant artificial cells will represent valuable tools for both modern biotechnology and improving our understanding of the mechanisms that underlie the origins of life on Earth (Figure 7.5A).

In the early 1990s, our laboratory at ETH Zurich embarked upon an experimental research programme investigating construction of what we were the first to term 'minimal

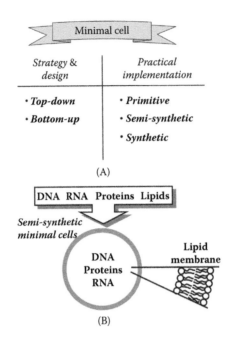

FIGURE 7.5

Experimental approaches to minimal cells. (A) Based on the experimental strategy and design, minimal cells can be built by following a top-down or a bottom-up approach. Depending on the nature of molecules used, one could tentatively distinguish between minimal cells composed of primitive-like compounds (fatty acids, ribozymes, short catalytic peptides, etc.), those formed by a synthetic assembly of modern biocompounds (DNA, RNA, proteins, ribosomes, lipids, etc.), and minimal cells constructed from totally synthetic building blocks (polymers, peptide-nucleic acids, synthetic surfactants, etc.). (B) Semi-synthetic minimal cells are defined as those constructions based on the encapsulation of the minimal number of DNA, RNA and protein components inside lipid vesicles (liposomes).

cells' (Schmidli et al. 1991). Clearly linked to autopoietic theory and to self-reproducing vesicles, minimal cell studies are typically concerned with the encapsulation, within lipid vesicles, of the minimal and sufficient number of molecules that would react between each other in order to reproduce the autopoietic phenomenology, that is, continuous production of a reaction network whose products are the elements of the network and of the boundary.

As we have explored in further articles that complement this chapter (Luisi 2002; Luisi et al. 2006), despite the understandable significance of a minimal cell construction using only primitive molecules (aimed at observing the onset of primitive metabolism), such a route presents very challenging practical and conceptual difficulties. Our proposal is that we can instead proceed in constructing minimal cells by using DNA, RNA and protein taken from modern organisms, which are available in laboratory, and focus on unveiling the constraints of autopoietic organization rather than be firmly bound to the constraints of primitive chemistry (Figure 7.5B). We called this approach 'semi-synthetic' to emphasise the hybrid nature of the enterprise: make synthetic cells by using 'natural' molecules. In this context, a mention should be made on the minimal number of genes that might characterise the minimal cell. Comparative genomic analyses have revealed that a small number (ca. 200) of genes are present in all smallest microorganisms. This set of genes is known as the minimal genome (Table 7.1), and its connection with the construction of synthetic cells is briefly discussed in Box 7.1.

TABLE 7.1

Comparison between Minimal Gene Sets

	Mushegian and Koonin (1999)	Gil et al. (2004)	Forster and Church (2006)
Amino acids and derivatives	3	0	0
Carbohydrate metabolism	21	14	0
Cell division and cell cycle	7	9	0
Cell wall and capsule	6	0	0
Cofactors, vitamins, prosthetic groups, pigments	8	15	0
DNA recombination and repair	5	3	0
DNA replication machinery	17	17	4
Electron transport chain	10	9	0
Fatty acids and lipids	7	7	0
Nucleoside and nucleotides	21	15	0
Protein biosynthesis	76	67	71
Protein folding	9	5	2
Protein turnover	1	2	0
RNA processing and modification	2	3	2
Transcription machinery	7	7	4
Transmembrane transport	19	4	0
tRNA synthesis and modification	26	27	25
Unclear/uncharacterised functions	6	4	1
TOTAL	**251**	**208**	**109**

BOX 7.1   THE MINIMAL GENOME

What is the minimal number of genes of a living cell? The concept of the minimal genome is widely discussed in biology (Mushegian and Koonin 1996; Gil et al. 2004; Islas et al. 2004; Forster and Church 2006; Fehér et al. 2007; Moya et al. 2009; Henry et al. 2010). It is known that *Mycoplasma genitalium* is the smallest organism that can be grown in pure culture (Fraser et al. 1995). Its genome (482 genes) is already quite reduced when compared with free-living prokaryotes (compare with the *E. coli* genome consisting of about 4,500 genes). Thanks to a comparative genomic approach, based on the comparison between the genomes of *M. genitalium* and other endosymbionts (such as *Buchnera*) or parasites (*Haemophilus influenzae*), it has been possible to sketch the minimal number of common genes.

Mushegian and Koonin (1996) proposed a minimal set of about 250 genes. In 2004, Moya and co-workers reduced this number to about 200 genes (Gil et al. 2004). Forster and Church (2006) searched for the minimal biochemical description of essential functions, arriving to the conclusion that the minimal genome should include about 100 genes. A comparison among these three minimal gene sets is shown in Table 7.1. Note that the largest group of genes refers to protein biosynthesis. The minimal genome essentially encodes the instruction for translating and duplicating the genetic information in very permissive conditions (biochemically rich environment). The essential part of the genome is dedicated to the production of internal components that reciprocally produce each other (from DNA to protein, from protein to DNA), plus those dedicated for boundary production, as demanded by the autopoietic theory. Further simplifications of the minimal genome would be possible only by a strong alteration of the known biochemistry. For example, one could imagine a 'primitive' minimal cell that relies on proteins containing (hypothetically) a reduced number of amino acids, or the loss of enzyme specialization, in favour of few enzymes that are able to carry out multiple functions (but less efficiently), or by reducing the complexity of ribosomes.

Even considering the very small number of ca. 100 genes (Forster and Church 2006), it is difficult to imagine the construction of such a minimal cell. To date, the figure of 100 genes is still far beyond the current experimental approaches, which are described in Sections 7.3.3 through 7.3.5.

Therefore, the relationship between the theoretical minimal genome and the experimental minimal cell is rather weak. It remains unknown whether or not a synthetic cell controlled by a minimal genome can ever be constructed.

7.3.1 Biochemical Reactions inside Liposomes – The Pioneer Phase (ETH Zurich 1990–1999)

It is over 40 years since it was first shown that active enzymes can be encapsulated inside liposomes (Sessa and Weissmann 1970) and a wide range of enzymes have since been encapsulated in this way (Walde and Ichikawa 2001). The application of liposome-encapsulated biochemical reactions as experimental models of primitive cells was pioneered by our ETH Zurich laboratory. Many of the methodologies developed by our laboratory in the 1990s are still widely used in current research, such as those described for the encapsulation of PNPase (Walde et al. 1994a), DNA polymerase (Oberholzer et al. 1995a), Qbeta replicase (Oberholzer et al. 1995b) and ribosomal machinery (Oberholzer et al. 1999). Methods developed more recently include the 'water-in-oil droplet' transfer procedure for the preparation of liposomes (Pautot et al. 2003a,b). Box 7.2 details how liposomes are typically prepared for intra-liposome enzyme activity experiments.

BOX 7.2 ENZYME REACTIONS INSIDE LIPOSOMES

In order to carry out enzyme reactions inside liposomes, the first step consists of the choice of liposome preparation method. This should be selected from the various available methods (Walde 2004), taking into account what lipid(s) are going to be used and the sensitivity of the molecules that need to be encapsulated. PCs, like POPC or DOPC have been largely used because of their very good vesicle-forming capacity in a wide pH range, their chemical compatibility with almost all proteins and nucleic acids and their low melting temperature (about −4 and −20°C, respectively).

Common methods to form lipid vesicles for encapsulation water-soluble enzymes and macromolecules are shown in Chart 7.1A. The most general and well-established method is based on the hydration of dry lipids with the solution of interest. The molecules contained in that solution will be encapsulated within liposomes during their formation. The 'thin-film hydration method' consists of adding an aqueous solution to a lipid film formed on the surface of a glass container. Generally, the lipid film is

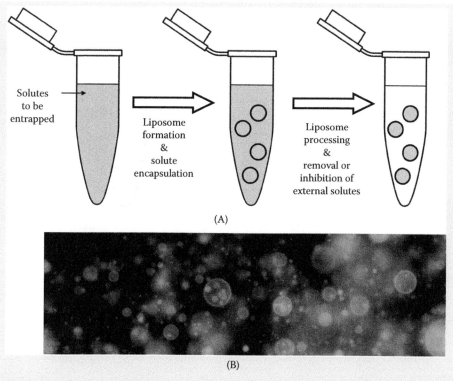

(A)

(B)

CHART 7.1

(A) The typical strategy for carrying out reaction inside liposomes starts from the solution containing the solutes that must be entrapped inside liposomes, for example, an enzyme(s) solution. Liposome formation brings about the formation of a population of compartments containing a variable number of solutes inside, according to the liposome size as well as to recently discovered stochastic effects (Luisi et al. 2010; Souza et al. 2011, 2012). After liposome processing, for example freezing in liquid nitrogen and thawing for reducing the lamellarity and increasing the entrapment yield, or extrusion through pores with defined size for obtaining a monodisperse population, the free unentrapped solutes are removed by gel filtration chromatography on Sepharose 4B (or similar gels), or their activity is blocked by adding killing agents or inhibitors. In all steps, liposomes should be maintained in isotonic conditions and possibly above the lipid melting temperature (T_m; note that for POPC in water, $T_m \sim -4°C$). In the case of protein synthesis, a cell-free transcription/translation kit (e.g. the PURE system or cell extracts), including also DNA, is used to prepare liposomes. The external reaction can be blocked by addition of RNase, EDTA, proteinase K, or just by dilution. In some cases, one or more chemicals needed for the reaction can be added later, allowing the intravesicle reaction only after their permeation through the lipid membrane. (B) Fluorescence micrographs of rhodamine 6G-stained giant vesicles (5–50 μm).

formed by evaporation of a chloroform or methanol lipid solution, by means of a rotavapor or by nitrogen flushing. Liposomes of various size and morphology are formed by gentle shaking and/or vortexing (large unilamellar vesicles [LUVs], multi-lamellar vesicles [MLVs], multivesicular vesicles [MVVs], GVs). If the thin film is left in contact with the aqueous solution without any mechanical perturbation (in the so-called 'natural swelling' protocol), giant lipid vesicles (GVs) are preferentially produced after several hours. The nature of lipids/buffer strongly affects the efficiency of the natural swelling method. The long time required for GV production by the natural swelling methods can be shortened to about 1 hour by applying alternate electric

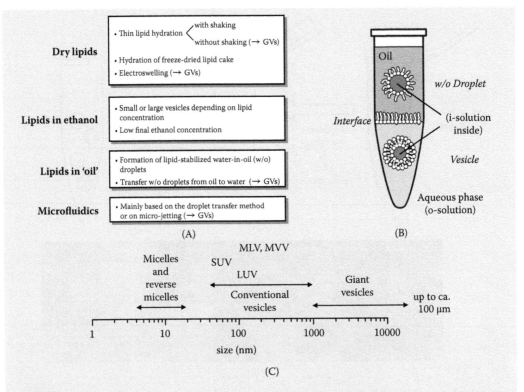

CHART 7.2

Methods of liposome preparation and liposome size. (A) Liposomes of various size and morphology can be prepared by hydration of dry lipids, by diluting (injection) an alcoholic lipid solution in water, from lipid-stabilised water-in-oil droplets, or by microfluidic technology (only a few examples, for a review see Stano et al. 2010; Matosevic 2012; Swaay and deMello 2013). See text for details. (B) The droplet-transfer method consists in the transfer of a lipid-stabilised water-in-oil droplet from the oil phase to an aqueous phase. The droplet converts to vesicle when crossing a lipid-containing interface. This method permits the formation of good yields of solute-filled giant vesicles. The vesicle content is defined by the inner content of the water-in-oil droplet. (C) Whereas micelles and reverse micelles are very tiny supra-molecular structures, whose size ranges from about 4 to 20 nm, conventional vesicles typically have a size below 1 μm. The smallest vesicle diameter can be around 30–40 nm (depending on the lipids). Those are called small unilamellar vesicles. Large unilamellar vesicles are instead typically 100–400 nm large. Multilamellar vesicles and multivesicular vesicles do not have defined sizes but are defined as containing several concentric or non-concentric vesicles inside the outermost vesicle, respectively. Giant vesicles (GVs) and giant unilamellar vesicles are very large vesicles that can be visualised by optical microscopy; their size is typically below 50 μm, but it can occasionally also reach 100 μm. GVs are generally prepared according to special methods (natural swelling, electroswelling, droplet transfer, microfluidics).

field on the lipid film, which is deposited over platinum or indium tin oxide electrodes, the so-called 'electroswelling method' (Angelova and Dimitrov 1986). Instead of hydrating a thin lipid film, liposomes with heterogeneous size and morphology can be obtained from the hydration of freeze-dried liposomes (lipid 'cake') (Chart 7.2).

The 'ethanol injection method' consists of the injection of an alcoholic lipid solution into a buffer. Injection is typically done using a syringe, in order to provide shearing forces. Ethanol mixes with water and the lipids self-assemble to give vesicles. The lipid concentration in ethanol, rather than their final concentration in the aqueous phase, affects the size and the morphology of the resultant liposomes (Domazou and Luisi 2002; Stano et al. 2005). Low lipid concentrations favour the formation of small liposomes, so that this method can be applied to form small unilamellar vesicles (SUVs). A drawback is that a certain amount of ethanol is left in the liposome preparation (typically <5%), but it can be removed by dialysis or gel filtration chromatography.

A recently introduced method, known as 'emulsion inversion', 'droplet inversion' or 'droplet transfer' (Pautot et al. 2003a,b), consists of the transformation of lipid-stabilised water-in-oil droplets into lipid vesicles (Chart 7.2B). This method is typically used to produce GVs. Starting from a lipid-containing oil ('mineral oil', squalane or dodecane), a solution of interest is used to produce a water-in-oil emulsion by mechanical dispersion. The solution of interest (inner solution, 'i-solution') typically contains sucrose. All solutes contained in the i-solution are therefore compartmentalised inside each droplet. The emulsion droplets are then stratified over a previously formed lipid-containing interface. The lipid-covered droplets cross the macroscopic oil-water interface and become covered by a second lipid layer, so that GVs are assembled stepwise. Such droplet transfer is facilitated by applying a centrifugal force. The external solution (outer solution, 'o-solution'), that typically contains glucose, must be isotonic with the i-solution.

Finally, new emerging microfluidic technologies have been already used to produce GVs by adaptation of the droplet transfer method or by jetting strategies (Matosewic 2012; Swaay and deMello 2013).

Chart 7.2C shows the typical size of conventional (<1 μm) and GVs (>1 μm), and a comparison with micelles and reverse micelles, two other typical surfactant supramolecular aggregates.

We have seen that it is possible to self-reproduce fatty acid vesicles by simply adding fatty acid precursors (e.g. fatty acid anhydrides or fatty acid vesicles) to the vesicles (Sections 7.2.4 and 7.2.5), giving rise to a growth-division mechanism, but the vesicle core is not involved in this dynamics. The next question therefore becomes: is it possible to incorporate a prebiotically relevant enzyme reaction *inside* self-reproducing vesicles so to mimic simple primitive cell behaviour?

Due to the central role of nucleic acids in origin-of-life hypotheses such as the 'RNA world' (Gilbert 1986) many early investigations into minimal cells featured nucleic acid-based experimentation. Walde et al. (1994a) observed that PNPase can be incorporated within extruded, submicron-sized oleic acid/oleate vesicles. These vesicles were able to take up adenosine diphosphate (ADP) which, once present in the vesicle lumen, was converted into poly(A) strands by PNPase. Simultaneously oleic anhydride was added in the surrounding media, and the mechanism of anhydride hydrolysis and oleate incorporation in the existing vesicles triggered vesicle growth and division (Figure 7.6). Together these observations represented the first proof that it was possible to create self-reproducing vesicles capable of internally synthesising nucleic acids.

In these experiments, ADP permeated into vesicles because the membrane was composed of oleic acid/oleate. Diffusion across a phospholipid membrane is, in general, much more difficult. However, this has also been reported. In a similar study, PNPase-containing dimyristoyl phosphatidyl choline (DMPC) vesicles were employed to carry out the same reaction, but in this case an efficient ADP permeation was

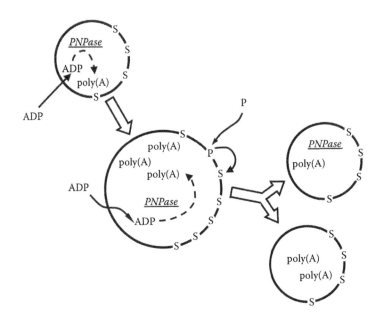

FIGURE 7.6
The 'Oparin reaction' inside liposomes (Walde et al. 1994a) as a paradigm of enzyme-reactions inside self-reproducing liposomes. The Oparin reaction consists in the polymerization of ADP in poly(adenilic) acid, that is, poly(A), catalysed by the polynucleotide phosphorylase (PNPase). The reaction takes its name by the fact that Oparin used it in his pioneer experiments of nucleic acid synthesis in coacervates. The mechanism consists in two independent reactions occurring simultaneously. First, PNPase-containing fatty acid vesicles are prepared and then they are treated with ADP and a fatty acid precursor, that is, fatty acid anhydride. ADP slowly permeates through the fatty acid membrane, enters the vesicle and it is polymerised to give poly(A). The latter cannot escape from the vesicle due to its size (as happens for PNPase). Simultaneously, fatty acid anhydride is absorbed in the vesicle membrane and hydrolysed to give more fatty acids, as described in Figure 7.3. Consequently, the vesicles grow and divide. The new vesicles will continue to synthesize poly(A) in their aqueous core. However, after a certain number of divisions, because PNPase is not produced by any process, some vesicles will certainly not contain any PNPase, so that they are dead-ends of the process. Such a scenario has also been called 'death-by-dilution' (Luisi et al. 2006) meaning that the catalyst(s) of the internal reaction network, if not reproduced, become(s) diluted among the several new compartments, which in turn will result inactive.

achieved only when DMPC vesicles were kept at 23°C – the melting temperature of DMPC. At this temperature, due to impaired packaging of DMPC in the membrane (i.e. membrane defects), solutes display a better permeability in terms of transversal diffusion (Chakrabarti et al. 1994).

A second example of RNA-producing vesicles, with a simultaneous vesicle self-reproduction, was achieved by entrapping the Qbeta replicase inside fatty acid vesicles (Oberholzer et al. 1995b). Qbeta replicase is a RNA-dependent RNA polymerase that synthesises an RNA strand on an RNA template. By introducing all reactants inside extruded submicron oleic acid/oleate vesicles, and providing oleic anhydride as feedstock, it was possible to detect RNA synthesis during vesicle self-reproduction.

Note that in both the earlier cases of enzyme-catalysed reactions, intravesicular reactions occur at the same time as vesicle self-reproduction. However, despite the increment of membrane molecules and of the products of the reactions, these schemes feature no mechanism for continued re-synthesis of the intravesicular enzymes and consequently the total mass of these enzymes remains constant. Given this situation, as the number of vesicles increases due to vesicle replication, the intravesicular enzyme concentration will decrease. After a given number of divisions, some vesicles would be predicted to contain zero enzyme molecules.

This scenario (and its extension to the general case) has been defined as 'death-by-dilution' (Luisi et al. 2006), meaning that in this condition it is expected that the encapsulated reaction network stops functioning after a certain number of cycles because at least one component is not produced. For a vesicle system to be fully self-sustaining and self-reproducing, all components of the vesicle boundary and lumen must be continually regenerated by the encapsulated reaction network (in autopoietic fashion).

Other examples of complex enzyme reactions encapsulated within extruded submicron phospholipid vesicles include (1) the PCR, based on DNA polymerase, and (2) the synthesis of a polypeptide (polyphenylalanine), based on ribosomes. Both these schemes were successfully achieved in 1-palmitoyl-2-oleoyl-*sn*-glycero-3-phosphatidylcholine (POPC) vesicles. In the first case, DNA polymerase, template DNA and deoxynucleotide triphosphates (dNTPs) were co-encapsulated inside vesicles and the expected DNA product was obtained after 25 thermal cycles, indicating the vesicles were robust to incubation at 95°C (Oberholzer et al. 1995a). No simultaneous vesicle self-reproduction was attempted in this case. In the second example, purified 70S ribosomes, poly(U)—acting as mRNA, phenylalanyl-tRNA synthetase, phenylalanyl-tRNA (tRNA[Phe]), phenylalanine, elongation factors EF-Tu, EF-G, EF-Ts, guanosine triphosphate (GTP), adenosine triphosphate (ATP), phosphoenolpyruvate and pyruvate kinase were all co-encapsulated inside liposomes. These components were able to successfully bring about intraliposomal translation of polyphenylalanine polypeptides (Oberholzer et al. 1999). This was the first example of ribosomal reaction in a synthetic cell-like microcompartment, and serves to demonstrate the feasibility of constructing a kind of protein-based artificial cell based on protein synthesis.

Next, Yu et al. (2001) and Oberholzer and Luisi (2002) extended the method of ribosomal machinery encapsulation going beyond poly(Phe) synthesis and independently reported the first studies on successful intraliposomal synthesis of functional proteins.

Specifically, a DNA sequence encoding green fluorescent protein (GFP) was mixed with a so-called 'transcription-translation' kit, a mixture including RNA polymerase, ribosomes, translation factors, tRNA, amino acyl tRNA synthetase, dNTPs and amino acids. This reaction mixture, when kept at low temperature, is almost inactive, whereas it can be triggered by increasing the temperature. POPC liposomes were prepared by the ethanol injection method (Oberholzer and Luisi 2002) or by the rehydration of previously freeze-dried liposomes (Yu et al. 2001) and the GFP synthesis was monitored by fluorescence measurements, microscopic evidences and flow cytometry. The last two techniques demonstrated that the protein synthesis was occurring only inside liposomes. Moreover, in order to abrogate the possibility of reactions occurring outside liposomes, ribonuclease (RNase), protease or EDTA were added after liposome formation. These substances cannot cross the liposome membrane and completely inhibit the protein synthesis by digesting RNA, digesting proteins or chelating Mg^{2+} ions (which are essential for transcription and translation).

These two latter reports mark symbolically the end of the 'pioneering' phase of realizing enzymatic reactions inside lipid vesicles; more and more experiments have been generated since 2001, with ever-increasing complexity.

It remains an unresolved question whether vesicle self-reproduction, which requires lipid synthesis and membrane incorporation, can be achieved more robustly by enzymatic synthesis or by enzyme-free chemical reactions.

In 1991, our group made attempts to use lipid-processing enzymes inside lipid-vesicles (Schmidli et al. 1991). In particular, the four enzymes of the salvage pathway for phosphatidylcholine synthesis, namely (1) *sn*-glycerol-3-phosphate acyltransferase,

(2) 1-acyl-*sn*-glycerol-3-phosphate acyltransferase, (3) phosphatidate phosphatase and (4) cytidinediphosphocholine phosphocholine transferase were simultaneously encapsulated (i.e. membrane-integrated) into soybean phosphatidylcholine liposomes. Our observations were consistent with the four enzymes of the salvage pathway being both associated with the proteoliposome membrane and in active and bound conformations.

Phosphatidylcholines with different alkyl groups and fatty acid moieties could be synthesised by these putatively encapsulated salvage pathway enzymes in a manner that coincided with changes in liposome size. However, direct evidence of liposome growth and division occurring as a result of the activity of liposome-associated salvage pathway enzymes has so far proven elusive.

The conceptual importance of this work lies in the fact that the lipids, which comprise the liposome boundary, can be synthesized by a catalyst enclosed within the liposome itself. This represents a major step towards establishing whether the self-contained boundary replication feature required by autopoietic theory can actually be achieved in reality.

7.3.2 Protein Synthesis and Cell-Free Technology

As has been mentioned, Yu et al. (2001) first reported the synthesis of a fully folded functional protein inside liposomes and the capability to achieve protein synthesis inside liposomes is now well established. The time is now ripe for tackling the design and the realization of more complex synthetic cells. Interested readers can find additional information in reviews by Luisi et al. (2006), Stano et al. (2011) and Stano and Luisi (2013).

Quite understandably, the gene expression (protein synthesis) process has a primary role in defining the functions of a minimal cell. By producing proteins, the minimal cell begins to establish its autopoietic organization. It is therefore not surprising that protein synthesis remains a key goal in the constructions of artificial cells.

The so-called 'semi-synthetic' approach to minimal cell construction typically involves entrapping a complete set of the pre-synthesised macromolecular machineries essential for gene expression within lipid-bounded compartments (Figure 7.7A). Non-entrapped material is typically either removed by dilution, by purification (liposome separation from non-entrapped chemicals via gel exclusion chromatography) or deactivated by proteinase or RNase addition. Minimal cell technology, therefore, can be seen as the convergence of liposome and cell-free technology, with a possible future involvement of microfluidic technology (Figure 7.7B). Thanks to the recent advancements of cell-free methods it is possible to synthesise several types of proteins, not only the water-soluble ones.

Typically, cell extracts or reconstituted cell-free systems are used. Commercial or lab-made cell extracts, often derived from bacteria, are supplemented with the DNA or RNA encoding for a target protein, and included during liposome preparation. Most of the work has been done on genes expressed under the control of the T7 promoter. Of particular interest is the so-called protein synthesis using recombinant elements (PURE) system, introduced in 2001 by Takuya Ueda (Shimizu et al. 2001, 2005). This is a reconstituted protein synthesis system that contains the minimal number of molecules necessary to express a gene from a DNA or RNA sequence. The PURE system contains, in addition to 70S ribosomes and tRNAs, 36 different *E. coli* protein components (individually purified), which are the 20 amino acyl t-RNA synthases, 11 translation factors (MTF, IF1, IF2, IF3, EF-G, EF-Tu, EF-Ts, RF1, RF2, RF3, RRF), T7 RNA polymerase and 4 energy-recycling factors (myokinase, creatine kinase, pyrophosphatase and nucleotide diphosphate kinase).

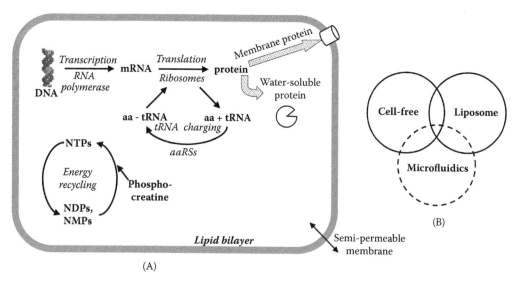

FIGURE 7.7

The state-of-the-art of *semi-synthetic minimal cell technology* focuses on the production of water-soluble and membrane proteins inside liposomes. (A) Cell-free systems are encapsulated within liposomes, producing one or more proteins starting from the corresponding DNA sequence. Note that cell-free systems actually consist in four interacting molecular reaction networks: transcription, translation, t-RNA charging (catalysed by aminoacyl-t-RNA synthases, aaRSs) and energy recycling (which ultimately requires a phosphate donor, e.g. phosphocreatine). The lipid membrane is semi-permeable, allowing the entrance of small molecules, possibly also thanks to channels incorporated in the membrane (Noireaux and Libchaber 2004). All macromolecular compounds required for protein synthesis, on the other hand, cannot leave the synthetic cell. (B) Semi-synthetic minimal cell technology as derived from the convergence of *cell-free systems*, like the PURE system (Shimizu et al. 2001), *liposome technology*, and possibly in the near future, *microfluidic technology*.

The PURE system is capable of synthesizing a functional protein thanks to a mixture of DNA plus 83 macromolecules (ribosomes + 36 protein factors + 46 *E. coli* tRNAs [Dong et al. 1996]). If the subcomponents of *E. coli* ribosomes are also counted (3 ribosomal nucleic acids [rRNAs] and 55 proteins), this makes a total of 140 different genetically encoded RNA or protein sequences required to specify all the PURE system components. This number accounts for about 70% of the Moya's minimal genome (Gil et al. 2004), further confirming that protein synthesis is really the keystone of a minimal cell. In order to sustain the reactions of (a) transcription, (b) translation, (c) amino acid charging and (d) energy recycling, the PURE system also contains low-molecular-weight compounds such as the nucleotide triphosphates, 20 amino acids, DTT, spermidine, formyl-tetrahydrofolate, salts and creatine phosphate (20 mM) as the ultimate phosphate donor. These low-molecular-weight compounds must be also co-encapsulated inside liposomes.

It is important to remark that the PURE system components are fully characterised, and their concentration is known and tunable, at least in principle. The PURE system can be considered as a standard cell-free chassis for synthetic biology (http://partsregistry.org).

Cell-free systems have been used to produce proteins inside liposomes. Generally simple water-soluble proteins are easily produced, with variable yields. Since the minimal cells should contain some membrane proteins, the encapsulation of any cell-free systems inside liposomes should be also able to produce membrane-inserted and correctly folded membrane proteins.

7.3.3 Production of Water-Soluble Proteins inside Liposomes

Following the first two papers (2001–2002), the research on protein synthesis inside liposomes, as well as on 'biopolymerization' reactions (DNA and RNA synthesis) expanded considerably. Here we summarise the main achievements, starting from the most relevant work on the synthesis of water-soluble proteins inside liposomes (for a review on the subjects, see Stano et al. 2011)

The group of Yoshikawa presented a study on the GFP synthesis in 1–100 μm GVs prepared by the so-called 'natural swelling' method (Nomura et al. 2003). Interestingly, it was shown that in the first 3 hours the reaction inside GVs proceeded faster than in the external phase. Two important works appeared in 2004. In the first one, protein synthesis inside liposomes was achieved by means of a two-step cascade genetic reaction (Ishikawa et al. 2004). In particular, SP6 RNA polymerase was added to a transcription-translation mixture (without T7 RNA polymerase). The mixture also contained a plasmid encoding two genes, the *t7rna* polymerase gene under SP6 promoter and the *gfp* gene under T7 promoter. The T7 RNA polymerase was effectively produced by the cell-free system inside liposomes; once produced, the T7 RNA polymerase allowed the expression of the *gfp* gene. GFP fluorescence was revealed in the liposome population via flow-cytometry. This showed for the first time that two genes can be expressed simultaneously inside liposomes (realizing a simple genetic circuit), paving the way to more complex design where gene expression can be activated according to a precise design.

The second important 2004 work was carried out by Noireaux and Libchaber (2004) who introduced two very interesting novel elements to the field. Their first innovation was the adoption of a novel GV preparation method, which was previously explored by Weitz and co-workers (Pautot et al. 2003a,b). This is the 'droplet transfer' or 'inversion of the emulsion' method (for a review, see Walde et al. 2010), which allows the entrapment of multisolutes mixtures in physiological conditions, with a good entrapment yield inside GVs (ca. 30%–40% according to our estimation; unpublished results). The second novel element was the formation of a pore on the phosphatidylcholine membrane allowing the material exchange between the GV and its environment. The pore was created by alpha-hemolysin, which was itself produced inside the GVs by cell-free expression. Alpha-hemolysin spontaneously integrates in the lipid membrane, forming a pore with 3 kDa molecular cut-off. This allowed the feeding of internal reactions by external addition of small molecules that can cross the pore (nucleotides, amino acids). In contrast, no macromolecules entrapped inside the GV could escape. In this way, it was possible to prolong the protein synthesis for 100 hours, as compared with the typical 1.5–3 hours when the alpha-hemolysin pore was not present.

These first studies were carried out by using commercial transcription-translation kits (from companies such as Promega or Roche) or laboratory-made cell extracts (e.g. *E. coli* S30 cell extracts). The Yomo group (Sunami et al. 2006) and our group (Murtas et al. 2007) independently reported the first examples of intravesicle GFP production by the PURE system (see Section 7.3.2), which was encapsulated inside phospholipid vesicles. Currently, the PURE system is considered as a synthetic biology standard for synthetic cell construction.

T. Yomo's group in Osaka, Japan, have investigated several quantitative aspects of intraliposomal protein expression by synthesizing not only GFP but also active enzymes, such as beta-galactosidase and beta-glucuronidase (Hosoda et al. 2008; Nishimura et al. 2009; Sunami et al. 2010b). Flow cytometry has also been used to dissect the influence of lipids on the intravesicular protein synthesis by Bui et al. (2008, 2010), Umakoshi et al. (2009) and Suga et al. (2013). Yoshikawa and co-workers (Saito et al. 2009) followed the course of GFP production in individual GVs and reported the intervesicle 'diversity' (Stano 2007) of GFP production kinetics and yield, due to variation in intravesicle solute concentration and surface-to-volume ratio geometric effects.

T. Yomo's group also managed the remarkable achievement of nucleic acid self-amplification inside lipid vesicles, using a combination of protein synthesis and RNA polymerization (Kita et al. 2008). The system was designed as follows: mRNA molecules encoding the RNA-dependent RNA replicase, Qbeta-replicase, were encapsulated inside liposomes together with essential transcription-translation components. The Qbeta-replicase coding sequence was successfully translated and the resultant Qbeta-replicase enzyme was indeed functional. It therefore replicated the RNA gene, producing its complementary strand. In turn, the complementary RNA strand also encoded for beta-galactosidase, translation of which was detected by a conversion of fluorogenic substrate.

Further attempts to influence the course of protein synthesis inside liposomes by tuning the membrane permeability were recently reported by the group of C. Danelon. According to classical strategy, the transcription-translation mixture was encapsulated inside a 4:1 mixture of dimyristoyl phosphatidylcholine/dimyristoyl phosphatidylglycerol (DMPC/DMPG) liposomes. Myristoyl chains (in DMPC/DMPG) are shorter that palmitoyl-oleoyl chains (in POPC), so that the permeability of small molecules is greatly increased (Nourian et al. 2012). In this way, it was possible to exploit the natural passive diffusion of small molecules inside liposomes in order to feed the reaction by fresh 'nutrients'.

In order to achieve a level of control over rates of cell-free protein synthesis, Martini and Mansy (2011) have explored the application of the theophylline riboswitch. The theophylline riboswitch is a synthetic untranslated region of mRNA, which increases the translation efficiency upon ligand binding (Lynch and Gallivan 2009).

Noireaux and co-workers recently reported the development of a 'toolbox' for realizing gene circuits inside artificial cells, based on *E. coli* sigma factors affecting the transcription of *E. coli* RNA polymerase (Shin and Noireaux 2012). Elementary circuit motifs, such as multiple stage cascades, AND-gates and negative feedback loops can be constructed with the six sigma factors, two bacteriophage RNA polymerases and a set of repressors. The GFP synthesis inside liposomes, prepared by the droplet transfer method, was regulated by the arabinose system (transcription is repressed by the protein AraC in the absence of arabinose and activated by AraC in the presence of arabinose). A study on intravesicle positive-feedback circuits has been published very recently (Kobori et al. 2013).

7.3.4 Production of Membrane Proteins inside Liposomes

With the exception of alpha-hemolysin, whose synthesis was first reported by Noireaux and Libchaber (2004), all cases illustrated in the previous paragraph refer to water-soluble proteins.

However, very important cellular processes, which are required for the assembly of a synthetic minimal living cell, depend on membrane proteins. This is not only true in modern cells, but hydrophobic functional proteins might have had a key role also in primitive cells. It is then clear that one of the most challenging goals of synthetic cell methodology deals with cell-free membrane protein synthesis inside liposomes. This is one of the current frontiers in biotechnology because it offers a complementary way to understand and use membrane proteins. In fact, the difficulty of isolating and manipulating active membrane proteins, and their sensitivity to extraction methods, is well known.

In the field of cell-free methods, several authors already managed to *produce* membrane proteins, instead of isolating them from cells, by transcription-translation reactions in the presence of lipid matrices, like micelles, bicelles or vesicles. Several groups are active in the field, such as the groups of Bernhard (Klammt et al. 2004), Swartz (Wuu and Swartz 2008), Yokohama (Shimono et al. 2009), Endo (Nozawa et al. 2011) and others.

In 2009, Akiyoshi, Morita and co-workers (Kaneda et al. 2009) reported the synthesis of functional (i.e. oligomer-forming) connexin 43 (Cx43) in the presence of egg yolk phosphatidylcholine (PC) liposomes using a DNA template and a rabbit reticulocyte extract. Protein synthesis was detected inside and outside the vesicles.

The so-obtained connexin-containing liposomes were able to form gap junctions between liposomes and cells, and passive diffusion-driven transport of a probe was observed. This research was continued by the same group, who demonstrated that Cx43 can be also synthesised, in its functional form, by the PURE system in the presence of 1,2-dioleoyl-*sn*-glycero-3-phosphatidylcholine (DOPC) liposomes. The presence of liposomes during the translation allowed a spontaneous insertion of Cx43 in the lipid membrane and then the formation of the connexon channel (Moritani et al. 2010).

We recently applied synthetic biology approaches in our attempts to achieve expression of lipid synthase enzymes inside liposomes (Kuruma et al. 2009).

As discussed in Section 7.3.1, the lipid salvage pathway consists of a four-step conversion of glycerol-3-phosphate (G3P), a water-soluble molecule, into diacyl phosphatidylcholine, a membrane-forming compound. We focused only on the first two steps of the pathway, namely the attachment of two acyl chains to G3P, to give a diacyl phosphatidic acid (PA), which is itself capable of integrating in lipid membrane and forming stable lipid vesicles.

Acyl transferases catalyse the two sequential steps in PA production: first the integral membrane protein, G3P-acyl transferase (GPAT), EC number 2.3.1.15, and then the peripheral membrane protein, lysophosphatidic acid acyl transferase (LPAAT), EC number 2.3.1.51. Our general strategy was based on the co-encapsulation of two plasmids together with the PURE system and the substrates of the lipid salvage reaction inside liposomes. The target molecule was 1-palmitoyl-2-oleoyl-*sn*-glycero-3-phosphatidic acid (POPA), and its substrates are G3P (the backbone), and palmitoyl-CoA and oleoyl-CoA as acyl donors. Three conditions need to be satisfied in order to produce POPA inside the vesicles: (1) the liposome membrane should not chemically interfere with the PURE system, (2) the lipids used should form liposomes of sufficient stability and lumen volume and (3) the lipid composition should allow the correct insertion and folding of GPAT and LPAAT. Clearly, phosphatidylcholine liposomes fit with the first two conditions, but unfortunately do not with the third one. The best suitable combination of lipids was a synthetic lipid mixture that mimics the *E. coli* membrane, further stabilised by phosphatidylcholine. In particular, POPC:POPE:POPG:cardiolipin 50.8 : 35.6 : 11.5 : 2.1 was effectively used. Enzyme activity was monitored by following a radioactive tracer. Activity studies revealed that the synthesis of GPAT inside liposomes correctly produced a membrane-localised catalytically active enzyme. Although the final goal was the simultaneous synthesis of GPAT and LPAAT, so that their concerted catalytic activity would have brought about the synthesis of membrane lipids, this was only partially achieved. In fact, the optimal redox conditions for the activities of the two enzymes were different. GPAT had the highest activity in reducing conditions, whereas the opposite was true for LPAAT (Luci 2003). The two-step lipid synthesis was achieved by pre-synthesizing enzyme-containing vesicles in different redox conditions, and mixed afterward. POPA synthesis was monitored by radioactive labelling, demonstrating that it is in principle possible to reconstruct the lipid salvage pathway in synthetic cells. The low yield, however, did not allow the observation of any morphological transformation due to the production of membrane lipids.

An alternative way to produce lipids exploits fatty acid synthase (FAS). An attempt has been recently reported that features entrapping partially purified FAS inside liposomes (Murtas 2010) in the presence of fatty acid precursors (acetyl-CoA, malonyl-CoA, NADPH). This attempt was complicated by the very large size of FAS.

Another recently reported case of membrane protein synthesis inside liposomes deals with the reconstruction of MreB/MreC cytoskeletal system (Maeda et al. 2012). MreB is an actin-like polymerisable protein, which forms filaments. MreC is a membrane-associated protein that interacts with MreB filaments, providing an anchor point to the membrane. The plasmids encoding from MreB and MreC proteins were co-encapsulated within giant lipid vesicles, formed by the droplet transfer method, together with *E. coli* crude extracts. MreB filaments were shown to form, after 12 hours, only on the inner surface of <15 μm GVs. In the absence of MreC, MreB polymerised as clusters, suggesting that correctly synthesised and membrane-inserted MreC plays a role in the formation of MreB filaments. It has been suggested, however, that MreB can interact directly with lipid membranes via an *N*-terminal amphipathic helix (Salje et al. 2011).

Lipid-free experiments performed with water/oil/water (w/o/w) double emulsions (Martino et al. 2012a), and water/oil (w/o) droplets (Chanasakulniyom et al. 2012), showed that a freshly synthesised fusion protein, comprised of MreB joined to red fluorescent protein (RFP), accumulated at the surfactant-based emulsion interface.

When MreB-RFP protein was synthesised inside microfluidic-produced polymersomes, its distribution was partially on the membrane, partially localised in the polymersome core (Martino et al. 2012a).

The olfactory receptor complex from the silkmoth *Bombyx mori* has been also synthesised inside GVs prepared by the droplet transfer method (Hamada et al. 2014). Interestingly, the receptor was able to respond to the presence of the ligand bombykol, suggesting that future developments of synthetic cell systems capable of communicating via chemical signals and surface receptors (Stano et al. 2012; Rampioni et al. 2014) are indeed possible.

Very recently, Ueda and collaborators proposed a strategy for constructing synthetic cells with membrane proteins embedded in the membrane, based on SecYEG translocon. The idea is to endow liposomes with the SecYEG translocon, a heterotrimer complex capable of facilitating the correct insertion of membrane proteins in the membrane. The authors have shown that the PURE system is capable of synthesizing a SecYEG complex in functional conformation (Matsubayashi et al. 2014). Coexpression of SecYEG genes and a membrane protein by a liposome-encapsulated PURE system would, if properly developed, in effect constitute a synthetic cell decorated with correctly folded membrane proteins.

7.3.5 Additional Examples of Nucleic Acid Synthesis inside Liposomes

Achieving nucleic acid synthesis inside liposomes represents another major technological milestone for both origin of life studies and synthetic biologists seeking ultimately to design and construct synthetic cells. The performance of intravesicular PCR has been recently revisited by Sugawara and co-workers. They were able to successfully port the PCR experiments described by Oberholzer et al. (1995a) from vesicles to giant lipid vesicles and to follow the reaction using flow cytometry. Intra-GV PCR gave higher amplicon yields then vesicular PCR, presumably due to the greater volume of trapped solute (Shohda et al. 2011).

Takakura et al. (2003) and Toyota et al. (2008) have previously demonstrated that GVs prepared from a synthetic surfactant can subsequently grow and divide by incorporating surfactant precursors that react *in situ* with a membrane-bound catalyst to form the membrane-forming surfactant. When Kurihara et al. (2011) combined these strategies for GV replication and intra-GV DNA replication, they found that the efficiency in vesicle reproduction was influenced (improved) by the DNA content. This result suggested there might be a functional coupling between internal and external reactions within even simple vesicles. Such a phenomenon could be mediated by DNA/membrane interaction, with DNA duplication influencing vesicle dynamics by an as yet undefined mechanism.

Additional examples of DNA synthesis inside liposomes have been reported (Shohda and Sugawara 2006; Torino et al. 2011). In a recent study, Kim, Kwon and collaborators reported the amplification of 1.6 kb DNA sequences by intraliposomal PCR (Lee et al. 2014) as a tool for preparing gene-containing neutral liposomes for non-toxic transfection (due to the absence of cationic lipids).

DNA transcription has been often used as a way of testing nucleic acid production inside liposomes. Clearly, when a DNA sequence is used as a template for synthesizing proteins inside liposomes, this implicitly means that RNA has also successfully been produced. Some studies have been specifically designed to verify the transcription efficiency and discover the possible role of lipids. Examples can be found in the literature (Tsumoto et al. 2001; Fischer et al. 2002; Monnard and Deamer 2002; Monnard et al. 2007).

7.3.6 Role of Mathematical Modelling in Synthetic Cell and Primitive Cell Research

The work presented until now deals with experiments on biochemical transformations inside liposomes. These studies triggered the interest of several research groups using mathematical modelling to understand the behaviour and the constraints of micro-compartmentalised reactive systems in terms of compartment transformation (e.g. growth, division) and internalised reactions.

Several studies on modelling have been inspired by the autopoietic vesicles, for example see Bozic and Svetina (2004, 2007), Fanelli and McKane (2008), the Wattis group (Coveney and Wattis 1998; Bolton and Wattis 2003a,b), Solé (2009) and Van Santen (Markvoort et al. 2007, 2010).

The ENVIRONMENT platform developed by Mavelli and Ruiz-Mirazo (2007, 2008, 2010, 2011, 2013) enables stochastic simulation of vesicle systems capable of growing at the expense of externally supplied feedstock and it has been specifically designed to deal with micro-compartmentalised reactions.

ENVIRONMENT simulates vesicle growth and autopoietic self-reproduction by using experimentally derived kinetic constants for each elementary step of the supposed mechanism and in this way can function as an *in silico* vesicle laboratory (Ruiz-Mirazo and Mavelli 2008; Ruiz-Mirazo et al. 2011).

When conditions for reproduction are reached (e.g. doubling the vesicle surface), the vesicle divides and solutes are redistributed to the two 'daughter' vesicles. By using ENVIRONMENT, it has been possible to simulate several real and hypothetical systems. For example, Mavelli and collaborators were able to extract kinetic constants of elementary vesicle processes from experimental data (mostly published by the Szostak group). Grounded on these experiment-related values, *in silico* modelling can be used to model the dynamics and the transformation of a hypothetical minimal RNA cell (Mavelli 2012), which was described in theoretical terms by Szostak et al. (2001).

Very recently, we proposed a detailed stochastic simulation of the PURE system inside a lipid vesicle, in collaboration with Marangoni (Lazzerini-Ospri et al. 2012; Calviello et al. 2013). Deterministic analyses of protein production by the PURE system were published recently (Karzbrun et al. 2011; Stögbauer et al. 2012), whereas Yomo and co-workers proposed a theoretical/experimental systematic study on protein synthesis yield based on combinatorial variations of PURE system components, revealing a kind of synthetic 'epistatic interactions' (Matsuura et al. 2009).

7.3.7 Open Questions and Perspectives

We have seen the origin of and the recent developments in the research on biosynthetic reaction networks inside liposomes. In addition to nucleic acid synthesis, mainly based on polymerases (DNA-dependent DNA and RNA polymerases or RNA-dependent RNA polymerase), protein synthesis plays a key role in the field. The reason is very clear. Cell horsepower resides in structural and catalytic proteins and therefore any attempt to construct a DNA/RNA/protein minimal synthetic cell must face the issue of autopoietic protein production. Moreover, the protein synthesis 'module' represents a very significant part (>50%) of the functions encoded in the minimal genome. Despite the recent technical progress, however, several questions remain still unanswered. For example, it has not yet been shown that a synthetic cell can grow and divide purely through the action of internal enzyme-based lipid synthesis. Advancements in this direction have been hampered by the chemical nature of the required enzymes (i.e. they are membrane integral enzymes, like the GPAT; or extremely large and complex water soluble enzymes like FAS). Moreover, the production of lipids should be followed by a very precise series of events, that is, vesicle growth, destabilization and division. Simultaneously, internal components should also be (re)produced (and possibly at a similar rate), for instance thanks to DNA duplication, ribosome synthesis and enzyme synthesis.

For example, ribosome synthesis certainly represents a very critical issue for a fully self-reproducing synthetic cell. Although it is well known that ribosomes might be reconstituted *in vitro* from the isolated components (Nierhaus 1990), their reconstitution procedure consists in a very well defined set of ordered operations, and it is not clear whether it might occur spontaneously inside vesicles. Moreover the ribosome components should be synthesised *in situ* by transcription/translation processes. Preliminary data on the *in vitro* preparation of ribosomal subunits, involving contacting polypeptides with rRNA, incubating polypeptides and rRNAs at constant temperature and magnesium ion concentration, and assembling ribosomal subunits have been first presented by M. Jewett and G. Church at the fourth Synthetic Biology conference, Hong Kong 2008 and recently published (Jewett et al. 2013).

Another interesting question is whether it is possible to perform vesicle self-reproduction experiments with protein-synthesizing vesicles. It has been previously shown that oleate vesicles encapsulating the Oparin reaction (polymerization of ADP catalysed by PNPase) or the Qbeta-replicase system were capable of simultaneous growth and division in a way that consumed oleic anhydride (Walde et al. 1994a; Oberholzer et al. 1995b). Growth and division capabilities have also been observed in vesicle studies with water-in-oil droplets (Fiordemondo and Stano 2007) but no equivalent work has been performed with protein-synthesizing vesicles.

An open question for such experimental approaches is whether, and to what extent, the internalised solutes are evenly distributed among the daughter compartments, enabling a continuation of the internal reaction in the first or second generation of newly formed compartments.

All these aspects also point to the need of moving from the current approach based on 'burst' bioreactors, where the reactions proceed till the consumption of the limiting reagent, to the 'continuous' bioreactor mode, so that the synthetic cellular system can continue its internal reactions homeostatically.

Other issues such as (1) the reconstruction of *cyclic* multi-enzyme networks, (2) the integration of regulatory/adaptation/responsive elements and (3) bioenergy production inside liposomes are still unanswered.

With regard to energy production, synthetic cells are currently totally dependent on the chemical energy provided by NTPs or other phosphate donors. Clearly, an autonomous system should be capable of producing sufficient bioenergy to sustain its internal metabolism. One approach to achieving this is to establish an electrochemical gradient that could be harnessed by ATP synthase. Preliminary work on a cell-free system that employs such an approach (production of some components of the F_0F_1-ATP synthase) has been reported by Kuruma et al. (2012). Studies concerned with generating energy within reconstituted systems (without protein synthesis) have been under way since the mid-1990s (Richard et al. 1995; Pitard et al. 1996; Choi and Montemagno 2005), showing that bacteriorhodopsin and ATP synthase proteins can be both reconstituted in liposomes or polymersome and function as expected. However, lack of correct orientation brings about external production of ATP, whereas the internal production is the target of synthetic cell studies.

Moving to more speculative views, one might ask whether one could imagine a minimal cell with a minimal gene set much smaller than the about 200 genes indicated by comparative genomic analysis. A detailed analysis of some possibilities has been presented (Islas et al. 2004). One of the arguments for further genome reduction deals with the fact that the enzymes involved in primitive cells were certainly less complex and specialised than modern ones; therefore, the overall number of different macromolecular types in a minimal cell could be less than 200. Primitive ribosomes might be active with less ribosomal proteins, because the peptide synthase function resides in the rRNA (Nissen et al. 2000). Clearly, with these simplifications, the resulting self-maintenance capacity of such a hypothetical minimal cell should be quite low, yet it should be still capable of autopoietic self-sustainment. The term 'limping cells' has been used to describe such a situation (Luisi et al. 2006).

7.4 Entrapment of Solutes inside Liposomes

No less important than the biochemical constraints are the biophysical limitations that play a major role in origin of life research. Major questions remain regarding the mechanisms that lead to the material self-organization and to the formation of the first living cells from their non-living environment. These include, 'how do lipid boundaries spontaneously form?' and 'how are solutes entrapped within newly formed lipid boundaries?'

7.4.1 Co-Entrapment of Different Solutes: A Conundrum?

First, we will consider the entrapment of cell-free transcription/translation systems inside vesicles, and in particular inside submicron vesicles. Protein synthesis within 'conventional', as opposed to GVs, was first reported in a short technical study of enhanced GFP (eGFP) synthesis in vesicles prepared by ethanol injection (Oberholzer and Luisi 2002). Subsequently, we also reported eGFP expression in vesicles of diameter 200 nm as part of an effort to answer the question, 'how small can a viable cell be?' (Souza et al. 2009). In fact, this remains an open question, not least because of controversial reports on 'nanobacteria' (Kajander and Ciftçioglu 1998) and 'nanobes' (Uwins et al. 1998).

Researchers have debated the theoretical lower limits of cell size for some years, even organising dedicated symposia to examine the question, such as the 1999 workshop: 'Size Limits of Very Small Microorganisms'. We reason that the reactions necessary for cell-free protein synthesis reactions, although representing only a sub-section of a minimal metabolism, are still significantly complex. As such, we have proposed that the pathways required for protein synthesis could be considered in isolation as a simplified proxy for an entire minimal metabolism. In this context, we asked if it is possible to (1) efficiently co-entrap all different PURE system (transcription-translation machinery) macromolecules within 200 nm diameter lipid compartments and then use these to (2) synthesise a well-folded and functionally active protein. Towards this aim, we applied largely standard procedures for liposome formation, solute entrapment, and liposome manipulation (see Box 7.2) and measured eGFP fluorescence.

The data obtained revealed that although the overall eGFP amount was very small, almost negligible, eGFP was indeed synthesised inside POPC vesicles. The average eGFP yield was estimated at about 5 ± 2 eGFP molecule every 100 lipid vesicles. Clearly, not all vesicles could synthesise eGFP, probably due to their very small internal volume. Such a small volume might not be able to co-entrap, on average, the about 80 different macromolecular components and the low molecular weight ones necessary to carry out protein synthesis.

This suggested that only some of the 200-nm vesicles were metabolically active, whereas the vast majority was not, as a consequence of the small vesicle volume. We indeed reasoned that a more realistic interpretation of the low average yield would be that one 'lucky vesicle' out of a hundred was able to co-entrap all required molecules, and then produce about 5 eGFP molecules (or one out of a thousand, producing 50 eGFP molecules). The presence of such a small fraction of metabolically active vesicles (let say, 1 out of 1,000) still correspond, in typical experimental conditions, to millions of vesicle per millilitre—not at all a negligible number.

Surprised and intrigued by these results, we asked whether such observations fit with the theoretical expectations, and then we started a direct investigation of macromolecular encapsulation inside lipid vesicles.

7.4.1.1 A Simple Statistical Model Based on Poisson Distribution

First, we faced the problem of finding a proper theoretical model against which to compare the data.

Let us consider the encapsulation of a single water-soluble chemical species (for example an enzyme), with concentration C_0, in liposomes all having the same volume V. The solutes are considered evenly distributed throughout the solution, and let us assume that suddenly (and randomly) a certain number of vesicles form, entrapping some of the solutes. Due to the concentration fluctuations at the molecular scale, some vesicles will entrap more molecules, other vesicles will entrap less molecules. Basic sampling theory says that the majority of vesicles will contain N_0 of entrapped molecules, where $N_0 = N_A C_0 V$ (where N_A is the Avogadro number). From a practical standpoint, the following formula holds: $N_0 = 2.52 \times 10^{-6} R^3_{nm} C_{0,\mu M}$ (R being the internal vesicle radius, in nanometres). It follows that for a population of monodispersed vesicles (all having the same volume), the mean number of entrapped molecules N_0 depends on the size (the volume) of the vesicles and on the concentration of the solute. Note that when $R = 100$ nm and $C_0 = 2$ μM, it follows that $N_0 = 5$, that is, a small number.

The probability p of finding a vesicle with N entrapped molecules, when N_0 are expected, can be calculated by the Poisson distribution (Boukobza et al. 2001; Sunami et al. 2006):

$$p(N, N_0) = e^{-N_0} \frac{N_0^{-N}}{N!} \tag{7.1}$$

The profile of the resulting Poisson distribution is shown in Figure 7.8, revealing that the majority of vesicles will actually contain about N_0 molecules, but it is also somehow expected that spontaneous fluctuations will lead to vesicles with N quite different from N_0. We conclude that it is not unexpected to have vesicles filled with solutes at local concentration (C) that differ from the average one (C_0), as far as the fluctuations are described by a Poisson curve.

The statistical meaning of a Poisson distribution is translated in physical terms by recalling the Poisson distribution tells us that there is a non-zero probability of sampling 0 molecule, 1 molecule, 3 molecules, 4 molecules ... when N_0 are expected. Since in our model the sampling event corresponds to a liposome entrapment event, in a large liposome population (e.g. a suspension of 10 mM POPC vesicles with diameter of 200 nm corresponds to about 10^{13} liposomes/mL) the fraction of empty vesicles will be $p(0, N_0)$, the fraction of vesicles containing 1 solute molecule will be $p(1, N_0)$ and so on. As shown in Figure 7.8, if $N_0 = 5$, it is expected that about 0.7% of vesicles are empty, whereas 3.4% of vesicles contain 1 solute, 8.4% contain 2 solutes and so on. The probability of finding a vesicle with 20 entrapped solutes is about 0.00003%.

By means of the Poisson statistics it is therefore possible to calculate the probability of solute entrapment and compare it with the experimentally determined one. As has been shown earlier, the theoretical calculus already includes 'natural fluctuations'. We will see, however, that natural fluctuations will not explain 'extremes' in the experimentally determined solute occupancy distribution.

7.4.1.2 Comparison between Theoretical Expectations and Results

We first used the Poisson model to compare the experimental evidence for eGFP synthesis inside conventional vesicles (Souza et al. 2009) with the ideal case of random co-entrapment of all PURE system components – a condition necessary to observe intra-vesicle eGFP synthesis.

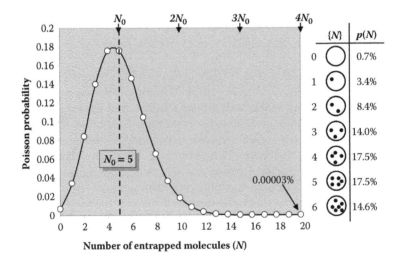

FIGURE 7.8
The Poisson distribution describes the expected solute occupancy distribution inside liposomes, as derived by a process of random entrapment. When the expected number of entrapped solutes (N_0) is 5, for example, it is expected that 0.7% of the vesicles are empty, 3.4% of the vesicles contain one solute molecule, and so on. Note that the probability of finding vesicles that contain many solutes rapidly decreases as N increases. Finding vesicles with four times more solutes than expected is a very improbable event (0.00003%).

Classic Poisson statistical analysis indicates there is a negligible probability of *simultaneous co-entrapment* of many different components occurring. The co-entrapment probability in the case of k different, independently entrapped solutes is given by the product of k Poisson terms. In simple terms, the probability that k different macromolecules are co-entrapped within the same liposome is essentially negligible, because it is the product of 83 small probability terms. The physical reason for this result is intuitively clear: it is very improbable that vesicles will encapsulate at least one copy of each molecule needed for protein synthesis, when already the individual entrapment of each of them is also not so probable. Note that small molecules are not included in the calculation because of their high C_0 concentration, that is, high number of molecules per unit of volume, which translates to $p \approx 1$ even for small vesicles.

We accurately calculated how the co-entrapment probability varies with vesicle size using the PURE system standard concentrations. This study revealed that the relationship between entrapment probability, solute concentration and vesicle volume could be described as a sigmoidal curve, with negligible values for small vesicles (diameter <250 nm) and with values equal to 1 for large vesicles (diameter >600 nm). The logical conclusion was that in order to achieve a measurable probability for the co-entrapment of the solutes of the PURE system inside 200 nm vesicles (and therefore explain the experimental data), each component of the PURE system should have been about 10–20 times more concentrated that its expected (nominal) concentration (Souza et al. 2009). This suggests that eGFP synthesis inside 200 nm diameter vesicles could be explained as a result of concentration of the PURE system components within the same vesicle, reaching a local intra-vesicle concentration at least 20 times higher than their external concentration.

Alternative explanations for the observed entrapment, such as it being driven by inter-molecular association, are not supported by direct electron microscopy or dynamic light scattering data (unpublished observations). Probability calculations, based on the assumption that the PURE components were associated within clusters of 50, 20, or 10 molecules, also did not explain the experimental data.

This 'entrapment conundrum' represents the most interesting aspects of synthetic biology research using liposomes as cell models. By comparing observations with theory, we concluded that liposomes could entrap solutes in a very efficient manner so that the internal concentration exceeds the bulk (C_0) expected value. Is it possible to demonstrate experimentally the supposed spontaneous over-concentration of solutes inside liposomes?

7.4.2 Macromolecular Crowding inside Vesicles: Exploring the Long Tail of the Power Law Distribution

As a result of the earlier observations from using eGFP production inside 200 nm vesicles, and their theoretical implications, we decided to investigate if (1) local concentration enrichment is indeed a possible event and (2) what the mechanism(s) might be.

Clearly, averaging techniques cannot be used for such study. Our approach was based on cryo-transmission electron microscopy (cryoTEM), one of the best techniques available to characterise liposomes populations at the level of single vesicles. Vesicle samples, in the form of thin layers (thickness <1 μm), are quickly frozen in liquid propane so that they are vitrified without the formation of ice crystals. In this way, the liposome structure is preserved. Thanks to cryoTEM it is possible to directly see the size, the shape and the lamellarity of lipid vesicles. In addition to the lipid membrane, electron-dense particles, if present, can be also directly visualised. Following our previous study on reactivity and transformation of fatty acid vesicles (Berclaz et al. 2001a,b), we employed ferritin – a large protein – that is clearly visible due to its cargo composed of thousands of iron atoms. Individual ferritin molecules can be easily seen and counted by cryoTEM (Singer and Schick 1960), as individual black spots. The study was designed in order to measure the actual ferritin distribution inside liposomes prepared by spontaneous methods such as film hydration or ethanol injection (Luisi et al. 2010).

Ferritin solutions of 4–16 μM were prepared with no homogenization (freeze/thaw) step and used to form ferritin-containing lipid vesicles. POPC, alone or mixed with cholesterol, was used as lipid. After imaging via cryoTEM, approximately 7,700 unilamellar liposomes were analysed in terms of size and internal ferritin concentration. It was then possible to build a 'solute occupancy distribution' curve (Figure 7.9A) to determine that, surprisingly, most liposomes were actually empty (ca. 85%), many contained an intermediate number of ferritin molecules (ca. 15%) and a small minority was full of ferritin molecules (Figure 7.9B).

Encapsulated ferritin was not aggregated and was not in contact with the lipid membrane. About 0.1% of vesicles contained a local intra-vesicle ferritin concentration at least one order of magnitude higher than expected, with some small vesicles encapsulating up to about 50 times more than expected; the ferritin encapsulation inside larger vesicles, on the other hand, followed more or less the statistical (Poisson law) expectations. In some cases, the local ferritin concentration was high and comparable in order of magnitude to intracellular media, so that the term 'crowding' can be applied to describe the physico-chemical conditions of these intraliposome solutions. These 'super-concentrated' or 'super-crowded' vesicles were totally unexpected from the viewpoint of Poisson distribution, as evident from Figure 7.9A.

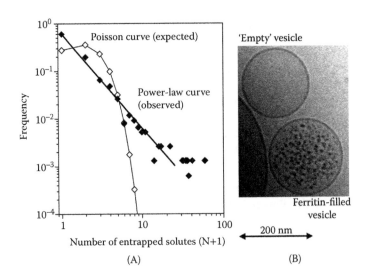

FIGURE 7.9
(A) Comparison between Poisson and power-law solute occupancy distribution in the case of ferritin encapsulation. The Poisson curve (empty symbols) refers to vesicles with radius of 100 nm, in the case of the entrapment of 8 μM ferritin. In contrast to the expectations, the analysis of about 8,000 vesicles reveals that the distribution of ferritin-containing POPC vesicles does not follow the Poisson distribution, being shaped instead as a power law (filled symbols). Note that the power-law probability of finding a lipid vesicle containing more than 10 solutes lies in the 0.001 probability range, whereas the Poisson curve rapidly decreases to very low values. (B) CryoTEM reveals the existence of 'super-filled' ferritin-containing vesicles together with empty vesicles. Note that the internal ferritin concentration can also be more than 10 times higher than the expected concentration. (Reprinted from Luisi, P. L., et al. *Chembiochem*, 11, 1989–1992, 2010.)

The observed solute occupancy distribution is not bell-shaped, like the Poisson one. Rather, it follows a power law, which is typically characterised by high probability values at low solute content, and by a long tail with probability values that *slowly* decrease when the number of encapsulated ferritin increases. In other words, events that are *de facto* impossible according to the random Poisson model are indeed not so rare according to the power law.

Interestingly, qualitatively and quantitatively similar observations were reported later in the case of ribosome encapsulation (Souza et al. 2011) as well as in the case of small RNA-peptidyl complexes (De Souza et al. 2012). The super-encapsulation of ribosomes is particularly relevant when this phenomenon is intended as a model for the origin of early cells.

What is the mechanism that generates the observed behaviour and in particular the power-law distribution as opposite to the Poisson distribution? Clearly, the model of random entrapment does not hold in this case. There must be an interplay between the mechanism of liposome formation and the solute capture (Figure 7.10). A possibility can be sketched for liposome formation mechanisms involving the closure of unstable lipid bilayer fragments (LBFs) (Lasic 1988). LBFs are considered the key intermediate in most liposome formation mechanisms. LBFs self-close to form spherical vesicles. Solutes can interact with LBFs, probably due to aspecific attractive forces, driven by the release of bound water. Solutes can therefore stabilise LBFs with respect to their closure, which is slowed down. This would further enhance the probability of solute capture. In other words, the two processes of LBFs

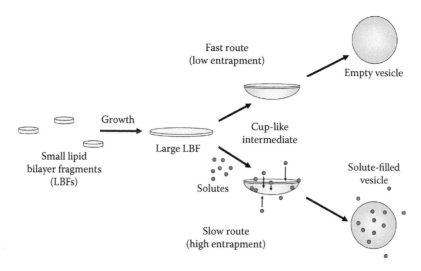

FIGURE 7.10

Hypothetical mechanism to explain the formation of super-filled vesicles (Luisi et al. 2010; Souza et al. 2011). The mechanism of liposome formation by the ethanol injection method has been supposed to occur via undulation and closure of highly unstable lipid bilayer fragments (LBFs) originating by fusion of small LBFs precursors (Lasic 1988). The rate of closure has been estimated to be ca. 1 ms⁻¹ (Hernández-Zapata et al. 2009). A LBF is very unstable due to the presence of the borders, which disappear when the disc becomes a spherical vesicle. Souza et al. (2011) reported a Smoluchowski encounter rate of 10 -20 ms-1 for solutes present in a vesicle-filling experiment. If naked LBFs close rapidly they may as a result form empty vesicles as solute molecules have insufficient time to diffuse into regions that becomes vesicle-enclosed. We considered that the presence of macromolecular solutes like ferritin or ribosomes might stabilize LBFs via non-specific, weak and/or transitory solute/membrane interactions. This could slow rates of LBF closure, resulting in a greater level of solute capture within closed LBFs. This would bring to solute-filled vesicles in cooperative way (solute-richer LBF will encapsulate more and more solutes). We account to local stochastic effects the existence of these different paths, with dominance of the 'empty vesicle' case (ca. 85%), whereas the formation of super-filled vesicles occurs in less than 1% of the instances.

closure and solute capture can be competitive, and whereas most of the LBFs will close to give empty vesicles, few LBFs can enter the 'solute-binding route', ending in a solute-filled vesicle. Stochastic local conditions might guide the LBFs to the first or second path.

We are currently investigating the consistency of this model using stochastic simulations to define the minimal conditions for generating the observed power law. On the other hand, a coarse grained model based on LBFs/solute interactions has been recently published (van Hoof et al. 2012), suggesting that the rate of LBF closure may not be greatly affected by the presence of solute molecules.

Although the mechanism at the roots of the formation of super-filled vesicles is still unknown, it is clear that experimental work in this field has revealed a new and important phenomena.

These observations open new perspectives to the interplay between the biophysics that rules vesicle formation and solute entrapment, and might shed light on the origin of primitive cells (note that additional interesting considerations refer to the fate of solutes after vesicle division) (see Box 7.3).

BOX 7.3 'RE-DISTRIBUTION' OF SOLUTES AFTER DIVISION

If the mechanism of vesicle division can be regarded as still unclear (see Section 7.2.4), the mechanisms by which solute molecules are lost or 're-distributed' between the daughter vesicles following division events remains almost totally unexplored. Currently no data are available on this process, despite its central importance to the self-reproduction of 'metabolically active' synthetic cells.

Ideally, experiments should be able to determine the distribution of each chemical species before and after the division. This could be approached by studying, in a population of self-dividing compartments, each division event individually, or trying to infer the details of solute redistribution from the average pattern measured on the whole population. A robust control of vesicle division is however required.

We recently reported the division of lipid-stabilised water-in-oil (w/o) droplets as a model of vesicle division (Fiordemondo and Stano 2007). In particular, we were able to divide pre-formed droplets by just increasing the amount of free lipids in the solution. Droplet division was followed in real time by video microscopy, revealing that after division the new daughter compartments were still capable of sustaining their own internal reaction.

We reasoned that w/o droplets can be used as simple models to investigate the fate of solute redistribution after division or fusion events. W/o droplets are typically generated by fragmentation of a large macroscopic aqueous drop in an apolar solvent, caused by mechanical shearing forces. Such mechanical forces can also trigger droplet encounter and coagulation (fusion), so that larger droplets derive from smaller ones. As the final outcome of these fragmentation/coagulation repetitive processes, a heterogeneous population of w/o droplet is obtained (a w/o emulsion). What happens to the solutes present in the droplets? Under typical circumstances the solute concentration lies in the micromolar range (e.g. 1 μM). Due to the large droplet size (generally ~1–50 μm), the average number of solute molecules in each droplet, even in the smallest ones, is very high (in 1 μm diameter droplet: $N_0 = 315$ molecules). It is known that the expected fluctuations should be in the order of $\sqrt{N_0}$, that is, around ±18 molecules (about ±6%).

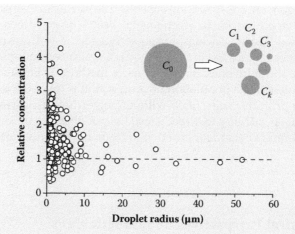

CHART 7.3
Normalised concentration (C_k/C_0; $C_0 = 0.5\ \mu M$) of phycoerythrin (PE, 240 kDa) inside lipid-stabilised water-in-oil droplets, as determined by confocal microscopy. 1-Palmitoyl-2-oleoyl-*sn*-glycero-3-phosphatidylcholine was used as lipid. The concentration of PE inside large droplets approximately corresponds to the PE concentration used to formulate the emulsion, whereas inside small droplets very high and very low PE concentrations can be observed. The spread around the average concentration value cannot be explained by simple stochastic fluctuations. Similar results have been obtained for other macromolecular compounds like dextran-FITC (150 kDa), allophycocyanin (104 kDa), albumin-FITC (66 kDa) and C-phycocyanin (110 kDa). The inset represents the mechanism of formation of microscopic small water-in-oil droplets (each one with its own internal solute concentration C_k) from the initial macroscopic droplet, whose internal solute concentration is C_0. The small droplets form during vortexing, following several coagulation-fragmentation elementary steps.

Actual experiments (carried out with fluorescent macromolecules, see Chart 7.3) show instead a very high heterogeneity in the solute concentration inside the droplets (up to ca. +400%), revealing that the redistribution of those solutes is far from being random, and that again the formation of compartments (in this case by repetitive fission-fusion) and the dynamics of their boundaries are intermingled with the nature and the composition of their aqueous contents. The heterogeneity associated with w/o droplets has also been reported by others (Kato et al. 2012), indicating that complex reactions might be particularly efficient inside the smallest droplets. We are currently investigating this phenomenon by combining experiments and numerical stochastic simulations.

If we assume the development of some kind of primitive metabolism in the bulk, the phenomenon of spontaneous concentration of solutes inside *in situ* formed compartments would explain the origin of cellular life. Consider a realistic scenario with molecules present in diluted form in primitive lagoons. The high dilution prevents any efficient reaction, so that the system remains static and might rapidly undergo inactivation. If, on the other hand, liposomes spontaneously form in that environment, some of them can encapsulate several molecules, bringing about a local (intravesicle) concentration that might trigger an efficient reaction inside vesicles (Luisi 2012). Moreover, the lipid bilayer might protect the encapsulated molecules. In a recent study, we demonstrated that liposomes can concentrate a diluted transcription-translation kit so that an efficient protein synthesis occurs inside a few vesicles, whereas the same reaction proceeds very slowly (at an undetectable rate) in the environment (Stano et al. 2013).

The second, more general, consideration concerns the intrinsic diversity (Stano et al. 2014) generated by microscopic stochastic phenomena. Vesicle populations are heterogeneous collections of particles. Every vesicle is a unique 'bioreactor' with respect to size and the composition of its inner space. Most of the methods of vesicle formation – with the exclusion of 'guided' ones like those involved in microfluidic methods (Matosevic 2012, van Swaay and deMello 2013) – generate a population of vesicles that is intended to mimic the spontaneous emergence of primitive cells. Only without neglecting this diversity can we have a correct view on (and a possibility to directly test) the possible dynamics related to competition/selection/cooperation behaviour based on differences in individual vesicle reactivity.

7.5 Biotechnological Perspectives of Synthetic Cells

Even if a substantial part of the work commented in this chapter has been carried out aiming at using liposomes as primitive cell models, the presented methods, theories and approaches can be tentatively defined as 'synthetic cell technology'. This technology is essentially based on the convergence of liposome technology and cell-free technology.

We have seen that as a cell-free system, the use of the PURE system (Shimizu et al. 2001) greatly advanced the field, and it is often considered as a synthetic biology standard for synthetic cell construction. The recently introduced 'droplet-transfer' method (Pautot et al. 2003a,b), discussed by Walde et al. (2010) and Stano et al. (2011), is probably the most important recent development in the field, and most of the recent work is based on this method (Noireaux and Libchaber 2004; Pontani et al. 2009; Nishimura et al. 2012; Carrara et al. 2012; Maeda et al. 2012; Hamada et al. 2014).

New studies have revealed routes by which microfluidic technology may enter this field in the near future (Matosevic 2012; Swaay and deMello 2013). Microfluidic methods could pave the way to scenarios where a synthetic cell can be constructed automatically (Martino et al. 2012b), in high-throughput manner, and with a well-defined internal composition. In this way, the physical diversity of liposome populations – commented on in Section 7.4 – could be strongly reduced to favour reproducibility. Consequently, the so-obtained synthetic cells can be used not only as primitive cell models but also for biotechnological applications. We discuss three possible applications of synthetic cell technology in the following sections.

7.5.1 Understanding Biological Mechanisms by Reconstructing Them in Artificial Minimal Systems

This approach could allow a simplification of the study of intracellular phenomena because background interferences – always present in real cells – would be eliminated (Liu and Fletcher 2009). Liposomes containing or expressing one or more proteins could be powerful tools to investigate the functions of intracellular machineries and their dynamic and reciprocal interactions in a cytomimetic environment. Interactions with small molecules (activators, inhibitors, etc.) and especially lipid membranes could also be studied. The possibility of synthesizing membrane proteins in surfactant-free liposomes (from inside) is quite attractive in this respect, especially when a proper orientation is needed (Kuruma et al. 2009; Sachse et al. 2014). Such studies will also benefit from the possibility of constructing liposomes with biphasic membranes such as domains, rafts and their boundaries (Dietrich et al. 2001; Baumgart et al. 2003).

This approach would be very valuable for testing hypotheses about intracellular mechanisms and could be really helpful if systems of increasing complexity could be prepared. The possibility of reconstructing in the lab a simplified model, with a known number and concentration of components, will also guide quantitative modelling. We might call this approach 'understanding by building'.

7.5.2 Synthetic Cell Technology to Build Tools for Research and Biotechnology

Here the focus is not only understanding, but also exploiting. Synthetic cells could be utilised as simplified cell models. For example, the effect of chemicals can be tested in a controlled and simplified cell-like matrix. For example, Zepik et al. (2008) have attempted to use liposomes for a toxicological assessment of xenobiotics. We have recently investigated the function of antioxidants on reactive oxygen species by means of synthetic cell models (Balducci et al. manuscript in preparation). In the field of drug discovery, especially at the beginning of large screening programs, synthetic cells can be used as substitutes of biological cells. For example, synthetic cells bearing protein membrane targets could be available at low costs and would not require specially controlled environments, conditions, and training. Tailor-made systems could be of interest for pharmaceutical companies for testing purposes, with the advantages, compared with natural cells, of being stable, inert, and readily storable.

Liposome display is another use of synthetic cells for biotechnological purposes that has been proposed by the group of Yomo (Fujii et al. 2013, 2014). Liposome display is a novel method for *in vitro* selection and directed evolution of membrane proteins, based on fluorescence-activated cell sorting; it allows the screening of DNA libraries of 10^7 mutants. Up to now it has been applied for the *in vitro* evolution of α-hemolysin from *Staphylococcus aureus*.

Interesting new directions can be found in micromanipulations and immobilization of liposomes for creating 2D 'liposome arrays' (Osaki et al. 2012) or liposomes embedded in gels (Ullrich et al. 2013 and references therein). Other pioneer studies focused on the construction of liposome networks (Jesorka et al. 2011) via lipid tubes or by DNA-mediated binding (Hadorn and Eggenberger Hotz 2010). In the latter case, protein synthesis in a sort of 'tissue' has been obtained (Hadorn et al. 2013). On the other hand, the development of multi-compartment vesicles (MCVs), small vesicles inside large vesicles, also called multi-vesicular vesicles or vesosomes, is also worth mentioning. MCVs could represent a model of superior (organelle-containing) cells that also could allow exploitation of multi-compartmentalization to enable, for example, intra-cellular gradients (Paleos et al. 2011; Chandrawati and Caruso 2012; Peters et al. 2014).

7.5.3 Construction of Synthetic Cells Capable of Interfacing with Natural Cells

This is probably the most fascinating application of synthetic cells, both for basic science and for applications in delivery of drugs, gene, biomarkers, and diagnostic molecules. The nascent arena of the biochemical-based information and communication technologies (biochem-ICTs) offers a new scenario for exploiting synthetic cell modularity, programmability and versatility (Nakano et al. 2011), and we have also proposed a realistic research program aimed at engineering a synthetic cell/natural cell chemical communication based on bacterial quorum sensing mechanisms (Stano et al. 2012; Rampioni et al. 2014). This might find application in nanomedicine, as pointed out by Le Duc et al. (2007), who devised a kind of synthetic cell (called a 'nanofactory') that encapsulates a triggerable metabolic or genetic/metabolic network that can produce a drug *in situ* (Figure 7.11). Injected in the bloodstream, the nanofactory could interact with living cells, by anchoring to receptors

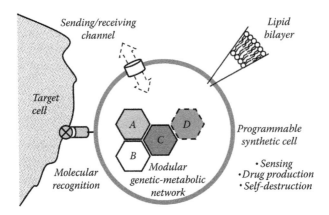

FIGURE 7.11

A potential use of programmable synthetic cells as drug nanofactories. The synthetic cell, injected in the body, might recognise specific target tissues, whereas its internal content consists of a modular genetic/metabolic network capable of detecting an externally present chemical trigger, and work co-ordinately to produce a chemical output, for example a drug. Entrance and release of compounds is achieved by controlling the membrane permeability by means of proteins or by changing lipid composition. (Redrawn on the basis of a published figure from Leduc, P. R. et al., *Nature Nanotechnology*, 2, 3–7, 2007).

presented on their surface, and release/produce a drug after triggering by specific chemical signals emitted by the target cell. Even if this scenario might appear very challenging and perhaps futuristic, the efforts put in attempting its realization will certainly help progress the field and further extend the knowledge and uses of synthetic cells.

7.6 Concluding Remarks

As kindly phrased in a recent review, *'we are much better at taking cells apart than putting them together'* (Liu and Fletcher 2009), and this has been true for most of modern biological investigation.

In the origin of life field, actually this is the only possible way to proceed, namely, constructing models of primitive cells, because we cannot observe how primitive cells are made. But, on the other hand, the same principles and methods developed in past years are going to converge in a powerful constructing methodology (possibly called synthetic cell technology).

While the theory of autopoiesis guides the efforts on the construction of minimal living systems by emphasizing that such systems must be composed of elements that produce each other thanks to reciprocal reactions, the advancements in reconstitution procedures allows the construction of cell-like systems of non-trivial complexity where compartmentalization and vectoriality are also taken into account.

In conclusion, the bottom-up construction of synthetic cells should be certainly considered one of the pillars of synthetic biology (de Lorenzo and Danchin 2008), which is under continuous expansion, and it is difficult not to share the vision that it will become increasingly successful in the coming years – both in theoretical and biotechnological fields. This progress will probably bring about the construction of the first artificial minimal living system constructed by a bottom-up approach (starting from non-living molecules), demonstrating at the same time that life is an emergent property resulting from a quite peculiar way of chemical organization (the autopoietic organization) and that its origin can be fully understood within the realm of physics and chemistry.

Acknowledgements

This work was derived from our recent involvement in studies on the construction of semi-synthetic minimal cells, funded by the Sixth Framework EU Program (SYNTHCELLS: Approaches to the Bioengineering of Synthetic Minimal Cells 043359), HFSP (RGP0033/2007-C), ASI (I/015/07/0), PRIN2008 (2008FY7RJ4); and further expanded thanks to networking initiatives SynBioNT (UK), the COST 'Systems Chemistry' action (CM0703), and the COST 'Emergence and Evolution of Complex Chemical Systems' (CM1304) action.

References

Akanuma, S., T. Kigawa, and S. Yokoyama. 2002. Combinatorial mutagenesis to restrict amino acid usage in an enzyme to a reduced set. *Proceedings of the National Academy of Sciences* 99: 13549–13553.

Anella, F., C. Chiarabelli, D. De Lucrezia, and P. L. Luisi. 2011. Stability studies on random folded RNAs ('Never Born RNAs'), implications for the RNA world. *Chemistry and Biodiversity* 8: 1422–1432.

Angelova, M., and D. Dimitrov. 1986. Liposome electroformation. *Faraday Discussions* 81: 303–311.

Bachmann, P., P. L. Luisi, and J. Lang. 1991a. Self-replicating reverse micelles. *Chimia* 45: 266–268.

Bachmann, P., P. L. Luisi, and J. Lang. 1992. Autocatalytic self-replicating micelles as models for prebiotic structures. *Nature* 357: 57–59.

Bachmann, P., P. Walde, P. L. Luisi, and J. Lang. 1990. Self-replicating reverse micelles and chemical autopoiesis. *Journal of the American Chemical Society* 112: 8200–8201.

Bachmann, P., P. Walde, P. L. Luisi, and J. Lang. 1991b. Self-replicating micelles—Aqueous micelles and enzymatically driven reactions in reverse micelles. *Journal of the American Chemical Society* 113: 8204–8209.

Baumgart, T., S. T. Hess, and W. W. Webb. 2003. Imaging coexisting fluid domains in biomembrane models coupling curvature and line tension. *Nature* 425: 821–824.

Benner, S. A., and A. M. Sismour. 2005. Synthetic biology. *Nature Reviews Genetics* 6: 533–543.

Berclaz, N., E. Blochliger, M. Muller, and P. L. Luisi. 2001a. Matrix effect of vesicle formation as investigated by cryotransmission electron microscopy. *Journal of Physical Chemistry B* 105: 1065–1071.

Berclaz, N., M. Muller, P. Walde, and P. L. Luisi. 2001b. Growth and transformation of vesicles studied by ferritin labeling and cryotransmission electron microscopy. *Journal of Physical Chemistry B* 105: 1056–1064.

Bich, L., and L. Damiano. 2012. Life, autonomy and cognition: An organizational approach to the definition of the universal properties of life. *Origins of Life and Evolution of the Biosphere* 42: 389–397.

Bitbol, M., and P. L. Luisi. 2004. Autopoiesis with or without cognition: Defining life at its edge. *Journal of the Royal Society, Interface* 1: 99–107.

Blochliger, E., M. Blocher, P. Walde, and P. L. Luisi. 1998. Matrix effect in the size distribution of fatty acid vesicles. *Journal of Physical Chemistry B* 102: 10383–10390.

Bolli, M., R. Micura, and A. Eschenmoser. 1997a. Pyranosyl-RNA: Chiroselective self-assembly of base sequences by ligative oligomerization of tetranucleotide-2′,3′-cyclophosphates (with a commentary concerning the origin of biomolecular homochirality). *Chemistry and Biology* 4: 309–320.

Bolli, M., R. Micura, S. Pitsch, and A. Eschenmoser. 1997b. Pyranosyl-RNA: Further observations on replication. *Helvetica Chimica Acta* 80: 1901–1951.

Bolton, C. D., and J. A. D. Wattis. 2003a. The size-templating matrix effect in vesicle formation I: A microscopic model and analysis. *Journal of Physical Chemistry B* 107: 7126–7134.

Bolton, C. D., and J. A. D. Wattis. 2003b. Size-templating matrix effect in vesicle formation. 2. Analysis of a macroscopic model. *Journal of Physical Chemistry B* 107: 14306–14318.

Boukobza, E., A. Sonnenfeld, and G. Haran. 2001. Immobilization in surface-tethered lipid vesicles as a new tool for single biomolecule spectroscopy. *The Journal of Physical Chemistry B* 105: 12165–12170.

Bourgine, P., and J. Stewart. 2004. Autopoiesis and cognition. *Artificial Life* 10: 327–345.

Bozic, B., and S. Svetina. 2004. A relationship between membrane properties forms the basis of a selectivity mechanism for vesicle self-reproduction. *European Biophysics Journal* 33: 565–571.

Bozic, B., and S. Svetina. 2007. Vesicle self-reproduction: The involvement of membrane hydraulic and solute permeabilities. *The European Physical Journal. E, Soft Matter* 24: 79–90.

Bui, H.T., H. Umakoshi, K. X. Ngo, M. Nishida, T. Shimanouchi, and R. Kuboi. 2008. Liposome membrane itself can affect gene expression in the *Escherichia coli* cell-free translation system. *Langmuir* 24: 10537–10542.

Bui, H. T., H. Umakoshi, K. Suga, T. Tanabe, K. X. Ngo, T. Shimanouchi, and R. Kuboi. 2010. Cationic liposome can interfere mRNA translation in an *E. coli* cell-free translation system. *Biochemical Engineering Journal* 52: 38–43.

Calviello, L., P. Stano, F. Mavelli, P. L. Luisi, and R. Marangoni. 2013. Quasi-cellular systems: Stochastic simulation analysis at nanoscale range. *BMC Bioinformatics* 14 (Suppl 7): S7.

Carrara, P., P. Stano, and P. L. Luisi. 2012. Giant cesicles "colonies": A model for primitive cell communities. *Chembiochem* 13: 1497–1502.

Caschera, F., P. Stano, and P. L. Luisi. 2010. Reactivity and fusion between cationic vesicles and fatty acid anionic vesicles. *Journal of Colloid and Interface Science* 345: 561–565.

Caschera, F., T. Sunami, T. Matsuura, H. Suzuki, M. M. Hanczyc, and T. Yomo. 2011. Programmed vesicle fusion triggers gene expression. *Langmuir* 27: 13082–13090.

Cevc, G., and H. Richardsen. 1999. Lipid vesicles and membrane fusion. *Advanced Drug Delivery Reviews* 38: 207–232.

Chakrabarti, A. C., R. R. Breaker, G. F. Joyce, and D. W. Deamer. 1994. Production of RNA by a polymerase protein encapsulated within phospholipid vesicles. *Journal of Molecular Evolution* 39: 555–559.

Chanasakulniyom, M., C. Martino, D. Paterson, L. Horsfall, S. Rosser, and J. M. Cooper. 2012. Expression of membrane-associated proteins within single emulsion cell facsimiles. *The Analyst* 137: 2939–2943.

Chandrawati, R., and F. Caruso. 2012. Biomimetic liposome- and polymersome-based multicompartmentalized assemblies. *Langmuir* 28: 13798–13807.

Chen, I. A., and J. W. Szostak. 2004. A kinetic study of the growth of fatty acid vesicles. *Biophysical Journal* 87: 988–998.

Chiarabelli, C., P. Stano, F. Anella, P. Carrara, and P. L. Luisi. 2012. Approaches to chemical synthetic biology. *FEBS Letters* 586: 2138–2145.

Chiarabelli, C., P. Stano, and P. L. Luisi. 2009. Chemical approaches to synthetic biology. *Current Opinion in Biotechnology* 20: 492–497.

Chiarabelli, C., J. W. Vrijbloed, D. De Lucrezia, R. M. Thomas, P. Stano, F. Polticelli, T. Ottone, E, et al. 2006a. Investigation of de novo totally random biosequences, part II: On the folding frequency in a totally random library of de novo proteins obtained by phage display. *Chemistry and Biodiversity* 3: 840–859.

Chiarabelli, C., J. W. Vrijbloed, R. M. Thomas, and P. L. Luisi. 2006b. Investigation of de novo totally random biosequences, part I: A general method for in vitro selection of folded domains from a random polypeptide library displayed on phage. *Chemistry and Biodiversity* 3: 827–839.

Choi, H.-J., and C. D. Montemagno. 2005. Artificial organelle: ATP synthesis from cellular mimetic polymersomes. *Nano Letters* 5: 2538–2542.

Coveney, P. V., and J. A. D. Wattis. 1998. Becker-Doring model of self-reproducing vesicles. *Journal of the Chemical Society-Faraday Transactions* 94: 233–246.

Damiano, L., and P. L. Luisi. 2010. Towards an autopoietic redefinition of life. *Origins of Life and Evolution of the Biosphere* 40: 145–149.

Deamer, D. 1985. Boundary structures are formed by organic-components of the murchison carbonaceous chondrite. *Nature* 317: 792–794.

DeClue, M. S., Monnard, P. A., Bailey, J. A., Maurer, S. E., Collis, G. E., Ziock, H. J., Rasmussen, S, et al. 2009. Nucleobase mediated, photocatalytic vesicle formation from an ester precursor. *Journal of American Chemical Society* 131: 931–933.

de Lorenzo, V., and A. Danchin. 2008. Synthetic biology: Discovering new worlds and new words. *EMBO Reports* 9: 822–827.

De Lucrezia, D., M. Franchi, C. Chiarabelli, E. Gallori, and P. L. Luisi. 2006a. Investigation of de novo totally random biosequences. Part III. RNA Foster: A novel assay to investigate RNA folding structural properties. *Chemistry and Biodiversity* 3: 860–868.

De Lucrezia, D., M. Franchi, C. Chiarabelli, E. Gallori, and P. L. Luisi. 2006b. Investigation of de novo totally random biosequences. Part IV. Folding properties of de novo, totally random RNAs. *Chemistry and Biodiversity* 3: 869–877.

De Souza, T. P., P. Stano, F. Steiniger, E. D'Aguanno, E. Altamura, A. Fahr, and P. L. Luisi. 2012. Encapsulation of ferritin, ribosomes, and ribo-peptidic complexes inside liposomes: Insights into the origin of metabolism. *Origins of Life and Evolution of the Biosphere* 42: 421–428.

Dietrich, C., L. A. Bagatolli, Z. N. Volovyk, N. L. Thompson, M. Levi, K. Jacobson, and E. Gratton. 2001. Lipid rafts reconstituted in model membranes. *Biophysical Journal* 80: 1417–1428.

Doi, N., K. Kakukawa, Y. Oishi, and H. Yanagawa. 2005. High solubility of random-sequence proteins consisting of five kinds of primitive amino acids. *Protein Engineering Design and Selection* 18: 279–284.

Domazou, A. S., and P. L. Luisi. 2002. Size distribution of spontaneously formed liposomes by the alcohol injection method. *Journal of Liposome Research* 12: 205–220.

Dong, H., L. Nilsson, and C. G. Kurland. 1996. Co-variation of tRNA abundance and codon usage in *Escherichia coli* at different growth rates. *Journal of Molecular Biology* 260: 649–663.

Duzgunes, N., J. Goldstein, D. Friend, and P. Felgner. 1989. Fusion of liposomes containing a novel cationic lipid, N-[2,3-(dioleyloxy)propyl]-N,N,N-trimethylammonium—Induction by multivalent anions and asymmetric fusion with acidic phospholipid-vesicles. *Biochemistry* 28: 9179–9184.

Endy, D. 2005. Foundations for engineering biology. *Nature* 438: 449–453.

Fanelli, D., and A. J. McKane. 2008. Thermodynamics of vesicle growth and instability. *Physical Review. E, Statistical, Nonlinear, and Soft Matter Physics* 78: 051406.

Fehér, T., B. Papp, C. Pal, and G. Pósfai. 2007. Systematic genome reductions: Theoretical and experimental approaches. *Chemical Reviews* 107: 3498–3513.

Fiordemondo, D., and P. Stano. 2007. Lecithin-based water-in-oil compartments as dividing bioreactors. *Chembiochem* 8: 1965–1973.

Fischer, A., A. Franco, and T. Oberholzer. 2002. Giant vesicles as microreactors for enzymatic mRNA synthesis. *Chembiochem* 3: 409–417.

Forster, A. C., and G. M. Church. 2006. Towards synthesis of a minimal cell. *Molecular Systems Biology* 2: 45.

Fraser, C., J. Gocayne, O. White, M. Adams, R. Clayton, R. Fleischmann, C. Bult, et al. 1995. The minimal gene complement of mycoplasma-genitalium. *Science* 270: 397–403.

Fujii, S., T. Matsuura, T. Sunami, T. Nishikawa, Y. Kazuta, and T. Yomo. 2014. Liposome display for in vitro selection and evolution of membrane proteins. *Nature Protocols* 9: 1578–1591.

Fujii, S., T. Matsuura, T. Sunami, Y. Kazuta, and T. Yomo. 2013. In vitro evolution of α-hemolysin using a liposome display. *Proceedings of the National Academy of Sciences USA* 110: 16796–16801.

Gánti, T. 1975. Organization of chemical reactions into dividing and metabolizing units: The chemotons. *Biosystems* 7: 15–21.

Gánti, T. 2003. *The Principles of Life*. New York, NY: Oxford University Press.

Gil, R., F. J. Silva, J. Peretó, and A. Moya. 2004. Determination of the core of a minimal bacterial gene set. *Microbiology and Molecular Biology Reviews* 68: 518–537.

Gilbert, W. 1986. Origin of life—The RNA world. *Nature* 319: 618.

Guiloff, G. D. 1981. Autopoiesis and neobiogenesis. In *Autopoiesis. A Theory of Living Organization*, ed. M. Zeleny, 118–124. New York, NY: North Holland.

Hadorn, M., E. Boenzli, K. T. Sørensen, D. De Lucrezia, M. M. Hanczyc, and T. Yomo. 2013. Defined DNA-mediated assemblies of gene-expressing giant unilamellar vesicles. *Langmuir* 29: 15309–15319.

Hadorn, M., and P. Eggenberger Hotz. 2010. DNA-mediated self-assembly of artificial vesicles. *PLoS One* 5: e9886.

Haines, T. 1983. Anionic lipid headgroups as a proton-conducting pathway along the surface of membranes—A hypothesis. *Proceedings of the National Academy of Sciences of the United States of America* 80: 160–164.

Hamada, S., M. Tabuchi, T. Toyota, T. Sakurai, T. Hosoi, T. Nomato, K. Nakatani, et al. 2014. Giant vesicles functionally expressing membrane receptors for an insect pheromone. *Chemical Communications* 50: 2958–2961.

Henry, C. S., R. Overbeek, and R. L. Stevens. 2010. Building the blueprint of life. *Biotechnology Journal* 5: 695–704.

Hernández-Zapata, E., L. Martínez-Balbuena, and I. Santamaría-Holek. 2009. Thermodynamics and dynamics of the formation of spherical lipid vesicles. *Journal of Biological Physics* 35: 297–308.

Hosoda, K., T. Sunami, Y. Kazuta, T. Matsuura, H. Suzuki, and T. Yomo. 2008. Quantitative study of the structure of multilamellar giant liposomes as a container of protein synthesis reaction. *Langmuir* 24: 13540–13548.

Ishikawa, K., K. Sato, Y. Shima, I. Urabe, and T. Yomo. 2004. Expression of a cascading genetic network within liposomes. *FEBS Letters* 576: 387–390.

Islas, S., A. Becerra, P. L. Luisi, and A. Lazcano. 2004. Comparative genomics and the gene complement of a minimal cell. *Origins of Life and Evolution of the Biosphere* 34: 243–256.

Jesorka, A., N. Stepanyants, H. Zhang, B. Ortmen, B. Hakonen, and O. Orwar. 2011. Generation of phospholipid vesicle-nanotube networks and transport of molecules therein. *Nature Protocols* 6: 791–805.

Jewett, M. C., Fritz, B. R., Timmerman, L. E., Church, G. M. 2013. In vitro integration of ribosomal RNA synthesis, ribosome assembly, and translation. *Molecular Systems Biology* 9: 678.

Kajander, E. O., and N. Ciftçioglu. 1998. Nanobacteria: An alternative mechanism for pathogenic intra- and extracellular calcification and stone formation. *Proceedings of the National Academy of Sciences of the United States of America* 95: 8274–8279.

Kaneda, M., S. M. Nomura, S. Ichinose, S. Kondo, K. Nakahama, K. Akiyoshi, and I. Morita. 2009. Direct formation of proteo-liposomes by in vitro synthesis and cellular cytosolic delivery with connexin-expressing liposomes. *Biomaterials* 30: 3971–3977.

Karzbrun, E., J. Shin, R. H. Bar-Ziv, and V. Noireaux. 2011. Coarse-grained dynamics of protein synthesis in a cell-free system. *Physical Review Letters* 106: 048104.

Kato, A., M. Yanagisawa, Y. T. Sato, K. Fujiwara, and K. Yoshikawa. 2012. Cell-sized confinement in microspheres accelerates the reaction of gene expression. *Scientific Reports* 2: 283.

Kita, H., T. Matsuura, T. Sunami, K. Hosoda, N. Ichihashi, K. Tsukada, I. Urabe, et al. 2008. Replication of genetic information with self-encoded replicase in liposomes. *Chembiochem* 9: 2403–2410.

Klammt, C., F. Lohr, B. Schafer, W. Haase, V. Dotsch, H. Ruterjans, C. Glaubitz, et al. 2004. High level cell-free expression and specific labeling of integral membrane proteins. *European Journal of Biochemistry* 271: 568–580.

Kobori, S., N. Ichihashi, Y. Kazuta, and T. Yomo. 2013. A controllable gene expression system in liposomes that includes a positive feedback loop. *Molecular Biosystems* 9: 1282–1285.

Kurihara, K., M. Tamura, K.-I. Shohda, T. Toyota, K. Suzuki, and T. Sugawara. 2011. Self-reproduction of supramolecular giant vesicles combined with the amplification of encapsulated DNA. *Nature Chemistry* 3: 775–781.

Kuruma, Y., P. Stano, T. Ueda, and P. L. Luisi. 2009. A synthetic biology approach to the construction of membrane proteins in semi-synthetic minimal cells. *Biochimica et Biophysica Acta* 1788: 567–574.

Kuruma, Y., T. Suzuki, S. Ono, M. Yoshida, and T. Ueda. 2012. Functional analysis of membranous Fo-a subunit of F1Fo-ATP synthase by in vitro protein synthesis. *The Biochemical Journal* 442: 631–638.

Lasic, D. D. 1988. The mechanism of vesicle formation. *The Biochemical Journal* 256: 1–11.

Lazzerini-Ospri, L., P. Stano, P. L. Luisi, and R. Marangoni. 2012. Characterization of the emergent properties of a synthetic quasi-cellular system. *BMC Bioinformatics* 13 (Suppl 4): S9.

Leduc, P. R., M. S. Wong, P. M. Ferreira, R. E. Groff, K. Haslinger, M. P. Koonce, W. Y. Lee, et al. 2007. Towards an in vivo biologically inspired nanofactory. *Nature Nanotechnology* 2: 3–7.

Leduc, S. 1912. *La Biologie Synthétique*. Paris: A. Poinat.

Lee, S., H. Koo, J. H. Na, K. E. Lee, S. Y. Jeong, K. Choi, S. H. Kim, et al. 2006. Organizational invariance and metabolic closure: Analysis in terms of (M,R) systems. *Journal of Theoretical Biology* 238: 949–961.

Lee, S., H. Koo, J. H. Na, K. E. Lee, S. Y. Jeong, K. Choi, S. H. Kim, et al. 2014. DNA amplification in neutral liposomes for safe and efficient gene delivery. *ACS Nano* 8: 4257–4267.

Liu, A. P., and D. A. Fletcher. 2009. Biology under construction: In vitro reconstitution of cellular function. *Nature Reviews Molecular Cell Biology* 10: 644–650.

Lonchin, S., P. L. Luisi, P. Walde, and B. H. Robinson. 1999. A matrix effect in mixed phospholipid/fatty acid vesicle formation. *Journal of Physical Chemistry B* 103: 10910–10916.

Luci, P. 2003. Gene cloning, expression and purification of membrane proteins. PhD Thesis Nr. 15108. Zurich: Swiss Federal Institute of Technology (ETH).

Luisi, P. L. 1985. Enzymes hosted in reverse micelles in hydrocarbon solution. *Angewandte Chemie-International Edition in English* 24: 439–450.

Luisi, P. L. 1998. About various definitions of life. *Origins of Life and Evolution of the Biosphere* 28: 613–622.

Luisi, P. L. 2002. Toward the engineering of minimal living cells. *The Anatomical Record* 268: 208–214.

Luisi P. L. 2003. Autopoiesis: A review and a reappraisal. *Die Naturwissenschaften* 90: 49–59.

Luisi, P. L. 2006. *The Emergence of Life: From Chemical Origins to Synthetic Biology*. Cambridge: Cambridge University Press.

Luisi, P. L. 2007. Chemical aspects of synthetic biology. *Chemistry and Biodiversity* 4: 603–621.

Luisi, P. L. 2011. The synthetic approach in biology: Epistemological notes for synthetic biology. In *Chemical Synthetic Biology*, ed. P. L. Luisi, and C. Chiarabelli, 343–362. Chichester: Wiley.

Luisi P. L. 2012. An open question on the origin of life: The first forms of metabolism. *Chemistry and Biodiversity* 9: 2635–2647.

Luisi, P. L., M. Allegretti, T. Souza, F. Steiniger, A. Fahr, and P. Stano. 2010. Spontaneous protein crowding in liposomes: A new vista for the origin of cellular metabolism. *Chembiochem* 11: 1989–1992.

Luisi, P. L., and C. Chiarabelli, eds. 2011. *Chemical Synthetic Biology*. Chichester: Wiley.

Luisi, P. L., F. Ferri, and P. Stano. 2006. Approaches to semi-synthetic minimal cells: A review. *Die Naturwissenschaften* 93: 1–13.

Luisi, P. L., M. Giomini, M. Pileni, and B. Robinson. 1988. Reverse micelles as hosts for proteins and small molecules. *Biochimica et Biophysica Acta* 947: 209–246.

Luisi, P. L., and P. Stano, eds. 2010. *The Minimal Cell: The Biophysics of Cell Compartment and the Origin of Cell Functionality*. New York, NY: Springer.

Luisi, P. L., and F. J. Varela. 1989. Self-replicating micelles—A chemical version of a minimal autopoietic system. *Origins of Life and Evolution of the Biosphere* 19: 633–643.

Lynch, S. A., and J. P. Gallivan. 2009. A flow cytometry-based screen for synthetic riboswitches. *Nucleic Acids Research* 37: 184–192.

Maeda, Y. T., T. Nakadai, J. Shin, K. Uryu, V. Noireaux, and A. Libchaber. 2012. Assembly of MreB filaments on liposome membranes: A synthetic biology approach. *ACS Synthetic Biology* 1: 53–59.

Markvoort, A. J., N. Pfleger, R. Staffhorst, P. A. J. Hilbers, R. A. van Santen, J. A. Killian, and B. de Kruijff. 2010. Self-reproduction of fatty acid vesicles: A combined experimental and simulation study. *Biophysical Journal* 99: 1520–1528.

Markvoort, A. J., A. F. Smeijers, K. Pieterse, R. A. van Santen, and P. A. J. Hilbers. 2007. Lipid-based mechanisms for vesicle fission. *The Journal of Physical Chemistry B* 111: 5719–5725.

Martini, L., and S. S. Mansy. 2011. Cell-like systems with riboswitch controlled gene expression. *Chemical Communications* 47: 10734–10736.

Martino, C., L. Horsfall, Y. Chen, M. Chanasakulniyom, D. Paterson, A. Brunet, S. Rosser, et al. 2012a. Cytoskeletal protein expression and its association within the hydrophobic membrane of artificial cell models. *Chembiochem* 13: 792–795.

Martino, C., S.-H. Kim, L. Horsfall, A. Abbaspourrad, S. J. Rosser, J. Cooper, and D. A. Weitz. 2012b. Protein expression, aggregation, and triggered release from polymersomes as artificial cell-like structures. *Angewandte Chemie-International Edition* 51: 6416–6420.

Matosevic, S. 2012. Synthesizing artificial cells from giant unilamellar vesicles: State-of-the art in the development of microfluidic technology. *BioEssays* 34: 992–1001.

Matsubayashi, H., Y. Kuruma, and T. Ueda. 2014. In vitro synthesis of the *E. coli* Sec translocon from DNA. *Angewandte Chemie International Edition*. DOI: 10.1002/anie.201403929.

Matsuura, T., Y. Kazuta, T. Aita, J. Adachi, and T. Yomo. 2009. Quantifying epistatic interactions among the components constituting the protein translation system. *Molecular Systems Biology* 5: 297.

Maturana, H. R., and F. J. Varela. 1972. *De Máquinas y Seres Vivos*. Santiago: Editorial Universitaria.

Maturana, H. R., and F. J. Varela. 1980. *Autopoiesis and Cognition: The Realization of the Living*. D. Reidel Publishing Company.

Maturana, H. R., and F. J. Varela. 1992. *Tree of Knowledge*. Rev Sub. Shambhala.

Mavelli, F. 2012. Stochastic simulations of minimal cells: The ribocell model. *BMC Bioinformatics* 13 (Suppl. 4): S10.

Mavelli, F., and K. Ruiz-Mirazo. 2007. Stochastic simulations of minimal self-reproducing cellular systems. *Philosophical Transactions of the Royal Society of London B* 362: 1789–1802.

Mavelli, F., and K. Ruiz-Mirazo. 2010. ENVIRONMENT: A computational platform to stochastically simulate reacting and self-reproducing lipid compartments. *Physical Biology* 7: 036002.

Mavelli, F., and K. Ruiz-Mirazo. 2013. Theoretical conditions for the stationary reproduction of model protocells. *Integrative Biology* 5: 324–341.

Mavelli, F., and P. Stano. 2010. Kinetic models for autopoietic chemical systems: The role of fluctuations in a homeostatic regime. *Physical Biology* 7: 16010.

Monnard, P.-A., and D. W. Deamer. 2002. Membrane self-assembly processes: Steps toward the first cellular life. *The Anatomical Record* 268: 196–207.

Monnard, P.-A., A. Luptak, and D.W. Deamer. 2007. Models of primitive cellular life: Polymerases and templates in liposomes. *Philosophical Transactions of the Royal Society of London B* 362: 1741–1750.

Moritani, Y., S. M. Nomura, I. Morita, and K. Akiyoshi. 2010. Direct integration of cell-free-synthesized connexin-43 into liposomes and hemichannel formation. *The FEBS Journal* 277: 3343–3352.

Moya, A., R. Gil, A. Latorre, J. Peretó, M. Pilar Garcillán-Barcia, and F. de la Cruz. 2009. Toward minimal bacterial cells: Evolution vs. design. *FEMS Microbiology Reviews* 33: 225–235.

Murtas, G. 2010. Internal lipid synthesis and vesicle growth as a step toward self-reproduction of the minimal cell. *Systems and Synthetic Biology* 4: 85–93.

Murtas, G., Y. Kuruma, P. Bianchini, A. Diaspro, and P. L. Luisi. 2007. Protein synthesis in liposomes with a minimal set of enzymes. *Biochemical and Biophysical Research Communications* 363: 12–17.

Mushegian, A. R., and E. V. Koonin. 1996. A minimal gene set for cellular life derived by comparison of complete bacterial genomes. *Proceedings of the National Academy of Sciences of the United States of America* 93: 10268–10273.

Nakano, T., M. Moore, A. Enomoto, and T. Suda. 2011. Molecular communication technology as a biological ICT. In *Biological Functions for Information and Communication Technologies*, ed. H. Sawai, 49–86. Springer Berlin Heidelberg.

Nierhaus, K. H. 1990. Reconstruction of ribosomes. In *Ribosomes and Protein Synthesis: A Practical Approach*, ed. G. Spedding, 161–189. Oxford, UK: IRL Press at Oxford University Press.

Nishimura, K., T. Hosoi, T. Sunami, T. Toyota, M. Fujinami, K. Oguma, T. Matsuura, et al. 2009. Population analysis of structural properties of giant liposomes by flow cytometry. *Langmuir* 25: 10439–10443.

Nishimura, K., H. Suzuki, T. Toyota, and T. Yomo. 2012. Size control of giant unilamellar vesicles prepared from inverted emulsion droplets. *Journal of Colloid and Interface Science* 376: 119–125.

Nissen, P., J. Hansen, N. Ban, P. B. Moore, and T. A. Steitz. 2000. The structural basis of ribosome activity in peptide bond synthesis. *Science* 289: 920–930.

Noireaux, V., and A. Libchaber. 2004. A vesicle bioreactor as a step toward an artificial cell assembly. *Proceedings of the National Academy of Sciences of the United States of America* 101: 17669–17674.

Nomura, S., K. Tsumoto, T. Hamada, K. Akiyoshi, Y. Nakatani, and K. Yoshikawa. 2003. Gene expression within cell-sized lipid vesicles. *Chembiochem* 4: 1172–1175.

Nourian, Z., W. Roelofsen, and C. Danelon. 2012. Triggered gene expression in fed-vesicle microreactors with a multifunctional membrane. *Angewandte Chemie (International Ed. in English)* 51: 3114–3118.

Nozawa, A., T. Ogasawara, S. Matsunaga, T. Iwasaki, T. Sawasaki, and Y. Endo. 2011. Production and partial purification of membrane proteins using a liposome-supplemented wheat cell-free translation system. *BMC Biotechnology* 11: 35.

Oberholzer, T., M. Albrizio, and P. L. Luisi. 1995a. Polymerase chain reaction in liposomes. *Chemistry and Biology* 2: 677–682.

Oberholzer, T., and P. L. Luisi. 2002. The use of liposomes for constructing cell models. *Journal of Biological Physics* 28: 733–744.

Oberholzer, T., K. H. Nierhaus, and P. L. Luisi. 1999. Protein expression in liposomes. *Biochemical and Biophysical Research Communications* 261: 238–241.

Oberholzer, T., R. Wick, P. L. Luisi, and C. K. Biebricher. 1995b. Enzymatic RNA replication in self-reproducing vesicles: An approach to a minimal cell. *Biochemical and Biophysical Research Communications* 207: 250–257.

Osaki, T., K. Kamiya, R. Kawano, H. Sasaki, and S. Takeuchi. 2012. Towards artificial cell array system: Encapsulation and hydration technologies integrated in liposome array. In *2012 IEEE 25th International Conference on Micro Electro Mechanical Systems (MEMS)*, 333–336.

Paleos, C. M., D. Tsiourvas, and Z. Sideratou. 2011. Interaction of vesicles: Adhesion, fusion and multicompartment systems formation. *Chembiochem* 12: 510–521.

Pantazatos, D. P., and R. C. MacDonald. 1999. Directly observed membrane fusion between oppositely charged phospholipid bilayers. *Journal of Membrane Biology* 170: 27–38.

Pantazatos, D. P., S. P. Pantazatos, and R. C. MacDonald. 2003. Bilayer mixing, fusion, and lysis following the interaction of populations of cationic and anionic phospholipid bilayer vesicles. *Journal of Membrane Biology* 194: 129–139.

Pautot, S., B. J. Frisken, and D. A. Weitz. 2003a. Production of unilamellar vesicles using an inverted emulsion. *Langmuir* 19: 2870–2879.

Pautot, S., B. J. Frisken, and D. A. Weitz. 2003b. Engineering asymmetric vesicles. *Proceedings of the National Academy of Sciences of the United States of America* 100: 10718–10721.

Peters, R. J. R. W., M. Marguet, S. Marais, M. W. Fraaije, J. C. M. van Hest, and S. Lecommandoux. 2014. Cascade reactions in multicompartmentalized polymersomes. *Angewandte Chemie International Edition* 53: 146–150.

Pitard, B., P. Richard, M. Duñach, G. Girault, and J. L. Rigaud. 1996. ATP synthesis by the F0F1 ATP synthase from Thermophilic bacillus PS3 reconstituted into liposomes with bacteriorhodopsin. 1. Factors defining the optimal reconstitution of ATP synthases with bacteriorhodopsin. *European Journal of Biochemistry* 235: 769–778.

Pontani, L.-L., J. van der Gucht, G. Salbreux, J. Heuvingh, J.-F. Joanny, and C. Sykes. 2009. Reconstitution of an actin cortex inside a liposome. *Biophysical Journal* 96: 192–198.

Rampioni, G., F. Mavelli, L. Damiano, F. D'Angelo, M. Messina, L. Leoni, and P. Stano. 2014. A synthetic biology approach to bio-chem-ICT: First moves towards chemical communication between synthetic and natural cells. *Natural Computing.* DOI: 10.1007/s11047-014-9425-x.

Rosen, R. 1958a. A relational theory of biological systems. *The Bulletin of Mathematical Biophysics.* 20: 245–341.

Rosen, R. 1958b. The representation of biological systems from standpoint of the theory of categories. *The Bulletin of Mathematical Biophysics.* 20: 317–341.

Souza, T., P. Stano, and P. L. Luisi. 2009. The minimal size of liposome-based model cells brings about a remarkably enhanced entrapment and protein synthesis. *Chembiochem* 10: 1056–1063.

Souza, T., F. Steiniger, P. Stano, A. Fahr, and P. L. Luisi. 2011. Spontaneous crowding of ribosomes and proteins inside vesicles: A possible mechanism for the origin of cell metabolism. *Chembiochem* 12: 2325–2330.

Rasi, S., F. Mavelli, and P. L. Luisi. 2003. Cooperative micelle binding and matrix effect in oleate vesicle formation. *Journal of Physical Chemistry B* 107: 14068–14076.

Richard, P., B. Pitard, and J.-L. Rigaud. 1995. ATP synthesis by the F0F1-ATPase from the Thermophilic bacillus PS3 co-reconstituted with bacteriorhodopsin into liposomes. Evidence for stimulation of ATP synthesis by ATP bound to a noncatalytic binding site. *Journal of Biological Chemistry* 270: 21571–21578.

Ruiz-Mirazo, K., and F. Mavelli. 2008. On the way towards 'basic autonomous agents': Stochastic simulations of minimal lipid-peptide cells. *BioSystems* 91: 374–387.

Ruiz-Mirazo, K., G. Piedrafita, F. Ciriaco, and F. Mavelli. 2011. Stochastic simulations of mixed-lipid compartments: From self-assembling vesicles to self-producing protocells. *Advances in Experimental Medicine and Biology* 696: 689–696.

Rushdi, A. I., and B. R. T. Simoneit. 2001. Lipid formation by aqueous fischer-tropsch-type synthesis over a temperature range of 100 to 400 degrees C. *Origins of Life and Evolution of the Biosphere* 31: 103–118.

Sachse, R., Dondapati, S. K., Fenz, S. F., Schmidt, T., and S. Kubick. 2014. Membrane protein synthesis in cell-free systems: From bio-mimetic systems to bio-membranes. *FEBS Letters.* DOI: 10.1016/j.febslet.2014.06.007.

Saito, H., Y. Kato, M. Le Berre, A. Yamada, T. Inoue, K. Yosikawa, and D. Baigl. 2009. Time-resolved tracking of a minimum gene expression system reconstituted in giant liposomes. *Chembiochem* 10: 1640–1643.

Salje, J., F. van den Ent, P. de Boer, and J. Löwe. 2011. Direct membrane binding by bacterial actin MreB. *Molecular Cell* 43: 478–487.

Schmidli, P., P. Schurtenberger, and P. Luisi. 1991. Liposome-mediated enzymatic-synthesis of phosphatidylcholine as an approach to self-replicating liposomes. *Journal of the American Chemical Society* 113: 8127–8130.

Sessa, G., and G. Weissmann. 1970. Incorporation of lysozyme into liposomes. A model for structure-linked latency. *The Journal of Biological Chemistry* 245: 3295–3301.

Shimizu, Y, A. Inoue, Y. Tomari, T. Suzuki, T. Yokogawa, K. Nishikawa, and T. Ueda. 2001. Cell-free translation reconstituted with purified components. *Nature Biotechnology* 19: 751–755.

Shimizu, Y., T. Kanamori, and T. Ueda. 2005. Protein synthesis by pure translation systems. *Methods* 36: 299–304.

Shimono, K., M. Goto, T. Kikukawa, S. Miyauchi, M. Shirouzu, N. Kamo, and S. Yokoyama. 2009. Production of functional bacteriorhodopsin by an *Escherichia coli* cell-free protein synthesis system supplemented with steroid detergent and lipid. *Protein Science* 18: 2160–2171.

Shin, J., and V. Noireaux. 2012. An *E. coli* cell-free expression toolbox: Application to synthetic gene circuits and artificial cells. *ACS Synthetic Biology* 1: 29–41.

Shohda, K., and T. Sugawara. 2006. DNA polymerization on the inner surface of a giant liposome for synthesizing an artificial cell model. *Soft Matter* 2: 402–408.

Shohda, K., M. Tamura, Y. Kageyama, K. Suzuki, A. Suyama, and T. Sugawara. 2011. Compartment size dependence of performance of polymerase chain reaction inside giant vesicles. *Soft Matter* 7: 3750–3753.

Simoneit, B. R. T. 2004. Prebiotic organic synthesis under hydrothermal conditions: An overview. In *Space Life Sciences: Steps Toward Origin(s) of Life*, eds. M. P. Bernstein, M. Kress, and R. NavarroGonzalez, 33:88–94. Kidlington: Pergamon-Elsevier Science Ltd.

Singer, S. J., and A. F. Schick. 1961. The properties of specific stains for electron microscopy prepared by the conjugation of antibody molecules with ferritin. *Journal of Biophysical and Biochemical Cytology* 9: 519–537.

Solé, R. V. 2009. Evolution and self-assembly of protocells. *The International Journal of Biochemistry and Cell Biology* 41: 274–284.

Stano, P. 2007. Question 7: New aspects of interactions among vesicles. *Origins of Life and Evolution of the Biosphere* 37: 439–444.

Stano, P., S. Bufali, A. S. Domazou, and P. L. Luisi. 2005. Effect of tryptophan oligopeptides on the size distribution of POPC liposomes: A dynamic light scattering and turbidimetric study. *Journal of Liposome Research* 15: 29–47.

Stano, P., P. Carrara, Y. Kuruma, T. Souza, and P. L. Luisi. 2011. Compartmentalized reactions as a case of soft-matter biotechnology: Synthesis of proteins and nucleic acids inside lipid vesicles. *Journal of Materials Chemistry* 21: 18887–18902.

Stano, P., E. D'Aguanno, J. Bolz, A. Fahr, and P. L. Luisi. 2013. A remarkable self-organization process as the origin of primitive functional cells. *Angewandte Chemie International Edition* 52: 13397–13400.

Stano, P., and P. L. Luisi. 2010. Achievements and open questions in the self-reproduction of vesicles and synthetic minimal cells. *Chemical Communications* 46: 3639–3653.

Stano, P., and P. L. Luisi. 2013. Semi-synthetic minimal cells: Origin and recent developments. *Current Opinion in Biotechnology*. DOI: 10.1016/j.copbio.2013.01.002.

Stano, P., G. Rampioni, P. Carrara, L. Damiano, L. Leoni, and P. L. Luisi. 2012. Semi-synthetic minimal cells as a tool for biochemical ICT. *Biosystems* 109: 24–34.

Stano, P., T. P. Souza, P. Carrara, E. Altamura, E. D'Aguanno, M. Caputo, P. L. Luisi, et al. 2014. Recent biophysical issues about the preparation of solute-filled lipid vesicles. *Mechanics of Advanced Materials and Structures*. DOI: 10.1080/15376494.2013.857743.

Stano, P., E. Wehrli, and P. L. Luisi. 2006. Insights into the self-reproduction of oleate vesicles. *Journal of Physics: Condensed Matter* 18: S2231.

Steering Group for the Workshop on Size Limits of Very Small Microorganisms, National Research Council. 1999. *Size Limits of Very Small Microorganisms: Proceedings of a Workshop*. Washington, DC: The National Academies Press.

Stögbauer, T., L. Windhager, R. Zimmer, and J. O. Rädler. 2012. Experiment and mathematical modeling of gene expression dynamics in a cell-free system. *Integrative Biology* 4: 494–501.

Suga, K., T. Tanabe, and H. Umakoshi. 2013. Heterogeneous cationic liposomes modified with 3β-{N-[(N',N'-dimethylamino)ethyl]carbamoyl}cholesterol can induce partial conformational changes in messenger RNA and regulate translation in an *Escherichia coli* cell-free translation system. *Langmuir* 29: 1899–1907.

Sunami, T., F. Caschera, Y. Morita, T. Toyota, K. Nishimura, T. Matsuura, H. Suzuki, et al. 2010a. Detection of association and fusion of giant vesicles using a fluorescence-activated cell sorter. *Langmuir* 26: 15098–15103.

Sunami, T., K. Hosoda, H. Suzuki, T. Matsuura, and T. Yomo. 2010b. Cellular compartment model for exploring the effect of the lipidic membrane on the kinetics of encapsulated biochemical reactions. *Langmuir* 26: 8544–8551.

Sunami, T., K. Sato, T. Matsuura, K. Tsukada, I. Urabe, and T. Yomo. 2006. Femtoliter compartment in liposomes for in vitro selection of proteins. *Analytical Biochemistry* 357: 128–136.

Swaay, D. van, and A. deMello. 2013. Microfluidic methods for forming liposomes. *Lab on a Chip* 13: 752–767.

Szostak, J. W. 2012. Attempts to define life do not help to understand the origin of life. *Journal of Biomolecular Structure and Dynamics*. 24: 599–600.

Szostak, J. W., D. P. Bartel, and P. L. Luisi. 2001. Synthesizing life. *Nature* 409: 387–390.

Takahashi, H., Y. Kageyama, K. Kurihara, K. Takakura, S. Murata, and T. Sugawara. 2010. Autocatalytic membrane-amplification on a pre-existing vesicular surface. *Chemical Communications* 46: 8791–8793.

Takakura, K., T. Toyota, and T. Sugawara. 2003. A novel system of self-reproducing giant vesicles. *Journal of the American Chemical Society* 125: 8134–8140.

Thomas, R. M., J. W. Vrijbloed, and P. L. Luisi. 2001. Towards random polypeptide synthesis. *Chimia* 55: 114–118.

Torino, D., C. Del Bianco, L.A. Ross, J. L. Ong, and S. S. Mansy. 2011. Intravesicle isothermal DNA replication. *BMC Research Notes* 4: 128.

Toyota, T., K. Takakura, Y. Kageyama, K. Kurihara, N. Maru, K. Ohnuma, K. Kaneko, et al. 2008. Population study of sizes and components of self-reproducing giant multilamellar vesicles. *Langmuir* 24: 3037–3044.

Tsumoto, K., K. Kamiya, and T. Yoshimura. 2006. Membrane fusion between a giant vesicle and small enveloped particles: Possibilities for the application to construct model cells. In *Proceedings of the International Symposium on Micro-NanoMechatronics and Human Science 2006*, 1–6.

Tsumoto, K., S. M. Nomura, Y. Nakatani, and K. Yoshikawa. 2001. Giant liposome as a biochemical reactor: Transcription of DNA and transportation by laser tweezers. *Langmuir* 17: 7225–7228.

Ullrich, M., J. Hanuš, J. Dohnal, F. Štěpánek. 2013. Encapsulation stability and temperature-dependent release kinetics from hydrogel-immobilised liposomes. *Journal of Colloid and Interface Sciences* 394: 380–385.

Umakoshi, H., K. Suga, H. T. Bui, M. Nishida, T. Shimanouchi, and R. Kuboi. 2009. Charged liposome affects the translation and folding steps of in vitro expression of green fluorescent protein. *Journal of Bioscience and Bioengineering* 108: 450–454.

Uwins, P. J. R., R. I. Webb, and A. P. Taylor. 1998. Novel nano-organisms from Australian sandstones. *American Mineralogist* 83: 1541–1550.

van Hoof, B., A. J. Markvoort, R. A. van Santen, and P. A. J. Hilbers. 2012. On protein crowding and bilayer bulging in spontaneous vesicle formation. *The Journal of Physical Chemistry. B* 116: 12677–12683.

van Swaay, D. and A. deMello. 2013. Microfluidic methods for forming liposomes. *Lab on a chip*. 13: 752–767.

Varela, F. J. 2000. *El Fenomeno de la Vida*. Dolmen.

Varela, F. J., H. R. Maturana, and R. Uribe. 1974. Autopoiesis: The organization of living systems, its characterization and a model. *Biosystems* 5: 187–196.

von Neumann, J. 1966. *Theory of Self-Reproducing Automata*. University of Illinois Press. Champaign, IL.

Walde, P. 2004. Preparation of vesicles (liposomes). In *ASP Encyclopedia of Nanoscience and Nanotechnology*, ed. H. S. Nalwa, 9:43–79. Valencia, CA: American Scientific Publishers.

Walde, P., K. Cosentino, H. Engel, and P. Stano. 2010. Giant vesicles: Preparations and applications. *Chembiochem* 11: 848–865.

Walde, P., A. Goto, P. Monnard, M. Wessicken, and P. L. Luisi. 1994a. Oparin's reactions revisited—Enzymatic-synthesis of poly(adenylic acid). *Journal of the American Chemical Society* 116: 7541–7547.

Walde, P, and S. Ichikawa. 2001. Enzymes inside lipid vesicles: Preparation, reactivity and applications. *Biomolecular Engineering* 18: 143–177.

Walde, P., R. Wick, M. Fresta, A. Mangone, and P. L. Luisi. 1994b. Autopoietic self-reproduction of fatty-acid vesicles. *Journal of the American Chemical Society* 116: 11649–11654.

Wuu, J. J., and J. R. Swartz. 2008. High yield cell-free production of integral membrane proteins without refolding or detergents. *Biochimica et Biophysica Acta* 1778: 1237–1250.

Yu, W., K. Sato, M. Wakabayashi, T. Nakaishi, E.P. Ko-Mitamura, Y. Shima, I. Urabe, et al. 2001. Synthesis of functional protein in liposome. *Journal of Bioscience and Bioengineering* 92: 590–593.

Yuen, G., and Kvenvolden, K. A. 1973. Monocarboxylic acids in Murray and Murchison carbonaceous meteorites. *Nature* 246: 301–302.

Zepik, H. H., E. Blochliger, and P. L. Luisi. 2001. A chemical model of homeostasis. *Angewandte Chemie-International Edition* 40: 199–202.

Zepik, H. H., P. Walde, E. L. Kostoryz, J. Code, and D. M. Yourtee. 2008. Lipid vesicles as membrane models for toxicological assessment of xenobiotics. *Critical Reviews in Toxicology* 38: 1–11.

Zhu, T. F., K. Adamala, N. Zhang, and J. W. Szostak. 2012. Photochemically driven redox chemistry induces protocell membrane pearling and division. *Proceedings of the National Academy of Sciences of the United States of America* 109: 9828–9832.

Zhu, T. F., and J. W. Szostak. 2009. Coupled growth and division of model protocell membranes. *Journal of the American Chemical Society* 131: 5705–5713.

8

Design Tools for Synthetic Biology

Philipp Boeing, Tanel Ozdemir and Chris P. Barnes

CONTENTS

8.1 Introduction

The *design process* can be defined as the set of steps an engineer takes in constructing a system to achieve a specified task. The first step is conceptualisation of the problem, including the objectives of the task and any other design constraints, such as safety or reliability. Once these have been established, an initial set of solutions is proposed. It is often not feasible to construct all the possible solutions, so the initial set of solutions is reduced by estimating their performance with respect to the objectives and constraints. The designs that are likely to offer the best performance are then implemented and tested to see how well their behaviour achieves the required goal in the real world. Information about the real-world performance feeds back to the design stage where the designs are modified accordingly. This process is repeated until a satisfactory solution is achieved. This is known as the *design-build-test* cycle. Computer aided design (CAD) tools that facilitate this process are integral to all engineering disciplines. Synthetic biology can be defined as the application of the engineering design process to construct novel biological systems. It is precisely this application of forward design that separates synthetic biology from traditional approaches such as molecular biology and genetic engineering.

Computational tools are revolutionising how we can manipulate biological systems. Representation in computer code of novel biological systems is commonplace. This allows us to easily create new designs (with editing tools), modify existing ones and disseminate them to colleagues and other members of the community in a globally understandable manner. This computational representation also allows us to *simulate* systems using mathematical models, thus allowing the exploration of competing designs for their ability to achieve the objective under different conditions. In principle, these *in silico* experiments allow us to estimate how well each design will perform when constructed *in vivo*, which has the potential to save a great deal of time and expense. Once the optimal design(s) are chosen, these are then taken forward to the wet lab to be constructed. CAD tools can help here too by generating and optimising the sequence implementation and instructing robots to automatically build the system. The implemented designs must be tested either through qualitative measurements (is the system functional?) or quantitative measurements (usually using flow cytometry or fluorescent time lapse imaging). Data collected at this stage are used for *characterisation* whereby the performance of the system is evaluated and more detailed mathematical models generated. These more detailed models are used to indicate specific aspects of the system that can be improved and aid in refining the design. The design-build-test cycle for synthetic biology is depicted in Figure 8.1.

The design-build-test cycle is an essential component of forward engineering. Two other concepts are *standards* and *abstraction*. Standards refer to the act of standardisation of parts and techniques. For example, prior to 1841, bolt manufacturers in Britain created screws of varying thread angle and depth, which meant that screws and bolts from one manufacturer could not be used interchangeably with another. In 1841, Joseph Whitworth created a standard which specified how thread angle and depth should change with the size of the screw. Conformation to the standard allowed interchangeable parts and eased the process of manufacturing, mostly for the railway industry. The concept spread to other areas of engineering and it is now hard to imagine electronics or construction without such standardised parts. One aim of synthetic biology is to create a similar parts standardisation for biological devices so that they can be used interchangeably within certain standardised organisms. Just as standardisation of screws enabled easier manufacturing of machinery, standardisation of synthetic biological parts, and chassis organisms, will allow systems to be constructed quickly and in a reliable and predictable manner [1].

Abstraction in engineering is best illustrated through the example of constructing a computer. A computer is formed from various modules such as the central processing unit (CPU), motherboard, memory, hard disk, graphics card and sound card. Each of these modules is constructed from microprocessors and chips of different types, which are in turn constructed from logic gates. Logic gates are constructed from transistors and there are billions of transistors in a single CPU alone. A computer gaming fanatic wants to build the fastest machine possible for the best graphics and gameplay responsiveness.

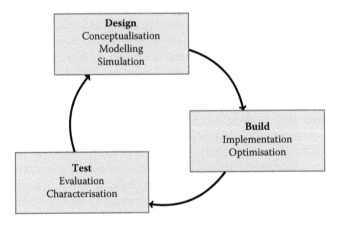

FIGURE 8.1
The design-build-test cycle in synthetic biology.

They can do this by choosing the component modules, ensuring that they work with each other (standardisation) and constructing their dream machine. Designing the computer by placing transistors on circuit boards is unimaginable, even though fundamentally the computer is constructed from them. This is abstraction: the designer is concerned only with the level that they require; the lower levels (logic gates, transistors) are abstracted away. Of course a CPU designer *is* concerned with placing logic gates on a circuit board but, even to someone working with logic gates, the transistor level is abstracted away. Only a designer of logic gates needs concern themselves with the transistor level. In order for synthetic biology to move beyond the transistor level of abstraction, we will require standardised parts that can be combined to create reusable modules (such as logic gates, oscillators, pulse generators, switches and sensors) that are reliable enough to form the basis of larger, more complex systems.

In many respects, systems biology and synthetic biology are complementary. One aim of systems biology is to generate mathematical and computational models of biological systems in order to understand their behaviour and to make testable predictions. Because modelling is a key component of the design process in synthetic biology, many modelling tools developed in the context of systems biology are applicable to synthetic biology. A major subfield of systems biology is *reverse engineering* which is the process of trying to understand a biological system, that is, by generating a mathematical model, through passive measurements (observing the natural system) or active experiments (e.g. gene knock-outs or knockdowns). Tools developed for reverse engineering in the context of systems biology can be used for the important process of characterisation in synthetic biology.

In a similar manner, bioinformatics offers an alternative set of tools to draw upon. Most of synthetic biology to date uses well characterised parts from different organisms. For example in prokaryotic systems, the transcription factors LacI, AraC, TetR and their associated promoters are extensively used. Bioinformatic mining of natural organisms for usable parts expands the available part libraries. One example where this was achieved was in the construction of orthogonal logic gates [2] and there have been other efforts to create orthogonal part libraries (reviewed in ref. [3]).

There have been a large number of tools developed for the design, abstraction and realisation of engineered biological systems [4]. There is now an annual conference on biodesign automation (www.iwbdaconf.org) with both the synthetic biology community and silicon chip design automation community interested in this area. While there are parallels with silicon chip design and manufacture, it should be noted that there are many significant and technologically challenging differences between the two engineering disciplines. These include the inherent stochasticity of transcriptional and translational processes, the changing cellular context and selective forces on engineered systems [3, 5, 6].

In this chapter, we aim to present a current snapshot of the tools available for the design of synthetic biological systems. We have tried to split the tools into logical groups based on their functionality and approach. However, as is often the case, some tools will span multiple sections and we try to point this out in the text. Our grouping, by no means unique, takes the following form:

- Part registries
- Device languages
- Modelling languages
- Network optimisation

- Part engineering
- DNA editing tools
- CAD tools and workflow systems

Since mathematical modelling is integral to the design process, before moving on to describe these tools, we present a very brief overview of mathematical modelling in synthetic biology.

8.2 Mathematical Modelling in Synthetic Biology

Mathematical modelling is an integral part of the engineering design process. The mathematical formalism used in a given situation will depend on the nature and size of the system, the essential details to be captured and the problem to be solved [7]. A model's behaviour is governed by both its *structure* and the *parameters* contained within it. For example, often in synthetic biology the model consists of a set of biochemical reactions (such as transcription, translation, binding) each of which has an associated rate constant. The reactions constitute the model structure and the rate constants constitute its parameters.

An example of a model for the expression of a gene, X, is shown in Figure 8.2. Gene expression is a complex process[*] involving the uncoiling of DNA, binding of RNA polymerase to the promoter region, transcriptional elongation, binding of the transcript to the ribosome, binding of transfer RNAs, formation of the peptide chain and folding of the peptide into the mature protein. Here our model of this complex process contains just two species X_{RNA} and X that represent the amount of transcript and protein product, respectively, and four reactions: production of the transcript, translation of the transcript to the protein product, decay of the transcript and decay of the protein. These

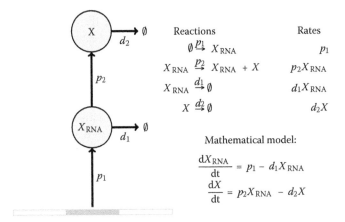

FIGURE 8.2
A model for the expression of a gene. The complex process of gene expression is approximated by two species and four reactions. The system of two ordinary differential equations form a representation of this model.

---

[*] Here, we will concentrate on gene expression in prokaryotes. Eukaryotic gene expression is more complex although it could be modelled by similar approaches.

four reactions have associated parameter rate constants which are denoted by p_1, p_2, d_1, d_2, respectively. This may seem like a vast oversimplification of the process, and indeed it is for many purposes, but models such as these can be extremely useful for gaining insight into biological processes, and are used extensively.

Generally speaking, mathematical models come in two main types. Deterministic models produce the same output on repeated simulation with the same parameters. Stochastic models contain an inherent degree of randomness such that repeated simulations with the same parameters give rise to different outcomes. The reason for choosing a stochastic model is that at low numbers of molecules, such as when transcription factors bind to a promoter, probabilistic effects become important.

Figure 8.2 shows an example of a deterministic mathematical model for the gene expression system. The mathematical formalism used is that of *ordinary differential equations*, which represent the change in the species amounts[†] over time ($\frac{d}{dt}$ can be read in words as 'rate of change of' and denotes the time derivative). In order to derive this model some assumptions must be made on how the rates of a reaction depend on the amounts of the reactants. A common assumption is the law of mass action, which states that the rate of a reaction is proportional to the product of the reactant amounts. For example, the reaction

$$X_{RNA} \xrightarrow{d1} \phi,$$

which denotes the decay (degradation or dilution) of transcript X_{RNA}, has a rate of reaction of $d_1 X_{RNA}\ Mh^{-1}$ (assuming X_{RNA} is measured in molar concentration). On the other hand, the reaction

$$\phi \xrightarrow{p1} X_{RNA},$$

which represents the production of RNA transcript does not have any reactants (in this simplified model) and has a reaction rate of $p_1\ Mh^{-1}$. The differential equation for each species is formed by balancing the production and decay rates due to each reaction it is involved in. For example, the rate of change of X_{RNA} is given by the rate of its production, p_1, minus the rate of its decay $d_1 X_{RNA}$.

Once the mathematical model has been constructed it can be analysed or simulated to try to understand the system behaviour that it represents. This generally requires that the parameters must be known at least approximately. An example simulation using MATLAB® (Section 8.5) is shown in Figure 8.3. To generate this simulation, we consider an *E coli* cell expressing a gene of around 3,000 nucleotides and 1,000 codons/amino acids. Approximate numbers for different processes can be found using the BioNumbers database (bionumbers.hms.harvard.edu). Here, we assume that the production rate of RNA transcript is $5 \times 10^{-8}\ Mh^{-1}$ and the decay rate of transcript is $20\ h^{-1}$. Translation is assumed to occur at 50 proteins h^{-1} per transcript. Decay of the protein is assumed to be due purely to the dilution through growth of the cell. Assuming $d_2 = ln2/T$, where T is the doubling time of the cells, and that that $T = 1\ h$, gives a value $d_2 = 0.69$. A more thorough treatment of the parameterisation of models in synthetic biology can be found in ref. [8]. In general, simulations can provide insights into synthetic systems *in silico*, before they are constructed in the laboratory. For example in our simple model, we can see that the protein and mRNA levels reach a constant value (steady state), with the protein concentration around 0.17 µM. In more complex systems, the behaviour can be unexpected, in a way

---

[†] Note that by convention in deterministic equations, the amount of a substance is measured in molar concentration or moles per litre. In stochastic models the convention is to use number of molecules.

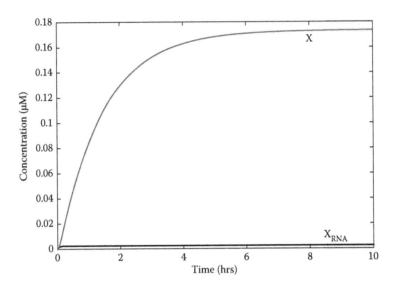

FIGURE 8.3
An example simulation of the system given in Figure 8.2, with parameters $p_1 = 5 \times 10^{-8}\ Mh^{-1}$, $d_1 = 20\ h^{-1}$, $p_2 = 50\ h^{-1}$ and $d_2 = 0.693\ h^{-1}$.

that is unpredictable from the model structure. There are other methods to gain insights into biological systems from mathematical models such as bifurcation analysis where the global behaviour of the system is studied, and sensitivity analysis where a model's parameters are examined for their impact on the model output [7].

Estimating all the parameters within a complex model (and also the model structure) is a non-trivial task. Often the only way to do this for biological systems is to use statistical inference techniques to estimate the parameters from observations (reverse engineering). Parameterised models can be used to make predictions, which can then be tested by experimental methods. If the model and data disagree, we re-examine the assumptions underlying the model, including the structure, the parameter values or the mathematical formalism. Constructing mathematical models in this way is the most accurate way to represent our knowledge of the system under study.

A full treatment of this subject is beyond the scope of this chapter but further reading on systems biology in general can be found in Alon's *Introduction to Systems Biology* [9]. A more technical – but still thoroughly readable – introduction to biochemical modelling, both deterministic and stochastic, can be found in Wilkinson's *Stochastic Modelling for Systems Biology* [10], and a classic textbook on deterministic modelling is Strogatz's *Nonlinear Dynamics and Chaos* [11].

8.3 Part Registries

We begin with one of the most fundamental concepts of synthetic biology, that of standardised parts. The eventual goal of synthetic biology is to provide a set of reusable standard parts that are well characterised and behave in a reproducible manner when placed together in a specified chassis or added to a native system.

Often known simply as 'the Registry', the Registry of Standard Biological Parts co-evolved with the International Genetically Engineered Machine (iGEM) competition at MIT in 2003 and was originally inspired by electronics catalogues. The registry is hosted at parts.igem.org and is currently in version 6.0. It was created to facilitate the standardisation and sharing of biological parts and currently contains just under 10,000 parts plus plasmid backbones. These have been mostly added by iGEM teams over the years and each year more are submitted as a result of the growing iGEM competition. For example, around 1,500 new parts were added in 2013.

In the Registry, parts are categorised by type (promoter, ribosome binding site [RBS], protein domains, etc.), function (cell signalling, chemotaxis, biosafety, etc.) and chassis (such as *E. coli*, yeast or *B. subtilis*) with each part having its own information page. Parts are of varying quality ranging from excellent, with lots of data submitted by different research groups from all over the world, to parts for which no DNA material or characterisation data has ever been deposited in the Registry. Since many of the parts are submitted by students as part of the iGEM competition, for a long time, the focus has been more on creating new parts and less on quality assurance. However, there are some indications that this is changing, such as the 2013 redesign of the registry, the introduction of postgraduate iGEM and the pruning of low quality parts. The part information includes sequence, a wiki for user reviews and experience, authorship and assembly compatibility. The data are also machine readable via extensible markup language (XML).

The Inventory for Composable Elements (ICE) is an open-source registry software for biological parts, developed by the Joint BioEnergy Institute (JBEI) [12]. This tool allows laboratories to track, annotate, share and search constructs with the aim to create a distributed infrastructure for the sharing, cataloguing and re-use of parts. Users can access the tool either via a web interface for user-friendly access or an application programming interface (API) that allows the development of software tools to interface with ICE. Individual labs can set up their own installation on a local server and the latest release includes features for distributed hosting of parts, so that parts can be hosted on different servers belonging to different labs, but still be searchable in the wider community, if security and privacy settings allow it. JBEI runs a public registry at public-registry.jbei.org.

ICE can store a wide range of properties for each part, including author and principal investigator names, biological information such as the part's backbone, origin of replication and selection markers, and other information such as funding sources, intellectual property status and biosafety level. Constructs and sequences can be visualised within ICE using Pigeon (Section 8.9) and VectorEditor (Section 8.9). Data can be imported and exported using a wide variety of file formats, including FASTA—a standard DNA sequence text format—and synthetic biology open language (SBOL) (Section 8.5).

The BioFab project (www.biofab.org) aims to create a registry of professionally characterised biological parts and is an allusion to the fabrication plants used for engineering semiconductor chips. The aim is to create standardised characterised parts in large quantities with a major emphasis on production and characterisation. A data access client and an API exist; GeneDesigner (discussed in Section 8.8) can import BioFab parts directly.

Most of the existing registries iteratively add parts and information over time as they are used to contract novel systems, or as part mining and engineering are performed. An alternative to this approach is provided in the database GenoLIB [13]. The authors of this database examined around 2,000 plasmids already in use and deposited in the SnapGene resources library (SnapGene is a DNA editing tool, discussed in Section 8.8). From this data set, they built a non-redundant database of plasmid features and derived sets of parts that are species specific. The database, complete with part, sequence and host chassis information, is available to download in various formats including SBOL (Section 8.5) from the online supplementary information on the Nucleic Acids Research website.

8.4 Device Languages

The concept behind device languages in synthetic biological design is to allow the expression of parts and the interactions between them in the form of a formal language. There are two main objectives that this can facilitate. The first is to ensure that parts are combined together in a biologically meaningful manner; for example a terminator can be placed downstream of an open reading frame but a promoter cannot. In this case, a device language serves as a grammar checker for biological constructs. The second objective is to facilitate the design of novel genetic systems. In essence, the goal is to develop programming languages whose outputs are biological devices. Many of these languages are based on the idea of an *Abstract Gene Regulatory Network* [14]. These are gene regulatory networks in which some of the genetic parts and transcription factors are unspecified, but constrained by parameters and by the interactions with other elements in the network. Such a synthetic biology programming language must then provide a mechanism to translate abstract, high-level specification into biologically viable devices that match the desired behaviour. Often, multiple biological implementations will be possible, in which case it is desirable to rank them according to a useful score, such as a performance metric, the robustness of the system or the cost-efficiency of building it.

The GenoCAD language aids the designer by specifying a 'context free grammar' that formalises rules for the composition of parts. It thus enforces combinations of parts that will generate a biologically valid construct [15, 16]. GenoCAD has been implemented as an online web application, enabling users to graphically design a DNA construct from parts that are either user uploaded or from a pre-loaded library (genocad.com). A number of grammars are available, and new grammars can be designed by the user. The resulting DNA sequence, validated by the chosen grammar, can then be downloaded by the user. Grammars can include simulation information, so GenoCAD constructs based on these grammars can generate simulations, which can be downloaded in synthetic biology markup language (SBML, Section 8.5) or run online using the complex pathway simulator (COPASI, Section 8.5).

Genetic engineering of cells (GEC) was developed at Microsoft Research in Cambridge, UK and allows the specification of genetic circuits involving abstract parts, which can then be compiled into potentially multiple suitable implementations [17]. GEC uses a database of genetic parts each with specified properties. For example, a promoter has properties such as the identities of the proteins responsible for transcription, the nature of its transcription (constitutive or inducible) and the associated binding rate constants. Other parts include RBS, terminator and protein coding regions. There is also a database of reactions and this, together with the parts database, is used to validate and constrain a device specification and return a list of valid implementations. GEC is currently closed-source but accessible via a web interface (research.microsoft.com/en-us/projects/gec/).

Eugene is a language for specifying devices made of synthetic biology parts, as well as creating permutations of parts constrained by user defined rules [18]. Its purpose is to speed up and innovate the design of novel biological devices and it aims to become a human-readable user-level design specification language for synthetic biology. Its workflow begins with device specification, followed by an exploration of design space using valid permutations of logically identical parts. These designs can then be simulated in an external tool (for example Synthetic Biology Software Suite [SynBioSS], discussed in Section 8.9, can read Eugene XML). Eugene is an open-source application available on SourceForge (eugenecad.org). There is also now a web-based application called miniEugene [19].

Proto is an open-source functional programming language for spatially distributed computation developed at MIT [20]. It is not designed specifically for synthetic biology but has been developed to write programs for spatial computers, which are composed of a collection of devices distributed across a space. Examples include swarms of robots, sensor networks or a collection of biological cells, such as a colony or biofilm. The unique feature of Proto is that with the Proto environment space becomes an abstraction and as such, Proto is not used to describe the behaviour of individual devices in the space, but instead describes the behaviour of regions of space, which are then approximated by the network of devices. This makes Proto useful as a high-level language for synthetic biology. Instead of specifying a biology system based on its genetic network, which is the case with GenoCAD, GEC and Eugene, in Proto a programmer specifies only the computation that is desired by the synthetic biology system. Using the dedicated BioCompiler, the program can then be compiled into a genetic network using biological motifs. In general, the abstract devices will be very complex, but can be optimised in the same way that computer code can be optimised. There is a web implementation available at synbiotools.bbn.com

8.5 Modelling Languages and Tools

There exists a close relationship between systems and synthetic biology and modelling tools developed in the former field are readily applied to modelling and design of synthetic biological systems. In this section, we review some of the most widely used modelling tools in systems and synthetic biology.

SBML is a widely used standard for the representation of models in systems biology implemented in XML [21]. It is supported by many applications through the libSBML library [22]. It is primarily an exchange format, that is, machine but not human readable and enables the use of models in different tools without rewriting. Therefore models can be easily shared and results verified and replicated. In SBML, models are represented by collections of molecules, reactions, parameters and compartments, together with relational functions, rules, events and constraints. The format has been used to represent a huge range of systems and synthetic biology models such as cell signalling networks, metabolic pathways, biochemical reactions, gene regulation networks and gene circuits. LibSBML is open source and features support for different SBML versions and levels, and object-oriented representation of models. The standard is under active development and currently at level 3, which provides a modular structure such that developers can add packages on top of core functionality. One important point to note is that since a level generally contains a whole set of new features, levels are not upwards compatible; a valid level 1 document is not a valid level 2 or 3 document.

CellML is another XML-based model representation language to facilitate the sharing and reuse of models [23, 24]. It differs from SBML primarily in the manner in which models are encoded. In CellML, models are represented as a network of components that allows for the editing, reuse and sharing of sub-models and all biological information is stored as metadata. This makes the language more general than SBML but the types of models are less constrained, placing more emphasis on the tools that read and interpret the CellML models. While there exist converters from CellML to SBML, there is not a one-to-one correspondence between the two modelling representations and they should be viewed as complimentary approaches.

SBOL is an open-source standard for the electronic representation of synthetic biology designs (sbolstandard.org). It aims to facilitate the exchange of designs between different researchers and different CAD tools and serve as a standard file format that can be sent to fabrication facilities, uploaded to online registries and embedded in publications. SBOL is based on an object-oriented model of DNA components that can be composed of sub-components, mirroring BioBrick parts making up a larger composite BioBrick. DNA components can be annotated with sequence information. SBOL designs can be exported as an XML file. Libraries implementing SBOL exist in Java, C and Python. Extensions to the SBOL core have been proposed and are under development. These include data on the performance of components, contextual information regarding the host organism, mathematical modelling, assembly data and annotation of regulatory constraints. Furthermore, SBOL Visual is an open-source graphical notation that establishes a common set of symbols to describe genetic designs. Many software tools mentioned in this chapter support SBOL to some extent. SBOL Designer‡ is a graphical tool to build SBOL designs. The interface uses the symbols defined by SBOL Visual (Figure 8.4), and designs can be exported as SBOL XML. Furthermore, DNA components can be imported from the synthetic biology parts knowledgebase (SBPkb), a knowledge database based on the Registry of Parts (Section 8.3). BioNetGen is a rule-based modelling language developed to tackle the combinatorial complexity of signalling pathways involving proteins with multiple binding sites and post-translational modifications. Rather than specifying reactions between every element in the system, which can reach huge numbers, models are specified via types of reactions between types of elements. Thus a small number of rules can equate to a large numbers of possible reactions. The rules can be processed to generate a large network of biochemical reactions (in SBML format) that can in turn be simulated via standard methods. Alternatively, there are stochastic simulation methods that can simulate the rules directly. It is open source, stable and has a wide user and developer community [25–27].

Kappa defines a process algebra approach to modelling biological systems, again developed to handle combinatorial complexity of protein interaction networks [28, 29]. Models defined in Kappa can be simulated using KaSim (Kappa Simulator), which executes one or more Kappa files using stochastic simulation. The open-source project is under active development and available from www.kappalanguage.org.

MATLAB is a very popular commercial software/programming environment with additional functionality provided through external toolboxes (mathworks.com). One such toolbox of particular relevance is SimBiology. This is both a graphical and programmatic tool for modelling biological systems. It supports both deterministic and stochastic modelling, sensitivity analysis and the import/export of models in the SBML format. This tool has been used in many iGEM projects and was recently used for a whole cell model of *Mycoplasma genitalium* [30].

Mathematica is another popular commercial mathematical and statistical modelling framework (www.wolfram.com/mathematica). Support for modelling biological systems is enhanced through the SystemModeler package, which supports the import and export of SBML models and multiple simulation options.

---

‡ Available at clarkparsia.github.io/sbol/

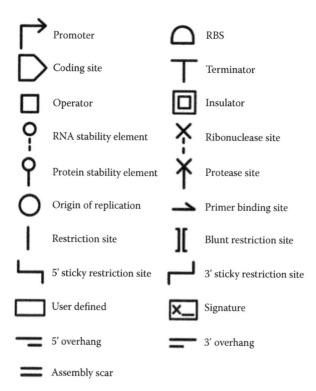

FIGURE 8.4
Symbols in SBOL Visual Version 1.0.0.

Python is an established programming language and is rapidly becoming the tool of choice for modelling in systems biology. The reasons for this are that it is open source, extremely flexible (it can be used for a vast range of applications), easy to learn and program (when compared with lower-level languages like C/C++) and has a huge range of external packages developed by its user base. Three essential packages for scientific computing are Numpy, SciPy and Matplotlib, which implement efficient matrix storage and manipulation, scientific algorithms (including differential equation solvers) and graphics, respectively. One disadvantage when compared with MATLAB or Mathematica for modelling applications is that developed code relies heavily on external packages which may be difficult to install or have version dependencies or varied support. This problem is alleviated by the availability of distributions, together with installers, that install a set of packages that are guaranteed to work together. Example include the SciPy Superpack for OS X (strong-inference.com/ScipySuperpack), or Enthought Canopy, which provides a fully integrated development environment (www.enthought.com/products/canopy).

COPASI is a tool for modelling biochemical reaction systems [31]. Models are entered as sets of biochemical reactions, specified via a text-based language. It contains an inbuilt library of reaction rate laws (plus the ability to specify user-defined rate equations) and supports reading and writing of models described in SBML. There are many analysis features including deterministic and stochastic simulation, steady state analysis, parameter estimation and sensitivity analysis. It has a wide user base in the systems biology community and is highly cited.

Developed in 2009, ProMOT was one of the first modular modelling tools for systems biology. It provides both a graphical environment and a programmatic language (model description language [MDL]) for the construction of models of modular biological systems [32]. This tool, together with the rule-based language BioNetGen, was used to design eukaryotic synthetic gene circuits [33].

Little b is a flexible, modular computational modelling framework where models are expressed programmatically in Lisp (LispWorks environment) and output as MATLAB code [34]. Biochemical processes can be expressed as reaction systems or using the BioNetGen rule-based framework. The framework implements modularity without encapsulation, which the authors reason is unique to biological systems (as opposed to engineering and physical science) because only membranes can provide true encapsulation. Thus in this framework, modularity is a modelling convenience allowing for large models to be built incrementally and avoiding the need to re-implement monolithic models after a change in a few species or reactions. The advantages of this approach are that modules can be modified, for example by the addition of species or reactions, and the downstream effects on other modules inferred automatically, together with any new entities required or equivalent entities removed.

Antimony is another modular modelling framework that aims to tackle the problem of building large models in an incremental fashion [35]. Biochemical reactions can be specified in a simple syntax. This has the advantage that the user does not have to learn a programming language to generate models. Modularity is implemented via a namespace mechanism and SBML models can be imported directly into module objects. Identical elements contained across modules can be declared as being the same and modules can have defined inputs and outputs. There is also syntax for the specification of genetic constructs.

PySB differs from many of the existing modelling languages by specifying models via an extension to the Python language rather than defining a new language [36]. The advantages to this are that established principles from software development – such as encapsulation, abstraction, modularity and inheritance – can be applied to the reuse, distribution and modification of large and complex models. The authors also envisage an open-source community of modellers (not unlike GitHub) that will collaborate on large and complex modelling tasks. Models in PySB are described through core classes such as Model, Monomer, Parameter, Compartment and Observable and can be simulated using ODEs. However the emphasis is on rule-based models and it incorporates the BioNetGen and Kappa language definitions.

NFSim is another rule-based modelling tool that allows efficient (exact) simulations of stochastic systems specified via an extension of the BioNetGen language [37]. The tool uses an agent-based representation of a reaction system and can be used to simulate large reaction networks that would otherwise be computationally intractable. It incorporates methods for parameter estimation and coarse-graining where sets of reactions can be simulated via simplified phenomenological models.

8.6 Network Optimisation

Whereas tools such as GEC and Eugene can output many different implementations of an abstract genetic regulatory network, they rely on the designer understanding the possible network structure (or topology) that can give rise to their desired behaviour. For example to build a pulse generator from a network of three interconnected genes, one must understand the positive and negative regulation structure that can give rise to such a behaviour. Models of gene regulatory networks are generally nonlinear and in some cases should be treated stochastically for an accurate description of the dynamics. These two aspects make intuition about genetic systems very difficult and a further complication is that an abstract gene regulatory network is likely to give a large range of different behaviours depending on the component parts (and hence the biochemical parameters). This section reviews some of the tools available for searching the space underlying models of small gene networks. Their function is to find or compare regulatory models that satisfy some desired behaviour often specified through an input-output mapping.

AutoBioCAD takes as an input some specified target dynamics and outputs a gene network capable of achieving it, assembled from a library of parts represented by mathematical models [38]. It outputs values for the associated kinetic parameters and the sequences required to construct the network. The optimisation is performed using a type of evolutionary algorithm and is constrained by allowing only certain biologically relevant combinations of parts. The optimal model is output in SBML format. AutoBioCAD is available online at jaramillolab.issb.genopole.fr/display/sbsite/Download

RoVerGeNe is a tool written in MATLAB that takes a genetic network as input and examines whether some specified dynamical behaviour is achievable. The dynamical behaviour is expressed through constraints known as linear temporal logic (LTL). If the desired behaviour is possible, the tool outputs parameters that give rise to the specified dynamics. The code is available online (iasi.bu.edu/~batt/rovergene/rovergene.htm).

ABC-SysBio implements an approximate Bayesian computation algorithm for parameter fitting and model selection in systems biology models [39]. It can read models in SBML format and simulate them both deterministically (ordinary differential equations) and stochastically (stochastic differential equations, chemical master equation). The software can equally be used in a synthetic biology design setting for ranking a set of designs according to how robustly they achieve a desired behaviour [40, 41]. The software is available to download from sourceforge.net/projects/abc-sysbio/.

The MatchMaker tool is not a network optimisation tool *per se* but rather attempts the conversion of an optimised abstract gene regulatory network into real parts and DNA sequence [42]. It does this in three steps. The first is feature matching where the general relationships (activation, repression) are determined that preserve the pattern of regulation in the abstract genetic regulatory network. The next stage is signal matching where parts are matched using characterisation data (transfer curves) such that the desired digital behaviour is output. Finally, parts matching is performed where a sequence is obtained from a parts database to match the specification generated from the first two stages. MatchMaker has been implemented as a Clotho plugin (Section 8.9).

8.7 Part Engineering

In this section, we look at a small number of widely used tools that can estimate the properties of nucleic acids and proteins based purely on their sequences. These tools can be used to construct parts with rationally designed properties and aid the predictability of synthetic circuits [43].

The Mfold web server is a free tool developed to predict the secondary structure of DNA and RNA[§]. Through established thermodynamic methods, the software is able to calculate

---

[§] Available at mfold.rna.albany.edu

the folding and hybridization of a given nucleic acid sequence [44]. Due to increasing use of oligonucleotides within molecular biology and medicine, Mfold is growing to become a vital tool in the correct design and execution of many experiments.

Sfold is another free online tool designed to look at statistical folding and rational design of nucleic acids [45]. In addition to general RNA folding, the Sfold algorithm has been developed to create several application modules that can aid in the design of siRNA, antisense oligonucleotides, ribozymes and microRNA binding sites. The package is free to access on sfold.wadsworth.org.

The ability to accurately control protein expression is vital for the creation of complex genetic circuits. In bacteria, protein expression is effectively controlled by the RBS that determines the translation initiation rate. RBS Calculator is a software implementation that builds on a thermodynamic predictive method to either reverse or forward engineer RBS [46]. Through reverse engineering, it is able to predict the translation initiation rate of a given codon on an mRNA sequence. Through forward engineering, it can edit an RBS to rationally control the translation initiation rate and in turn the protein expression levels. The software is free to access on salis.psu.edu/software/forward.

The Rosetta software suite (www.rosettacommons.org) is a large-scale collaborative project between a number of institutes and universities. It is free to use and includes several powerful algorithms for computationally modelling and analysing protein structures. Through a variety of protocols, the widely used tool enables users to carry out *de novo* protein design (RosettaAbinitio), enzyme design (RosettaEnzdes), DNA-protein interactions (RosettaDNA) and small molecule–protein interactions (RosettaLigand).

8.8 DNA Editing Tools

Often the design of a synthetic genetic circuit comes down to the editing of DNA sequence by placing parts into a plasmid backbone, rearranging and inserting sequences, or planning cloning and assembly protocols. The tools presented in this section have various features that allow for the ease of designing and editing DNA sequence.

A plasmid Editor (ApE) is a cross-platform free to use DNA editing tool developed by Wayne David from the University of Utah. The package consists of several plasmid visualization and editing tools. The tool has a relatively strong level of uptake from academia but lacks several DNA editing features available from more recently developed tools. It is free to download from biologylabs.utah.edu/jorgensen/wayned/ape.

Genome Compiler is an integrated genetic design software that allows users to manipulate all abstraction layers of a synthetic biology project. Users can zoom in and edit right down to DNA, parts or protein levels. The extensive software includes a library of genomes, vectors, restriction sites and parts from the iGEM registry to enable drag and drop construction. In addition to this, genes can be simply imported into the built-in library from the NCBI database [www.ncbi.nlm.nih.gov]. Projects can be stored and shared on the cloud and the software is also capable of seamlessly quoting and ordering constructs or sequences from a number of vendors. While the wide variety of features can be cumbersome to navigate initially, online tutorials help to unlock the true capability of the software. Academic users can request a free download at www.genomecompiler.com/enterprise.

Stemming from an 2012 MIT iGEM Entrepreneurship entry, Benchling is another free browser-based tool for plasmid visualization and editing. In addition to the traditional

DNA editing tools such as restriction analysis and primer design, it incorporates a library of protocols that users can apply, edit and share. Benchling includes tools for gel visualization and analysis, as well as CRISPR genome engineering. The simple interface is easy to navigate and welcoming to users who have little or no experience in molecular biology. It is free to sign up for academic users at www.benchling.com.

SnapGene is a molecular biology software for editing, visualizing and sharing plasmid maps (www.snapgene.com). Extensive features and tutorials provide a platform to enable users to quickly design primers, restriction digests and cloning strategies. The 'cut and paste' interface provides a welcoming introduction to novice users in both microbiology and synthetic biology. The software is available as a free 30-day trial but will then require a license to be purchased. The requirement of a license could become quite restrictive for labs with a large number of members or users with multiple machines. In addition to the free trial, a free version is also available but with more limited features that mainly allow for visualization.

Gene Designer is a free web-based tool that also aims to have an ergonomic GUI for a range of molecular and synthetic biology techniques (www.dna20.com/resources/gene-designer). In addition to construct visualization, primer design and 'drag and drop' editing, the tool uses codon optimization algorithms for recombinant protein expression in different organisms [47]. Due to its free web-based format, several academic institutions have incorporated the tool within their Synthetic Biology curriculum and produced a variety of free online tutorials.

8.9 CAD Tools and Workflow Systems

Up until recently, synthetic biological constructs have been relatively small. For these systems, designers only had to choose between a small set of possible parts, which they could do by hand. However, as the complexity of the designs and the number of available parts increase, new scalable workflows need to be found that can handle the complexity of the design, build characterisation and debugging stages. The tools described in this section offer aspects of this scalability.

Pigeon (pigeoncad.org) is an online design visualizer for synthetic biology, with the desire to replace the manual drawing of visual synthetic biology designs by an automatic tool [48]. Pigeon defines a simple textual language which is then used to automatically render a visualization, using SBOL Visual symbols. An example of how circuits are defined, together with the graphical output, is shown in Figure 8.5.

The iBioSim tool, based on the principles of electronic design automation (EDA), provides a project-based framework for the design and modelling of genetic circuits together with model reduction techniques for more efficient simulation. Additionally, it implements an inferential framework that allows for the determination of genetic part connectivity using time series data. It incorporates an SBML model editor together with a genetic circuit model editor that allows for the abstraction and combination of genetic parts [49]. Recently, functionality was also added that can bridge SBML models with SBOL system descriptions [50].

TinkerCell is a free open-source CAD tool for synthetic biology [51]. Within it, biological networks can be constructed using a visual editor. These can then be analysed with user-defined code that can be written with C/C++ or Python using the TinkerCell API.

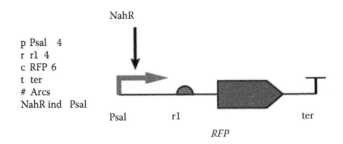

p Psal   4
r r1 4
c RFP 6
t ter
Arcs
NahR ind   Psal

FIGURE 8.5
Pigeon graphical output of example code, creating a visualization of a simple genetic circuit.

Note that while it is still available to download and use, at the time of publishing TinkerCell is no longer being actively developed.

Clotho is an application platform for synthetic biology, originally emerging out of Berkeley's 2008 iGEM team [52], but now part of Autodesk Research (www.clothocad.org). Clotho provides a platform for applications and defines a data layer and an application layer. The data layer is based on ideas in this chapter, such as SBOL, parts registries, grammars, design calculators, as well as automated assembly methods. Applications running on the Clotho platform can work with these data objects. Thus the data become standardised, allowing many tools to work together as 'Apps' or plug-ins, similar to software on smartphones. New applications can be added and removed from the Clotho platform, so that each Clotho installation can be customised to an individual user or lab. Clotho manages data in a private database and data is linked, enabling tools to create workflows such as finding all references belonging to all the parts of a genetic circuit. Current applications include Sledrunner, a sequence viewer and editor for parts, oligos, plasmids and features; BglBrick Digester, which calculates fragment sizes resulting from a digest of plasmids in the BglBricks format; Registry Importer, which allows for the import of parts into the Clotho data layer from the iGEM Registry; and a script editor, where scripts based on the Groovy language, a dynamic language related to Java, can access the Clotho API to quickly program new functionality. The current version available is Clotho 2, but Clotho 3.0, presented in 2013 and currently still under development, adds a great deal more functionality, flexibility and extensibility to the data model.

SynBioSS is a collection of applications for each of the steps involved in the development of a synthetic biology model [53]. There are three separate applications:

- SynBioSS Designer covers the creation of a kinetic model. It is a web application that can be accessed at www.synbioss.org/designer/. Based on a network of genetic parts and a set of rules, the software generates a set of reactions representing the transcription, translation, regulation, induction and degradation of the parts in the network. It outputs an SBML file that can be used to simulate the system.

- SynBioSS Wiki stores kinetic data in a web accessible database curated by users, in a similar way to Wikipedia (www.synbioss.org/wiki). The aim is to create a better collection of kinetic parameters that are necessary for models, for example reaction rates and binding affinities. It is implemented in an extension of MediaWiki, the software behind Wikipedia. Extensions include pages for database entries and structures, so that species and reaction data can be stored and edited.

- SynBioSS DS (Desktop Simulator) allows simulation of the models. There are two simulator implementations, one for a graphical desktop PC and one for a 'supercomputer', accessed by a command line. Both run simulations based on Hy3S, a hybrid jump-continuous Markov process integrator [54]. This achieves a similar accuracy to a stochastic simulation, whilst cutting time on high-concentration reactions. SynBioSS DS can be obtained from synbioss.sourceforge.net.

SBROME [55] and related work [56, 57] provide methods for the scalable design of larger synthetic circuits, mostly concentrating on logic gate implementations. Incorporated within the software is a part and module library called PAMLib, which includes experimentally determined characterisation data and model parameters. A user defines an abstract circuit together with a desired input–output relationship. The parts database is then searched for sets of parts and models that can achieve the desired behaviour. Initially a search based on logic relationships is performed to reduce the space of possible models, and then a more fine-grained dynamical simulation is used to measure the performance of the potential solutions.

The aim of Toolchain to Accelerate Synthetic Biological Engineering (TASBE) is to speed up and reduce errors in the synthetic biology design flow from high-level descriptions, through specific genetic network designs and finally to assembly instructions [14]. The workflow is designed around four stages: specification, compilation, part assignment and assembly and incorporates prior projects that have addressed each particular stage. The specification stage uses the Proto language to specify the desired behaviour of the organism and the desired target platform, such as *E. coli* or mammalian cells. In the compilation stage, the specifications are turned into an abstract genetic regulatory network, using a dictionary that maps specification elements to genetic motifs (Proto, Section 8.4). In the part assignment stage, the abstract genetic regulatory network is mapped onto one or more sequences of parts using MatchMaker (Section 8.6). This process is constrained by a database of parts and corresponding characterization data. Finally in the assembly stage, the generated sequences can be assembled using the tools available to the laboratory (using liquid handling robots within Clotho for example). TASBE is free, open source and allows researchers to develop their own customised tools.

DeviceEditor, j5 and VectorEditor are three closely integrated web tools available at j5.jbei.org after a user registration. DeviceEditor provides a graphical design tool for synthetic biology circuits [58]. Circuits can be designed using SBOL Visual symbols and each abstract part can be given multiple possible assignments. Eugene rules (Section 8.4) can be used to constrain the derived circuits. DNA parts can be copied in from VectorEditor, a tool to visualise and annotate sequences, with import and export features to SBOL and GenBank file formats [12]. After circuits have been designed, DeviceEditor acts as an interface to j5, which automatically creates assembly protocols for scar-free methods such as Gibson and Golden Gate [59]. Many j5 parameters can be customised, including prices for synthesis, so that the tool is able to estimate the constructs that should be ordered from a synthesis company and which should be assembled in house.

8.10 Summary

In this chapter, we have covered a range of tools available for the design, simulation and manipulation of synthetic biological systems. As synthetic biology matures these types of tools will become ever more crucial to the design-build-test cycle. The future success of synthetic biology will rely on many different parallel efforts in novel biochemistry, rational part design, chassis optimisation and the part mining of organisms. The design tools for synthetic biology must also mirror this development. One area of increasing importance is context dependence, which can manifest at the sequence, parts, host and environmental levels [3, 5, 6]. Incorporation of context dependence into the design process, and doing this in a modular manner, will be challenging from a CAD perspective, but will be crucial to the forward design of cell-based synthetic systems.

References

1. Barry Canton, Anna Labno, and Drew Endy. Refinement and standardization of synthetic biological parts and devices. *Nature Biotechnology*, 26(7):787–793, July 2008.
2. Brynne C. Stanton, Alec A. Nielsen, Alvin Tamsir, Kevin Clancy, Todd Peterson, and Christopher A. Voigt. Genomic mining of prokaryotic repressors for orthogonal logic gates. *Nature Chemical Biology*, December 2013.
3. Jennifer A. N. Brophy and Christopher A. Voigt. Principles of genetic circuit design. *Nature Methods*, 11(5):508–520, May 2014.
4. Adrian L. Slusarczyk, Allen Lin, and Ron Weiss. Foundations for the design and implementation of synthetic genetic circuits. *Nature Reviews Genetics*, 13(6):406–420, 2012.
5. Stefano Cardinale and Adam P. Arkin. Contextualizing context for synthetic biology – Identifying causes of failure of synthetic biological systems. *Biotechnology Journal*, 7(7):856–866, July 2012.
6. Adam P. Arkin. A wise consistency: Engineering biology for conformity, reliability, predictability. *Current Opinion in Chemical Biology*, November 2013.
7. James T. Macdonald, Chris Barnes, Richard I. Kitney, Paul S. Freemont, and Guy-Bart V. Stan. Computational design approaches and tools for synthetic biology. *Integrative Biology: Quantitative Biosciences from Nano to Macro*, 3(2):97–108, February 2011.
8. Jason G. Lomnitz and Michael A. Savageau. Strategy revealing phenotypic differences among synthetic oscillator designs. *ACS Synthetic Biology*, 3(9):686–701, September 2014.
9. Uri Alon. *An Introduction to Systems Biology: Design Principles of Biological Circuits*, 1st edition. Chapman & Hall/CRC Mathematical and Computational Biology. Chapman and Hall/CRC, July 2006.
10. Darren J. Wilkinson. *Stochastic Modelling for Systems Biology*, 2nd Edition. Chapman and Hall/CRC Mathematical and Computational Biology. Boca Raton, Taylor & Francis, 2011.
11. Steven H. Strogatz. *Nonlinear Dynamics and Chaos: With Applications to Physics, Biology, Chemistry, and Engineering (Studies in Nonlinearity)*, 1st edition. New York, Westview Press, January 2001.
12. Timothy S. Ham, Zinovii Dmytriv, Hector Plahar, Joanna Chen, Nathan J. Hillson, and Jay D. Keasling. Design, implementation and practice of JBEI-ICE: An open source biological part registry platform and tools. *Nucleic Acids Research*, 40(18):e141, October 2012.
13. Neil R. Adames, Mandy L. Wilson, Gang Fang, Mathew W. Lux, Benjamin S. Glick, and Jean Peccoud. GenoLIB: A database of biological parts derived from a library of common plasmid features. *Nucleic Acids Research*, April 2015.
14. Jacob Beal, Ron Weiss, Douglas Densmore, Aaron Adler, Evan Appleton, Jonathan Babb, Swapnil Bhatia, Noah Davidsohn, Traci Haddock, Joseph Loyall, Richard Schantz, Viktor Vasilev, and Fusun Yaman. An end-to-end workflow for engineering of biological networks from high-level specifications. *ACS Synthetic Biology*, 1(8):317–331, August 2012.
15. Yizhi Cai, Brian Hartnett, Claes Gustafsson, and Jean Peccoud. A syntactic model to design and verify synthetic genetic constructs derived from standard biological parts. *Bioinformatics*, 23(20):2760–2767, October 2007.

16. Yizhi Cai, Mandy L. Wilson, and Jean Peccoud. GenoCAD for iGEM: A grammatical approach to the design of standard-compliant constructs. *Nucleic Acids Research*, 38(8):2637–2644, May 2010.

17. Michael Pedersen and Andrew Phillips. Towards programming languages for genetic engineering of living cells. *Journal of the Royal Society, Interface/the Royal Society*, 6 (Suppl 4):S437–S450, August 2009.

18. Lesia Bilitchenko, Adam Liu, Sherine Cheung, Emma Weeding, Bing Xia, Mariana Leguia, J Christopher Anderson, and Douglas Densmore. Eugene – A domain specific language for specifying and constraining synthetic biological parts, devices, and systems. *PloS One*, 6(4):e18882, 2011.

19. Ernst Oberortner and Douglas Densmore. Web-based software tool for constraint-based design specification of synthetic biological systems. *ACS Synthetic Biology*, December 2014.

20. Jacob Beal, Ting Lu, and Ron Weiss. Automatic compilation from high-level biologically-oriented programming language to genetic regulatory networks. *PloS One*, 6(8):e22490, 2011.

21. Michael Hucka, Andrew Finney, Herbert M. Sauro, Hamid Bolouri, John C. Doyle, Hiroaki Kitano, Adam P. Arkin, Ben J. Bornstein, Dennis Bray, Athel Cornish-Bowden, Autumn A. Cuellar, Serge Dronov, Ernst D. Gilles, Martin Ginkel, Victoria Gor, Igor I. Goryanin, Warren J. Hedley, Charles Hodgman, Jan-Hendrik Hofmeyr, Peter J. Hunter, Nick S. Juty, Jay L. Kasberger, Andreas Kremling, Ursula Kummer, Nicolas Le Novère, Les M. Loew, Daniel Lucio, Pedro Mendes, Eric Minch, Eric D. Mjolsness, Yoichi Nakayama, Melanie R. Nelson, Poul F. Nielsen, Takeshi Sakurada, James C. Schaff, Bruce E. Shapiro, Tom S. Shimizu, Huge D. Spence, Jorg Stelling, Kouichi Takahashi, Marasu Tomita, John Wagner, Jian Wang, and SBML Forum. The systems biology markup language (SBML): A medium for representation and exchange of biochemical network models. *Bioinformatics*, 19(4):524–531, March 2003.

22. Benjamin J. Bornstein, Sarah M. Keating, Akiya Jouraku, and Michael Hucka. LibSBML: An API library for SBML. *Bioinformatics*, 24(6):880–881, March 2008.

23. Autumn A. Cuellar, Catherine M. Lloyd, Poul F. Nielsen, David P. Bullivant, David P. Nickerson, and Peter J. Hunter. An overview of CellML 1.1, a biological model description language. *Simulation*, 2003.

24. Catherine M. Lloyd, Matt D. B. Halstead, and Poul F. Nielsen. CellML: Its future, present and past. *Progress in Biophysics and Molecular Biology*, 85(2–3):433–450, June 2004.

25. Michael L. Blinov, James R. Faeder, Byron Goldstein, and William S. Hlavacek. BioNetGen: Software for rule-based modeling of signal transduction based on the interactions of molecular domains. *Bioinformatics*, 20(17):3289–3291, November 2004.

26. James R. Faeder, Michael L. Blinov, and William S. Hlavacek. Rule-based modeling of biochemical systems with BioNetGen. *Methods in Molecular Biology (Clifton, N.J.)*, 500:113–167, 2009.

27. Stéphanie Rialle, Liza Felicori, Camila Dias-Lopes, Sabine Pérès, Sanaâ El Atia, Alain R. Thierry, Patrick Amar, and Franck Molina. BioNetCAD: Design, simulation and experimental validation of synthetic biochemical networks. *Bioinformatics*, 26(18):2298–2304, September 2010.

28. Vincent Danos and Cosimo Laneve. Formal molecular biology. *Theoretical Computer Science*, 325(1):69–110, September 2004.

29. Vincent Danos, Jérôme Feret, Walter Fontana, and Jean Krivine. Scalable simulation of cellular signaling networks. In *Programming Languages and ...*, pp. 139–157. Berlin, Heidelberg: Springer Berlin Heidelberg, 2007.

30. Jonathan R. Karr, Jayodita C. Sanghvi, Derek N. Macklin, Miriam V. Gutschow, Jared M. Jacobs, Benjamin Bolival, Nacyra Assad-Garcia, John I. Glass, and Markus W. Covert. A whole-cell computational model predicts phenotype from genotype. *Cell*, 150(2):389–401, July 2012.

31. Stefan Hoops, Sven Sahle, Ralph Gauges, Christine Lee, Jürgen Pahle, Natalia Simus, Mudita Singhal, Liang Xu, Pedro Mendes, and Ursula Kummer. COPASI – A COmplex PAthway SImulator. *Bioinformatics*, 22(24):3067–3074, December 2006.

32. Sebastian Mirschel, Katrin Steinmetz, Michael Rempel, Martin Ginkel, and Ernst Dieter Gilles. PROMOT: Modular modeling for systems biology. *Bioinformatics*, 25(5):687–689, March 2009.

33. Mario A. Marchisio and Jorg Stelling. Computational design of synthetic gene circuits with composable parts. *Bioinformatics*, 24(17):1903–1910, September 2008.

34. Aneil Mallavarapu, Matthew Thomson, Benjamin Ullian, and Jeremy Gunawardena. Programming with models: Modularity and abstraction provide powerful capabilities for systems biology. *Journal of the Royal Society, Interface/the Royal Society*, 6(32):257–270, March 2009.

35. Lucian P. Smith, Frank T. Bergmann, Deepak Chandran, and Herbert M. Sauro. Antimony: A modular model definition language. *Bioinformatics*, 25(18):2452–2454, August 2009.

36. Carlos F. Lopez, Jeremy L. Muhlich, John A. Bachman, and Peter K. Sorger. Programming biological models in Python using PySB. *Molecular Systems Biology*, 9:646, 2013.

37. Michael W. Sneddon, James R. Faeder, and Thierry Emonet. Efficient modeling, simulation and coarse-graining of biological complexity with NFsim. *Nature Methods*, 8(2):177–183, February 2011.

38. Guillermo Rodrigo and Alfonso Jaramillo. AutoBioCAD: Full biodesign automation of genetic circuits. *ACS Synthetic Biology*, 2(5):230–236, May 2013.

39. Juliane Liepe, Chris Barnes, Erika Cule, Kamil Erguler, Paul Kirk, Tina Toni, and Michael P. H. Stumpf. ABC-SysBio – Approximate Bayesian computation in Python with GPU support. *Bioinformatics*, 26(14):1797–1799, July 2010.

40. Chris P. Barnes, Daniel Silk, Xia Sheng, and Michael P. H. Stumpf. Bayesian design of synthetic biological systems. *Proceedings of the National Academy of Sciences of the United States of America*, 108(37):15190–15195, September 2011.

41. Chris P. Barnes, Daniel Silk, and Michael P. H. Stumpf. Bayesian design strategies for synthetic biology. *Interface Focus*, 1(6):895–908, December 2011.

42. Fusun Yaman, Swapnil Bhatia, Aaron Adler, Douglas Densmore, and Jacob Beal. Automated selection of synthetic biology parts for genetic regulatory networks. *ACS Synthetic Biology*, 1(8):332–344, August 2012.

43. Alec A. K. Nielsen, Thomas H. Segall-Shapiro, and Christopher A. Voigt. Advances in genetic circuit design: Novel biochemistries, deep part mining, and precision gene expression. *Current Opinion in Chemical Biology*, November 2013.

44. Michael Zuker. Mfold web server for nucleic acid folding and hybridization prediction. *Nucleic Acids Research*, 31(13):3406–3415, 2003.

45. Ye Ding, Chi Yu Chan, and Charles E. Lawrence. Sfold web server for statistical folding and rational design of nucleic acids. *Nucleic Acids Research*, 32(Web Server issue):W135–41, July 2004.

46. Howard M. Salis, Ethan A. Mirsky, and Christopher A. Voigt. Automated design of synthetic ribosome binding sites to control protein expression. *Nature Biotechnology*, 27(10):946–950, 2009.

47. Alan Villalobos, Jon E. Ness, Claes Gustafsson, Jeremy Minshull, and Sridhar Govindarajan. Gene designer: A synthetic biology tool for constructing artificial DNA segments. *BMC Bioinformatics*, 7(1):285, 2006.

48. Swapnil Bhatia and Douglas Densmore. Pigeon: A design visualizer for synthetic biology. *ACS Synthetic Biology*, 2(6):348–350, June 2013.

49. Chris J. Myers, Nathan Barker, Kevin Jones, Hiroyuki Kuwahara, Curtis Madsen, and Nam-Phuong D. Nguyen. iBioSim: A tool for the analysis and design of genetic circuits. *Bioinformatics*, 25(21):2848–2849, November 2009.

50. Nicholas Roehner and Chris J. Myers. A methodology to annotate systems biology markup language models with the synthetic biology open language. *ACS Synthetic Biology*, 3(2):57–66, February 2014.

51. Deepak Chandran, Frank T. Bergmann, and Herbert M. Sauro. TinkerCell: Modular CAD tool for synthetic biology. *Journal of Biological Engineering*, 3:19, 2009.

52. Douglas Densmore, Anne Van Devender, Matthew Johnson, and Nade Sritanyaratana. A platform-based design environment for synthetic biological systems. In *TAPIA'09: The Fifth Richard Tapia Celebration of Diversity in Computing Conference: Intellect, Initiatives, Insight, and Innovations*. ACM Request Permissions, April 2009.

53. Anthony D. Hill, Jonathan R. Tomshine, Emma M. B. Weeding, Vassilios Sotiropoulos, and Yiannis N. Kaznessis. SynBioSS: The synthetic biology modeling suite. *Bioinformatics*, 24(21):2551–2553, November 2008.

54. Howard Salis, Vassilios Sotiropoulos, and Yiannis N. Kaznessis. Multiscale Hy3S: Hybrid stochastic simulation for supercomputers. *BMC Bioinformatics*, 7:93, 2006.

55. Linh Huynh, Athanasios Tsoukalas, Matthias Köppe, and Ilias Tagkopoulos. SBROME: A scalable optimization and module matching framework for automated biosystems design. *ACS Synthetic Biology*, 2(5):263–273, May 2013.

56. Linh Huynh and Ilias Tagkopoulos. Optimal part and module selection for synthetic gene circuit design automation. *ACS Synthetic Biology*, 3(8):556–564, August 2014.

57. Linh Huynh and Ilias Tagkopoulos. Fast and accurate circuit design automation through hierarchical model switching. *ACS Synthetic Biology*, April 2015.

58. Joanna Chen, Douglas Densmore, Timothy S. Ham, Jay D. Keasling, and Nathan J. Hillson. DeviceEditor visual biological CAD canvas. *Journal of Biological Engineering*, 6(1):1, 2012.

59. Nathan J. Hillson, Rafael D. Rosengarten, and Jay D. Keasling. j5 DNA assembly design automation software – ACS Synthetic Biology (ACS Publications). *ACS Synthetic Biology*, 2011.

9

New Genetic Codes

Heinz Neumann

CONTENTS

9.1 Introduction

The genetic code is the key to the translation of the information stored in the form of DNA into the functional world of proteins. DNA stores information in the form of a linear sequence of the four bases: adenine (A), cytosine (C), guanine (G) and thymine (T). In protein coding regions, combined in triplets, these bases encode a corresponding amino acid eventually present in a protein. Both polymers are co-linear and a sophisticated machinery has evolved that communicates between these two regimes that are so different in their chemical nature. The flow of information from nucleic acids to proteins is irreversible, which was summarised by Francis Crick as the central dogma of molecular biology. The central components of the genetic code, the ribosomes, tRNAs and aminoacyl-tRNA synthetases (AARSs), are common to all domains of life on Earth, suggesting that our last common ancestor already possessed essentially the same decoding machinery as present day organisms. It is possible that the possession of this machinery represented such an enormous advantage that all remaining forms of life were outcompeted. Subsequently, the genetic code remained unaltered since any changes to its nature would lead to pleiotropic effects on essentially the entire proteome, thus freezing the state of the genetic code once and for all. The past 15 years have changed this picture dramatically. Many organisms have acquired (by human intervention) additional coding capacities, thereby facilitating the production of proteins with a vast range of new chemical and physical properties.

The genetic code is set by (i) the complementary interaction of mRNA codons and tRNA anticodons and (ii) the high fidelity recognition between AARSs and their substrate amino acid with which they charge cognate tRNAs. In this way, all of the 64 possible triplet codons are assigned to one of the 20 canonical amino acids or decode the termination of translation. Genetic code expansion exploits the degeneracy of this assignment by introducing an additional, novel pair of tRNA and AARS usually targeted towards the relatively rare amber stop codon, UAG. Under the prerequisite that the novel tRNA/AARS pair does not cross-react with the host's tRNA and AARS repertoire (i.e. 'orthogonal' to the endogenous tRNAs and AARSs), its presence will alter the decoding properties of the host cell by reassigning the amber codon to the amino acid recognised by the AARS (Figure 9.1). This strategy was first successfully demonstrated in *E. coli* by Furter in 1998 using the yeast PheRS/tRNA$^{Phe}_{CUA}$ pair [1] (see Figure 9.2 for an overview of conventional tRNA notation). Soon after the Schultz laboratory reported the successful evolution of the TyrRS/tRNA$^{Tyr}_{CUA}$ pair derived from *M. jannaschii* [2] towards recognition of new amino acids in *E. coli*. Key to success was the development of a powerful selection system that enabled the group to identify AARS variants with the desired specificity from a large library of variants (usually >10^9) containing a set of targeted mutations in active site residues (Figure 9.3).

In the first round of positive selection the mutant library is used to transform an *E. coli* strain harbouring an antibiotic resistance gene (usually chloramphenicol acetyltransferase)

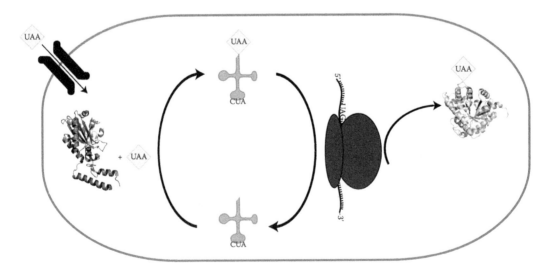

FIGURE 9.1
Engineering translation. Genetic code expansion is achieved by the heterologous expression of an evolved orthogonal tRNA/aminoacyl-tRNA synthetase pair in the host organism. The unnatural amino acid (UAA) is provided with growth medium, taken up by endogenous transport systems and activated by the AARS. The tRNA mediates the incorporation of the UAA at the genetically determined position (in most cases amber codons).

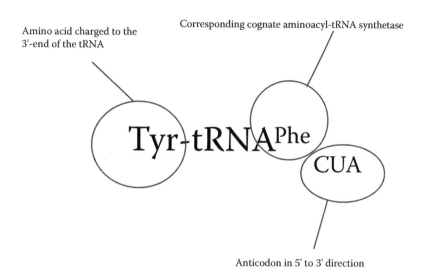

FIGURE 9.2
Conventional tRNA notation.

interrupted by an amber codon. In the presence of an active amber suppressor tRNA and synthetase pair, recognizing the unnatural amino acid (UAA) or a natural amino acid, this codon is suppressed, conferring resistance to the antibiotic to the host cell. Thus, by growing the library of cells on media containing the antibiotic and the UAA, only those will survive that contain an active AARS mutant. AARSs that recognise natural amino acids are eliminated in a subsequent round of negative selection in the absence of the UAA. Here the amber codon is placed into a toxic gene that will kill the cell in the presence of an active suppressor pair. In reiterative rounds of positive and negative selection, the AARS variant of interest is eventually isolated from the library. Almost one hundred unique UAAs have been added to the genetic code of various organisms by this approach [3] (Figure 9.4). As detailed further in the following sections, the addition of UAAs to the genetic code offers a vast potential for applications in protein chemistry, biochemistry and cell biology. UAAs can be used as spectroscopic probes, to photo-cage proteins, for labelling in bioorthogonal reactions or to introduce post-translational modifications and much more.

9.2 Available Orthogonal Pairs and Their Host Systems

9.2.1 Prokaryotes

In 2001, Wang et al. published the first evolved version of the tyrosyl-tRNA synthetase/tRNA$_{CUA}$ pair from *Methanococcus jannaschii* using *E. coli* as the host system [2]. Since then this pair has been parent to numerous variants that incorporate diverse structural amino acid analogues, usually containing small aromatic ring systems. This system has proven very powerful for the expansion of the genetic code of *E. coli* but is unfortunately not orthogonal in most other host systems. The natural amber suppressor pair (pyrrolysyl-tRNA synthetase, PylS and its cognate tRNA, PylT) found in methanogenic archaea (Methanosarcinae) encoding the 22nd coding amino acid pyrrolysine [4, 5] is a more recent addition to the toolbox of synthetic biologists [6, 7]. Pyrrolysine is a lysine derivative

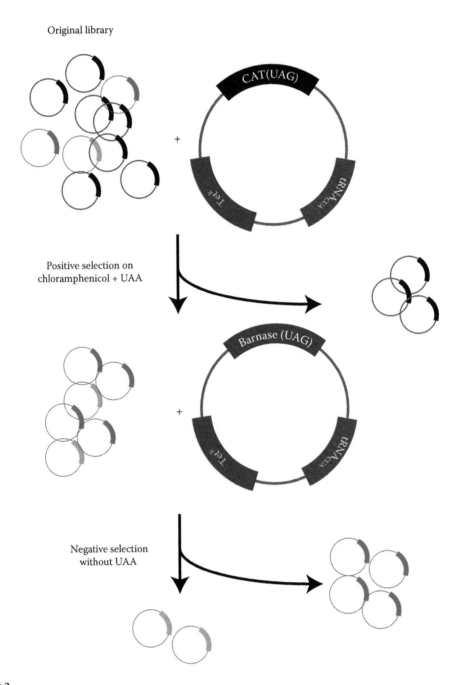

FIGURE 9.3

Selection strategy for aminoacyl-tRNA synthetase mutants with altered amino acid specificity. First, a library of AARSs with mutations in the active site is used to transform *E. coli* cells harbouring a reporter plasmid encoding an antibiotic resistance gene interrupted by an amber codon. Only cells obtaining an active AARS that activates the UAA provided with the growth medium (yellow) or a natural amino acid (green) will survive, while inactive AARSs (black) will be removed from the pool. In the second round of selection, AARSs activating natural amino acids will be removed using a reporter plasmid encoding a toxic gene (e.g. RNAse barnase) interrupted by amber codons. This selection procedure may be iterated until AARS clones of choice are isolated from the library.

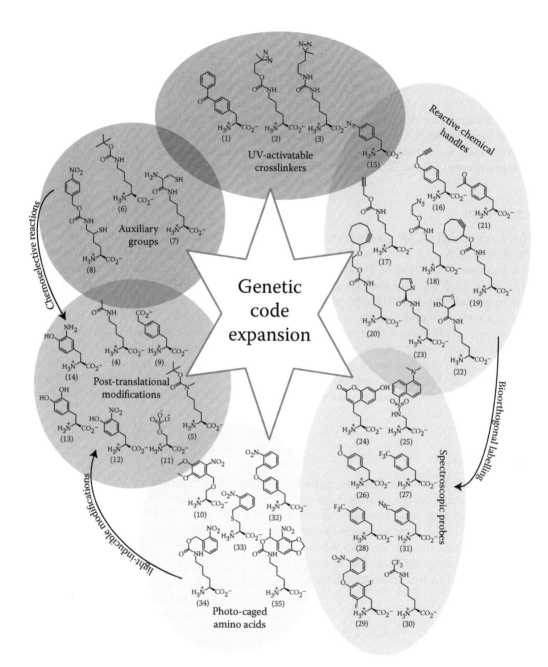

FIGURE 9.4
An immense chemical toolkit provided by genetically encoded unnatural amino acids (UAA). UAAs are numbered with respect to their appearance in the main text.

containing a five-membered pyrroline ring connected to the ε-amino group. The active site of the synthetase harbours a complementary hydrophobic cavity [8] which is malleable and able to adapt to various other functionalities (Figure 9.4). An additional advantage of the PylS/PylT system is the unusual structure of the tRNA which confers orthogonality in almost every host system tested so far.

Besides the well-established *Methanococcus jannaschii* TyrRS/tRNA$_{CUA}$ and *Methanosarcina barkeri/mazei* PylRS/PylT pairs several other options are available but have not been used extensively. The lysyl-tRNA synthetase/tRNA pair of *Pyrococcus horikoshii* has been used to create a frameshift suppressor encoding homoglutamine [9]. Multiple, mutually orthogonal prolyl-tRNA synthetase/tRNA pairs with different anticodons have been evolved from *P. horikoshii* ProRS, which may be used to evolve synthetases for the incorporation of proline derivatives [10]. Similarly, a tryptophanyl-tRNA synthetase/tRNA pair has been derived from *S. cerevisiae* TrpRS [11] which was recently improved and evolved to encode several different tryptophan analogues [12].

9.2.2 Yeasts

Three aminoacyl-tRNA synthetase/tRNA pairs have been used successfully to expand the genetic code of *S. cerevisiae* [13–15], *C. albicans* [16] and *P. pastoris* [17]. The first report of a successful addition of UAAs to the genetic code of *S. cerevisiae* was published by Chin et al. in 2003. The authors evolved the *E. coli* tyrosyl-tRNA synthetase/tRNA$_{CUA}$ pair to encode several different phenylalanine analogues. This pair has been used in numerous studies and proved orthogonal in mammalian cells. Therefore, it is possible to generate synthetase variants in this genetically easy-to-manipulate lower eukaryote and subsequently transplant the system to more complex cell types. Even more generally applicable (see Section 9.2.1) is the PylS/T system, which has also been adapted for *S. cerevisiae* [14]. The third alternative is derived from *E. coli* leucyl-tRNA synthetase/tRNA$_{CUA}$ and was used to encode photo-caged derivatives of serine and cysteine in *S. cerevisiae* [15, 18].

9.2.3 Metazoans

Directed evolution of AARSs has not been attempted in mammalian cell systems. The general approach for generating new genetic codes in mammalian systems is to transplant variants that were successfully evolved in *E. coli* or *S. cerevisiae* to higher eukaryotes. PylS/T is the most versatile system that can be readily modified to encode UAAs in mammalian cells [19, 20] and whole animals, such as *Caenorhabditis elegans* [21] and *Drosophila melanogaster* [22]. The *E. coli* TyrRS/tRNA$_{CUA}$ and LeuRS/tRNA$_{CUA}$ pairs have also been used in mammalian cells and *C. elegans* [23–25].

9.3 Using UAAs to Encode Post-Translational Modifications

Most proteins are modified chemically or enzymatically at different amino acid side chains often with dramatic effects on their activity, affinity for binding partners or subcellular localization. Many post-translational modifications have been mapped across the proteome of different cell types by mass spectrometry. However, for most of these modifications neither their impact on protein function nor the modifying enzymes have been identified. Genetic code expansion offers the possibility to directly encode many of these modifications and hence to produce modified proteins without the need to know the enzymes involved.

9.3.1 Lysine Acetylation

Lysine acetylation was initially observed on histones and soon correlated with transcriptional activity [26]. Lysine acetyltransferases were only identified three decades later and it took yet another decade until the enormous prevalence of this modification across the entire proteome was appreciated. The biological consequences of most of these acetylations have yet to be identified, which requires the ability to produce the modified proteins in sufficient quantities for biochemical analysis. The Chin lab has developed a method to install N(ϵ)-acetyl-lysine (compound 4) in recombinant proteins of *E. coli* using an evolved variant of pyrrolysyl-tRNA$_{CUA}$ synthetase from *Methanosarcina barkeri* [7]. Astonishingly, the parent tRNA$_{CUA}$/AARS pair is used by these methanogens to encode the 22nd canonical amino acid, pyrrolysine, in response to amber codons [4, 5]. Work by the Söll laboratory had shown that this pair is functional in *E. coli* [27, 28] while Chin and colleagues were subsequently able to establish a selection system for synthetase variants specific for alternative amino acids such as N(ϵ)-acetyl-lysine [7]. The first protein containing an encoded N(ϵ)-acetyl-lysine residue encoded by this approach showed signs of post-translational demodification by an endogenous lysine deacetylase of *E. coli*. After blocking the enzyme's activity with nicotinamide, cleanly acetylated proteins were obtained. Numerous acetylated proteins have been produced using this method (Figure 9.5). Chin and colleagues investigated the impact of acetylation of histone H3 at lysine 56 on the stability of nucleosomes [29]. This modification had been observed on the nascent protein and implicated in the modulation of chromatin structure during DNA replication and repair. Single-molecule experiments by the group of John van Noort at Leiden University provided a possible molecular explanation for these observations, showing that the acetylation indeed increased the dynamics of binding and dissociation of the DNA near the entry/exit site on the nucleosome. Robert Schneider and colleagues used this technology to produce histone H3 acetylated at Lys-122 towards the C-terminal end of the protein but still in the folded core of the nucleosome. They found that H3 K122ac stimulates transcriptional activation probably by directly affecting histone-DNA binding [30]. In another study, Lammers et al. investigated the influence of acetylation of K125 of cyclophilin A on its activity [31]. CypA is an abundant peptidylprolyl isomerase involved in HIV infection. Its acetylation modulates its enzymatic activity and binding to HIV capsid. This work illustrates the variability in the mechanisms by which the addition of a small acetyl group can influence different aspects of protein function.

This strategy to produce cleanly acetylated proteins is used now by many laboratories to study the impact of this modification on protein function. For example, Hsieh et al. found that acetylation of Ubc9 at Lys65, an E2 ligase of the SUMO pathway, modulates its substrate preference [32] and Ullmann et al. found that acetylation of SUMO itself affects binding to SUMO-interacting motifs [33].

9.3.2 UAAs as Synthetic Biology Tools to Direct Lysine Methylation and Ubiquitination

Lysine modifications are not restricted to acetylation. Methylation of the ϵ-amino group is a prominent modification on histone proteins but also on non-histone proteins. A sophisticated enzymatic machinery controls the addition and removal of one to three methyl groups from lysine side chains. Several domains of proteins have been identified that recognise the methylation state of their target proteins to affect downstream regulatory processes. Ubiquitin and related small proteins, such as SUMO, are attached to proteins via

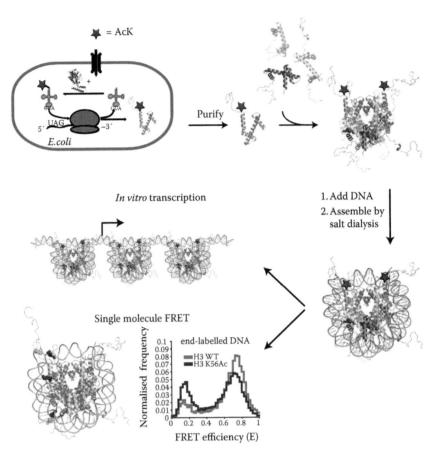

FIGURE 9.5
Genetically encoded acetyllysine has been used to study the consequences of this modification on chromatin. Recombinant histones containing the modification are produced in *E. coli*, purified and mixed with the remaining core histones to refold histone octamers. These are subsequently used to reconstitute nucleosomal core particles by spontaneous assembly with DNA upon gradual salt removal. These can be used in further experiments to study the role of the modification in nucleosome stability (Neumann et al., 2009) or chromatin dependent processes, for example transcription (Tropberger et al., 2013).

lysine side chains. Monomeric ubiquitylation is involved in many cellular processes, such as protein trafficking, transcription or DNA repair. Polymeric chains of ubiquitin (linked via Lysine 48 of ubiquitin) signal the degradation of the target protein, but the modification has many functions beyond this destructive task.

These post-translational modifications of lysine have so far escaped direct decoding by an evolved amber suppressor pair. Efforts to evolve pyrrolysyl-tRNA$_{CUA}$ synthetase towards N(ϵ)-methyl-lysine never produced a specific enzyme (personal observations). The synthetase recognises the carbonyl-oxygen of N(ϵ)-acetyl-lysine which is missing in N(ϵ)-methyl-lysine [8]. The residues involved in its recognition constitute a selectivity filter which, when mutated, loses the ability to discriminate pyrrolysine from other amino acids. The chemical difference between N(ϵ)-methyl-lysine and lysine is presumably too small to allow the creation of a specific enzyme. To circumvent this problem, Nguyen et al. incorporated N(ϵ)-*tert*.-butyl-oxycarbonyl-N(ϵ)-methyl-lysine (compound 5) into histone proteins [34]

(Figure 9.6A). The chemically labile *tert.*-butyl-oxycarbonyl group could then be removed post-translationally revealing the monomethylated lysine. Unfortunately, this strategy is not applicable to higher substituted N(ε)-methyl-lysines. Jason Chin and his co-workers therefore extended this protective group chemistry to all lysine residues within a protein (Figure 9.6B). Again using a histone protein as an example they encoded the lysine residue to be modified with a protection group (compound 6). After purifying the protein, they chemically protected the remaining lysine residues with a different group and subsequently identified conditions to specifically deprotect only the lysine of interest. The resulting protein contained a single reactive lysine residue which they could dimethylate chemically producing a site-specifically modified protein after removal of the remaining protection groups [35].

Using a similar strategy, combined with chemoselective ligation, the groups of Jason Chin and David Komander managed to produce diubiquitins connected by a native linkage via lysine residues other than the usual Lys48 or Lys63 (Figure 9.7). Screening a library of deubiquitinases with their diubiquitins as substrates, they could identify an enzyme specific for the Lys29 linkage [36]. An alternative strategy was followed by Chan and co-workers who incorporated N(ε)-Cys-lysine (compound 7) and used it as a handle in native chemical ligation. This allowed them to conjugate ubiquitin to calmodulin, which decreased its ability to activate phosphorylase kinase but not phosphatase 2B [37]. A disadvantage of installing (compound 7) is that ubiquitin is not attached via a native linkage, which may disturb the interaction with effector proteins or deubiquitinases. An elegant solution to this issue was developed by the Chin lab who incorporated a protected δ-thiol-lysine [8, 38]. The protection group is needed as a handle for the recognition by the synthetase but unstable under the reducing conditions of the *E. coli* cytosol, thus producing proteins containing δ-thiol-lysine at a genetically determined position. The thiol serves as an auxiliary group in native chemical ligation and is subsequently removed by desulfurization.

These studies demonstrate the enormous potential of this approach to investigate post-translational modifications of lysine. Still, the range of moieties found to be present on lysine side chains in nature goes far beyond synthetic biology's present abilities to produce proteins with such modifications directly. As such, synthetic biologists are currently lacking tools to produce ADP-ribosylated proteins, a modification that plays an important role in chromatin biology. Further ingenious combinations of genetic code expansion and chemoselective reactions are necessary to tackle these challenges.

9.3.3 Applications of UAAs as Synthetic Biology Tools to Investigate Serine and Tyrosine Phosphorylation

Phosphorylation of tyrosine, serine and threonine residues are widespread post-translational modifications of eukaryotic proteins with important roles in the regulation of protein function and signal transduction. Genetic code expansion has been used to incorporate *p*-carboxymethyl-phenylalanine (compound 9) into recombinant proteins of *E. coli* [39]. Functional analysis of a protein containing this UAA in place of a phosphorylatable tyrosine residue demonstrated its usefulness as a non-hydrolysable phosphotyrosine analogue. Another study by the Schultz lab installed a photo-caged derivative of serine (compound 10) in proteins of *S. cerevisiae* [18]. They could show that this approach can be used to photo-control the phosphorylation of a transcription factor, thereby enabling experiments that monitor nuclear transport processes in life cells.

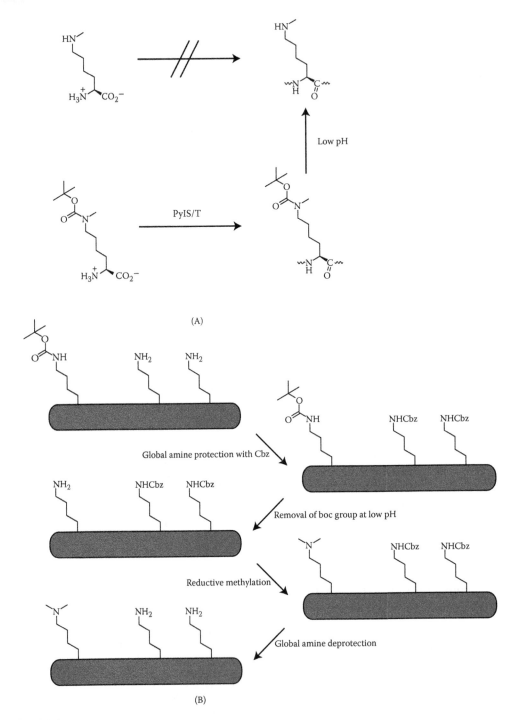

FIGURE 9.6
Installation of mono- and di-methyl-lysine on histone proteins. (A) Strategy to install N(ε)-methyl-lysine by genetically encoding N(ε)-*tert.*-butyl-oxycarbonyl-N(ε)-methyl-lysine (compound 5) and subsequent removal of the protection group (Nguyen et al., 2009). (B) Strategy to install N(ε)-dimethyl-lysine by encoding N(ε)-*tert.*-butyl-oxycarbonyl-lysine (compound 6) followed by several chemical modification steps (Nguyen et al., 2010).

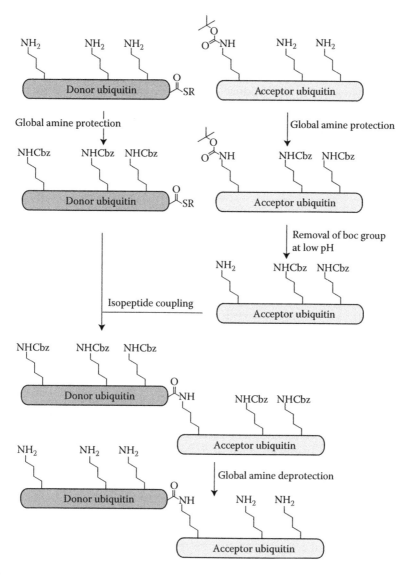

FIGURE 9.7
Strategy for the production of di-ubiquitins using a combination of chemical protection/deprotection of lysine residues and chemoselective ligation (Virdee et al., 2010).

Söll and colleagues succeeded in expanding the genetic code of *E. coli* with phosphoserine (compound 11) [40]. They made use of the fact that a phosphoseryl-tRNA synthetase (SepRS) is involved in cysteine production in archaea. Of course, this enzyme is a prime candidate to engineer bacteria with the ability to encode phosphoserine. However, the charged tRNA[Sep] is not a good substrate for bacterial EF-Tu, a problem that the authors overcame by engineering the elongation factor by directed evolution. The final system was able to introduce phosphoserine in response to the amber codon albeit with still rather low efficiency. Recently the system has been improved to produce recombinant histone H3 phosphorylated at serine-10 with high yields [41].

9.3.4 Synthetic Biology Tools for Investigation of Protein Oxidative Damage

In addition to numerous enzymatic modifications, proteins are also subject to chemical modifications, which often interfere with their function. Since chemical reactions lack the selectivity of enzymatic catalysis, it is nearly impossible to obtain cleanly modified protein for biochemical analysis. A general strategy to produce proteins containing 3-nitro-tyrosine (compound 12), a modification caused by nitrogen containing radicals and a marker of vascular disease, has been developed by the labs of Jason Chin and Ryan Mehl [42, 43]. Manganese superoxide dismutase with a genetically encoded 3-NT residue at position 34 showed dramatically reduced activity, confirming previous observations with chemically nitrated protein. Further oxidative damage modifications of tyrosine, for example 3-hydroxylation (compound 13) or 3-amination (compound 14), are also available using genetic code expansion [3].

9.4 UAAs as Synthetic Biology Tools to Detect Interactions Occurring within Cells

A very powerful application of genetic code expansion is the incorporation of UV-activatable photo-crosslinkers, most importantly because it allows the detection of interactions in living cells. Several alternative crosslinker UAAs have been genetically encoded.

9.4.1 *p*-Benzoyl-Phenylalanine

Judged by the number of publications, *p*-benzoyl-phenylalanine (pBPA, compound 1) is by far the most popular genetically encoded UV-crosslinker. This crosslinker has been installed on proteins in bacteria, yeast and mammalian cells [13, 23, 44] and used to investigate protein–protein and protein–DNA interactions *in vivo*. Among the first applications of pBPA in *E. coli* was to map the interaction surface between the translocon (SecYEG) and SecA, the ATPase which provides the energy required for protein translocation across the plasma membrane [45, 46]. These studies provided valuable information on their mutual interactions in solution, corroborating data obtained by X-ray crystallography [46]. By replacing a conserved tyrosine residue in the bacterial AAA+ chaperone ClpB, Schlieker et al. could show that the enzyme channels aggregated substrate proteins through its central pore like a molecular vacuum cleaner [47]. The interaction of trigger factor with nascent proteins emerging from the ribosomal exit tunnel was investigated by placing pBPA at various positions on the chaperone [48, 49]. These studies revealed the sequence of events during trigger factor binding and suggest a high degree of flexibility towards the nature of the nascent chain. In yeast, pBPA has been employed to analyse the interactions between subunits of the mitochondrial protein import machinery [50–52], between transcription factors and RNA polymerases [53, 54] and between a substrate protein and components of the ER associated degradation (ERAD) pathway [55], for instance. In all these cases, individual interactions were investigated that were predicted from prior *in vitro* studies. The *in vivo* experiments provide valuable insights into the biological relevance of these interactions and offer the opportunity to study the influence of mutations in the genetic background (e.g. by blocking post-translational modifications and up- or downstream pathways). The identification of novel interaction partners by this approach is a much more difficult task. A candidate approach employs gel shift assays in SDS–PAGE together with genetically tagged versions of potential target proteins. The fusion will increase the mass of the crosslink product compared with the same crosslink in an untagged strain. This results in a slower migration of the cross-linked protein in SDS–PAGE which can be detected by Western blot. Unfortunately, this approach requires the manual inspection of each individual candidate protein, limiting it to a small number of potential interaction partners. The identification of crosslink products by mass spectrometry may facilitate the identification of unexpected protein–protein interactions. However, such an unbiased mass spectrometric survey on protein–protein interactions trapped *in vivo* by pBPA crosslinking has not yet been performed. The reason is probably the low amount of material produced by UV-crosslinking and potentially the hydrophobicity of pBPA, which may lead to difficulties during MS analysis. Improvements to the incorporation efficiency, novel crosslinker UAAs and more sensitive mass spectrometers will certainly enable such experiments. An important tool to facilitate such an analysis has been developed by Ashton Cropp and colleagues by producing a perdeuterated version of pBPA for mass fingerprinting [56].

Identifying the specific peptides involved in crosslinking is extremely valuable when mapping protein–protein interaction surfaces and it is also experimentally challenging. Usually microgram quantities of the crosslinked proteins are required to facilitate its identification by mass spectrometry. Fragmentation of the crosslinked peptide may be impeded by the hydrophobic nature of pBPA. Also the size of the two peptides participating in the crosslink has a strong influence on its stability during MS–MS fragmentation. The larger the peptides are, the more complex the fragmentation patterns and the more difficult is their detection. Forné et al. have studied the conformation of the chromatin remodeler ISWI (Imitation *switch*) using pBPA and used the identified crosslinks as constraints to model its structure from homologous proteins of known structure [57]. To identify the crosslinked peptides, they developed an algorithm to mine the MS–MS data.

The techniques described earlier are discussed in the context of investigating static interactions between proteins or protein and nucleic acids. However, *in vivo* crosslinking with genetically encoded crosslinkers also has the potential to yield information on dynamic properties of such interactions. If a process can be blocked at different stages (e.g. by deletion of certain factors or inhibitors) or synchronised, the sequence of events can be interrogated. I will illustrate this in the following with two examples.

Gram-negative bacteria, such as *E. coli*, assemble lipopolysaccharide (LPS) molecules on their cell surface each time the cell divides. A multi-subunit protein complex transports LPS from the inner membrane, where it is synthesised, across the periplasm to its destination in the outer membrane. The organisation of the protein complex was recently investigated using pBPA to crosslink LPS with different subunits [58] (Figure 9.8A). Using anti-LPS antibodies the authors could identify several sites that interacted with LPS. Overexpression of certain components of the complex caused LPS to accumulate at different subunits. This revealed the order in which LPS was handed from subunit to subunit during translocation. Moreover, by blocking ATP hydrolysis in an *in vitro* transport assay they could show that translocation between the inner membrane proximal subunit and the next was energy dependent. These elegant experiments revealed the sequence of events during LPS transport and showed that multiple rounds of ATP hydrolysis are required to pump LPS against the concentration gradient between the inner and outer membrane.

The second example is from recent findings of my own laboratory. Cell division is characterised by the formation and separation of X-shaped metaphase chromosomes. While this process was first observed more than 120 years ago, our understanding of their structure and the forces driving their formation remains rudimentary. Using genetically

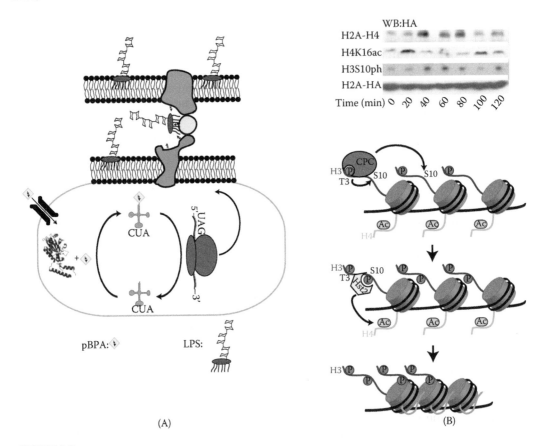

FIGURE 9.8

Genetically encoded crosslinkers trap protein–protein interactions *in vitro* and *in vivo*. (A) Transport of LPS molecules from the inner to the outer bacterial membrane is facilitated by a multi-subunit complex and driven by ATP hydrolysis. The individual steps in transport were investigated *in vivo* by trapping transport intermediates by UV-crosslinking to pBPA incorporated at different positions along the transport channel (Okuda et al., 2012). (B) Condensation of chromosomes in mitosis is induced by histone–histone interaction that is controlled by a cascade of modification events. The inter-nucleosomal interaction could be trapped *in vivo* using genetically encoded pBPA and was followed along the cell cycle in synchronised yeast. The sequence of upstream events was identified using mutant strains (Wilkins et al., 2014).

encoded UV-crosslinkers we identified a histone–histone interaction in living yeast, which we followed along the cell cycle (Figure 9.8B). We could show that this interaction takes place between histone H4 and histone H2A present in different nucleosomes and that it adds a driving force to chromosome condensation in mitosis. We could further reveal the events regulating this interaction by identifying a signalling cascade that involves phosphorylation of histone H3 Ser-10, a long-known hallmark of mitosis. This modification leads to the recruitment of a histone deacetylase which subsequently removes an acetyl group from histone H4 Lys-16. This frees the H4 tail to interact with the surface of neighbouring nucleosomes, thereby driving chromosome fibre condensation [59]. In this work, we combined for the first time crosslinking by genetically encoded crosslinkers with the synchronization of a cellular process. This approach could be extended to other synchronisable processes to investigate the sequence of protein–protein of protein–nucleic acid interaction events involved.

9.4.2 Other Phenylalanine-Based UAAs That Can Be Used as Crosslinkers

p-azido-phenylalanine (pAzF) can be used as a bioorthogonal reactive group to label proteins but also as an UV-activatable crosslinker. Irradiation with UV-light results in the cleavage of N_2 from the azide functional group yielding a highly reactive nitrene which can either react itself or after rearrangement to a ketenimine with C–H bonds in its close proximity. pAzF is smaller and less hydrophobic compared with pBPA, which can be an advantage at sensitive sites. The chemical decomposition of pAzF on the other hand limits the lifetime of the crosslinker, whereas pBPA simply relaxes to the ground state and can be excited again if no suitable reaction partner is available. Hence, different types of crosslinkers should be tried and may give rise to different crosslink patterns. Grunbeck et al. have compared pBPA and pAzF in their ability to trap ligand-GPCR interactions [60]. Interestingly, while certain sites produced crosslinks only when pAzF was incorporated, others required the presence of pBPA. Another alternative was developed by the Schultz lab who encoded a diazirine substituted phenylalanine using the MjTyrRS system [61]. UV excitation creates a carbene with similar reactivity to nitrenes.

9.4.3 Lysine-Based UAAs That Can Be Used as Crosslinkers

One significant drawback of pBPA and other phenylalanine-based crosslinkers is their rather bulky and hydrophobic nature. A slim and hydrophilic alternative is diazirine-modified lysines (compounds 2 & 3) that were successfully added to the genetic code using the PylS/T system [62, 63]. Chen and colleagues used compound 3 to profile the substrates of HdeA, a major acid-protection chaperone of the *E. coli* periplasm [64]. The same crosslinker has been used by this group in other pathogenic bacteria to investigate the client proteins of these chaperones [65]. A furan-based lysine was encoded by Schmidt and Summerer that is converted into a crosslinker upon reaction with singlet oxygen [66]. Conversion of triplet into singlet oxygen by irradiation with red light can be used to control its activity towards nucleic acids. This is an interesting alternative to UV-crosslinkers since longer wavelength light is potentially less cytotoxic and penetrates deeper into tissues.

In general, the great advantage of genetically encoded photo-crosslinkers over alternative crosslinking methods is the ability to obtain positional and therefore structural information on protein complexes in their natural environment.

9.5 Photo-Activatable UAAs

Current developments in the field of super resolution light microscopy allow us to visualise cellular processes with amazing precision. In order to better understand the dynamics of the processes, tools are needed to locally trigger these processes in order to observe for example the flow of information along a signalling cascade. Light is an ideal non-invasive tool to control protein function in a cellular environment potentially with nanometre scale precision. In order to trigger protein function by light, its activity must be masked by a light-sensitive functionality. To achieve this, protein domains that undergo a light-triggered conformational change, such as the LOV domain, have been fused to the protein of interest [67].

Alternatively, UAAs with a photo-cleavable group can be used to replace an important functional residue thereby inhibiting the function of the protein when in darkness. Irradiating the cell with light of appropriate wavelength frees the side chain thereby activating the protein. Photo-caged versions of tyrosine (compound 32), cysteine (compound 33), serine (compound 10) and lysine (compounds 34 and 35) have been genetically incorporated into proteins in bacteria, yeast and mammalian cells [14, 15, 18–20, 68, 69]. Numerous examples demonstrate the potential of this approach. As mentioned earlier, photo-caged serine has been used to control the phosphorylation state of a serine residue and with it the localisation of a transcription factor in yeast [18].

The Chin lab demonstrated that photo-caged lysine can be used to control the recognition of nuclear localisation signals by importins in mammalian cells [20]. Light-triggered activation of the NLS facilitated quantifying the kinetics of nuclear transport processes. By caging an active site lysine residue of MEK1 kinase, the same lab rendered this enzyme photo-activatable. This allowed them to monitor the kinetics of activation within the kinase subnetwork using time-lapse microscopy [70]. Recently, the Wang lab used photo-caged cysteine to control the conductivity of a potassium channel in mouse neurons and was thereby able to photocontrol neuronal activity *in vitro* and *in vivo* [71]. A photo-activatable tyrosine kinase effector was built using a PylS derivative for the incorporation of *o*-nitrobenzyl-*O*-tyrosine, which was used to replace the phosphorylation site of the effector protein [72].

Most of the studies published so far end with the demonstration of feasibility. It is obvious that this approach has great potential and awaits in-depth applications that address complex biological questions. For example, it would be interesting to combine photo-activatable calcium channels with genetically encoded biosensors, for example chameleon, to read out changes in space and time. Also, photo-activation could be combined with single-molecule spectroscopy to observe the consequences of activation.

The UAAs described earlier are 'single use' entities in the way that the activation state can only be changed once. After the caging group is removed the protein will remain in the activated state until it is inactivated by natural means. The ability to reversibly switch the activation state of the protein would facilitate additional experiments. A possible functionality that could be used in this regard is azobenzene, which exists in a *trans*- and *cis*-configuration that is inter-convertible by light. When placed into a suitable location this conformational change can be connected to the protein's activity, for example by blocking access of substrate molecules to the active site. Many studies have employed this approach using chemical means to introduce this functionality into biomolecules (reviewed in ref. [73]). A UAA based on azobenzene has been incorporated into proteins in *E. coli* [74], where it was used to trigger the activity of an endogenous transcription factor. Extending this approach to eukaryotic cells or animals would facilitate a number of exciting experiments.

9.6 Bioorthogonal Chemistry

Site-specific labelling of proteins is a prerequisite for many biochemical and biophysical experiments. The thiol group of cysteine is very nucleophilic and provides an ideal anchor point in cases when otherwise cysteine-free versions of a protein are available. This severe limitation can only be overcome by the use of bioorthogonal chemistries that are independent of the presence or absence of any of the 20 natural amino acids. Several of these 'click'-type reactions have been established for proteins and genetic code expansion provides the tools to introduce the relevant functional groups into the protein (Table 9.1).

TABLE 9.1

Protein Labelling Chemistries

	Remarks
Copper-catalysed Azide–Alkyne coupling (CuAAC)	Cytotoxicity of Cu(I) Difficult to remove Cu(I) Simple, flexible, polar linker UAA easily synthesized
Strain-promoted Azide–Alkyne coupling (SPAAC)	Efficient reaction, bulky linker Complex synthesis of UAA
Inverse-electron demand Diels–Alder reaction (IEDAR)	Very rapid reaction, suitable for pulse-chase Bulky linker Complex synthesis of UAA Limited stability of UAA
Oxime formation	Simple chemistry Labelling reaction requires low pH
Suzuki–Miyaura reaction	Reaction is not quantitative, requires elevated temp.

The paradigm of all click-reactions is the Cu(I)-catalysed cycloaddition between an azide and a terminal alkyne [75]. Azides and alkynes have been incorporated using the tyrosine- and pyrrolysine-derived amber suppressor systems [15–18, 76–79]. This approach becomes problematic in its requirement for Cu(I), which is cytotoxic and difficult to remove from the sample after the reaction.

An alternative to terminal alkynes is the use of cyclooctyne derivatives which react with azides independently of Cu(I) due to the additional ring strain [80, 81]. Cyclooctyne-modified lysines (compounds 19 and 20) have been incorporated using a mutant of PylS, facilitating copper-free click-reactions with azide-substituted ligands *in vivo* [82]. Similar pericyclic reactions with inverse-electron demand offer a very powerful alternative. Tetrazines react voraciously with electron-rich dienophiles, such as norbornene or bicyclo[6.1.0]nonyne. These functionalities have been installed on proteins using the PylS/T system, allowing highly efficient coupling of fluorescent dyes [82–86]. An additional advantage of this strategy is the fluorogenic nature of the reaction, because tetrazines quench the attached fluorophores, thereby reducing the background from uncoupled free dye.

Alternative bioorthogonal chemistries have been developed, expanding the repertoire of available labelling methods. For example, ketone-containing UAAs, such as *p*-acetylphenylalanine (compound 21), can be derivatised with hydroxylamines or hydrazides [87–89]. Unfortunately, the labelling reaction requires a pH < 5, which is not always compatible with protein stability. Nguyen et al. reported the conjugation of fluorophores to genetically encoded 1,2-aminothiols, such as 22, in a cyanobenzothiazole condensation [90]. This reaction is compatible with conventional labelling of cysteines with maleimide-conjugated dyes, hence allowing labelling of proteins with two distinct dyes at genetically predetermined positions. A very elegant strategy developed by the Geierstanger lab introduced pyrrolysine biosynthetic enzymes into the *E. coli* host [91, 92]. These converted D-ornithine into pyrroline-carboxy-lysine (compound 23), a suitable substrate of PylS. They established a conjugation reaction to 2-amino-benzaldehyde or 2-amino-acetophenone reagents to introduce a wide range of labels, biooligomers and small molecules.

With these methods at hand there are obviously many applications for the investigation of protein dynamics. The Lemke lab studied the properties of FG-repeat containing nucleoporins in single-molecule FRET experiments [93]. These proteins form the selectivity barrier of the nuclear pore that controls the flux in and out of the nucleus. The precise physical nature of this barrier is highly debated and the ability to observe its dynamics by this approach promises exciting new insights. Chakraborty et al. used the incorporation of pAzF to label two different subunits of the bacterial RNA polymerase clamp with a FRET pair. This allowed them to study the opening and closing of the polymerase during transcription initiation and elongation in single-molecule FRET experiments [94]. It is to be expected that these labelling methods will find more widespread applications in the future since few alternative methods for the site-specific attachment of fluorophores are available.

9.7 Orthogonal Ribosomes

Ribosomes are the site of all cellular protein synthesis and thus absolutely critical for cell survival. Changing the way ribosomes decode the genetic information would affect the entire proteome and lead to immediate cell death. To render this ancient assembler amenable to directed evolution, Jason Chin and Oliver Rackham identified a strategy to duplicate the prokaryotic translational apparatus. The recognition between mRNA and ribosome is mediated by an interaction between their Shine-Dalgarno and anti-Shine-Dalgarno sequences, mostly via Watson–Crick base pairing. By changing the respective sequences to all possible combinations they identified, in rounds of positive and negative selection, new functional 16S-rRNA/mRNA pairs that would no longer cross-react with their natural counterparts [95]. These orthogonal ribosomes form a parallel independent translational apparatus that may serve as a tool to investigate ribosome function without disturbing cellular processes.

The Weissman laboratory used orthogonal ribosomes to identify sites of ribosome stalling during translation [96]. They showed that normal ribosomes most frequently stall at Shine-Dalgarno-like sequences and that orthogonal ribosomes, due to the altered anti-Shine-Dalgarno sequence, stall at the corresponding sequences instead.

However, the most important feature of these orthogonal ribosomes is that they are dispensable for cellular survival and therefore free to be evolved towards new functions. Subsequent engineering work by the Chin lab identified mutations in the A-site that selectively interfered with the binding of release factor 1 (RF1) [97], rendering these ribosomes highly effective in the suppression of amber codons [98]. This study represents the first case of a directed evolution of the ribosome, which was previously deemed impossible. The selective increase in suppression efficiency, directed only to orthogonal mRNAs, minimises potential side effects caused by the suppression of genomic amber codons. Further work targeted the entire surface of the A-site belonging to the 16S-rRNA to identify ribosomal mutants with enhanced quadruplet decoding properties [99]. Indeed, just two mutations (A1196G, A1197G) enabled the ribosome to accept tRNAs with an extended anticodon with significantly increased efficiency. These ribosomes allow for the production of proteins with two distinct UAAs, which in this study was used to encode a genetically programmable intramolecular crosslink (Figure 9.9). Building on this original system, which encodes six exogenous components on four different plasmids, improved plasmid systems and additional PylT variants with altered anticodons have been developed [100–102]. Future work will certainly further expand the scope and applicability of orthogonal ribosomes enabling the production of proteins with multiple different UAAs or even entirely composed of UAAs.

One important requirement for the design of a cell with a parallel genetic code is the availability of a sufficient number of orthogonal tRNA/AARS pairs. Currently only two pairs – the *M. jannaschii* TyrRS/tRNA$_{CUA}$ and the *M. barkeri*/*M. mazei* PylS/PylT pairs – are being used extensively in the expansion of the genetic code of *E. coli*. Additional pairs need to be orthogonal to the endogenous tRNAs/AARSs and towards each other making the identification of such pairs increasingly more difficult. An alternative to the fortuitous isolation of these pairs from natural sources is the design of new pairs from existing ones. The feasibility of this approach has been demonstrated by the Chin lab, who duplicated the *Mj*TyrRS/tRNA$_{CUA}$ pair in several rounds of mutagenesis and selection to create a new *Mj*TyrRS/tRNA$_{UCCU}$ pair orthogonal to the parental pair [103].

9.8 Systemic Optimization of New Genetic Codes

There are, however, certain limitations towards the chemical nature of UAAs imposed by their cell permeability, the requirements of the ribosome and EF-TU and the size and structure of the AARS's active site. The efficiency of incorporation of UAAs depends on

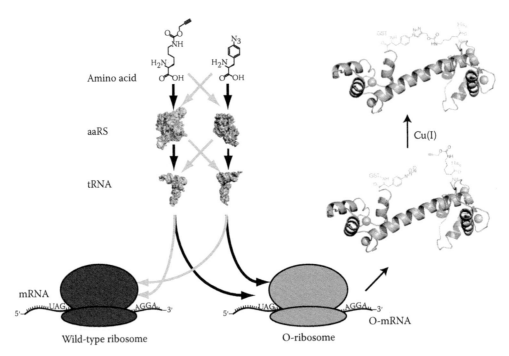

FIGURE 9.9
Orthogonal ribosomes facilitate the genetically encoded incorporation of a directional crosslink. Combination of two evolved orthogonal AARS/tRNA pairs with an evolved orthogonal ribosome and mRNA facilitated the installation of two bioorthogonal reactivities on the same protein (here calmodulin) (Neumann et al., 2010). Exposure to a Cu(I) catalyst induced the formation of a redox-insensitive directional crosslink that may be used to stabilise protein folds or lock conformational states.

several factors. A number of plasmid constructions have been explored, trying to maximise the yield of full-length UAA-containing protein [104–106]. The most convenient solutions are expressions of the AARS and tRNA from a single plasmid, which is compatible with most vectors for recombinant gene expression. A carefully balanced expression of all components is desirable since excessive production of a single component unnecessarily consumes biosynthetic power and is likely to produce toxic side effects [100].

In bacteria, the most significant limitation to protein yield is the competition between RF1 and the suppressor tRNA for binding to an amber codon in the A-site. This is accentuated when more than just a single UAA is to be incorporated into the same protein. Since deletion of RF1 from the *E. coli* genome is lethal [107], shifting this equilibrium to favour UAA incorporation could be achieved by growing strains containing a temperature-sensitive allele of RF1 at the restrictive temperature upon induction of protein expression [108]. This approach, however, is of limited use since many recombinant proteins would not tolerate the elevated temperature and prolonged induction times would lead to cell death. Liu and co-workers overexpressed the C-terminus of ribosomal protein L11 to reduce RF1 mediated termination [109]. This facilitated the incorporation of three identical UAAs into the same model protein in response to amber codons. Yokoyama and Sakamoto reported an RF1 deletion strain created in the background of *E. coli* cells harbouring an amber suppressor tRNA and a bacterial artificial chromosome encoding the seven essential amber-terminated genes of *E. coli* in which the termination signal had been replaced by the codon UAA [110]. They incorporated up to six iodo-tyrosine residues into the same polypeptide using this strain. This work clearly demonstrates the reassignment of the amber stop codon to a sense codon in this strain background. However, the genetic modifications reduce the fitness of the strain and it remains to be seen whether it proves to be a useful tool for genetic code expansion. Wang and colleagues have created a RF1-deletion strain by engineering RF2 (responsible for termination at opal and ochre codons). They propose that RF1 is not essential because of its function in termination at amber codons but rather because in its absence ochre codons (which are recognised by RF1 and RF2) are insufficiently terminated. Their *E. coli* strain is able to incorporate 10 and more identical UAAs into the same protein without showing tremendous growth defects [111]. Recently, the Church and Isaacs laboratory removed all amber codons from the genome of *E. coli* and subsequently deleted the gene for RF1 [112, 113] using multiplex automated genome engineering (MAGE) [114]. In their *E. coli* strain, all UAG codons are converted to UAA relieving the requirement for RF1. In the presence of an amber suppressor tRNA/AARS pair, efficient incorporation of multiple UAAs is observed. This strain may prove tremendously useful in the future.

References

1. Furter, R. (1998). Expansion of the genetic code: Site-directed p-fluoro-phenylalanine incorporation in *Escherichia coli*, *Protein Sci 7*, 419–426.
2. Wang, L., Brock, A., Herberich, B., and Schultz, P. G. (2001). Expanding the genetic code of *Escherichia coli*, *Science 292*, 498–500.
3. Liu, C. C., and Schultz, P. G. (2010). Adding new chemistries to the genetic code, *Annu Rev Biochem 79*, 413–444.
4. Hao, B., Gong, W., Ferguson, T. K., James, C. M., Krzycki, J. A., and Chan, M. K. (2002). A new UAG-encoded residue in the structure of a methanogen methyltransferase, *Science 296*, 1462–1466.
5. Srinivasan, G., James, C. M., and Krzycki, J. A. (2002). Pyrrolysine encoded by UAG in Archaea: Charging of a UAG-decoding specialized tRNA, *Science 296*, 1459–1462.
6. Namy, O., Zhou, Y., Gundllapalli, S., Polycarpo, C. R., Denise, A., Rousset, J. P., Soll, D., and Ambrogelly, A. (2007). Adding pyrrolysine to the *Escherichia coli* genetic code, *FEBS Lett 581*, 5282–5288.
7. Neumann, H., Peak-Chew, S. Y., and Chin, J. W. (2008). Genetically encoding N(epsilon)-acetyllysine in recombinant proteins, *Nat Chem Biol 4*, 232–234.
8. Kavran, J. M., Gundllapalli, S., O'Donoghue, P., Englert, M., Soll, D., and Steitz, T. A. (2007). Structure of pyrrolysyl-tRNA synthetase, an archaeal enzyme for genetic code innovation, *Proc Natl Acad Sci U S A 104*, 11268–11273.
9. Anderson, J. C., Wu, N., Santoro, S. W., Lakshman, V., King, D. S., and Schultz, P. G. (2004). An expanded genetic code with a functional quadruplet codon, *Proc Natl Acad Sci U S A 101*, 7566–7571.
10. Chatterjee, A., Xiao, H., and Schultz, P. G. (2012). Evolution of multiple, mutually orthogonal prolyl-tRNA synthetase/tRNA pairs for unnatural amino acid mutagenesis in *Escherichia coli*, *Proc Natl Acad Sci U S A 109*, 14841–14846.
11. Hughes, R. A., and Ellington, A. D. (2010). Rational design of an orthogonal tryptophanyl nonsense suppressor tRNA, *Nucleic Acids Res 38*, 6813–6830.
12. Chatterjee, A., Xiao, H., Yang, P. Y., Soundararajan, G., and Schultz, P. G. (2013). A tryptophanyl-tRNA synthetase/tRNA pair for unnatural amino acid mutagenesis in *E. coli*, *Angew Chem Int Ed Engl 52*, 5106–5109.
13. Chin, J. W., Cropp, T. A., Anderson, J. C., Mukherji, M., Zhang, Z., and Schultz, P. G. (2003). An expanded eukaryotic genetic code, *Science 301*, 964–967.

14. Hancock, S. M., Uprety, R., Deiters, A., and Chin, J. W. (2010). Expanding the genetic code of yeast for incorporation of diverse unnatural amino acids via a pyrrolysyl-tRNA synthetase/tRNA pair, *J Am Chem Soc 132*, 14819–14824.

15. Wu, N., Deiters, A., Cropp, T. A., King, D., and Schultz, P. G. (2004). A genetically encoded photocaged amino acid, *J Am Chem Soc 126*, 14306–14307.

16. Palzer, S., Bantel, Y., Kazenwadel, F., Berg, M., Rupp, S., and Sohn, K. (2013). An expanded genetic code in Candida albicans to study protein-protein interactions *in vivo*, *Eukaryot Cell 12*, 816–827.

17. Young, T. S., Ahmad, I., Brock, A., and Schultz, P. G. (2009). Expanding the genetic repertoire of the methylotrophic yeast *Pichia pastoris*, *Biochemistry 48*, 2643–2653.

18. Lemke, E. A., Summerer, D., Geierstanger, B. H., Brittain, S. M., and Schultz, P. G. (2007). Control of protein phosphorylation with a genetically encoded photocaged amino acid, *Nat Chem Biol 3*, 769–772.

19. Chen, P. R., Groff, D., Guo, J., Ou, W., Cellitti, S., Geierstanger, B. H., and Schultz, P. G. (2009). A facile system for encoding unnatural amino acids in mammalian cells, *Angew Chem Int Ed Engl 48*, 4052–4055.

20. Gautier, A., Nguyen, D. P., Lusic, H., An, W., Deiters, A., and Chin, J. W. (2010). Genetically encoded photocontrol of protein localization in mammalian cells, *J Am Chem Soc 132*, 4086–4088.

21. Greiss, S., and Chin, J. W. (2011). Expanding the genetic code of an animal, *J Am Chem Soc 133*, 14196–14199.

22. Bianco, A., Townsley, F. M., Greiss, S., Lang, K., and Chin, J. W. (2012). Expanding the genetic code of Drosophila melanogaster, *Nat Chem Biol 8*, 748–750.

23. Hino, N., Okazaki, Y., Kobayashi, T., Hayashi, A., Sakamoto, K., and Yokoyama, S. (2005). Protein photo-cross-linking in mammalian cells by site-specific incorporation of a photoreactive amino acid, *Nat Methods 2*, 201–206.

24. Liu, W. S., Brock, A., Chen, S., Chen, S. B., and Schultz, P. G. (2007). Genetic incorporation of unnatural amino acids into proteins in mammalian cells, *Nature Methods 4*, 239–244.

25. Parrish, A. R., She, X. Y., Xiang, Z., Coin, I., Shen, Z. X., Briggs, S. P., Dillin, A., and Wang, L. (2012). Expanding the genetic code of *Caenorhabditis elegans* using bacterial aminoacyl-tRNA synthetase/tRNA Pairs, *ACS Chemical Biology 7*, 1292–1302.

26. Allfrey, V. G., Faulkner, R., and Mirsky, A. E. (1964). Acetylation and methylation of histones and their possible role in the regulation of RNA synthesis, *Proc Natl Acad Sci U S A 51*, 786–794.

27. Ambrogelly, A., Gundllapalli, S., Herring, S., Polycarpo, C., Frauer, C., and Soll, D. (2007). Pyrrolysine is not hardwired for cotranslational insertion at UAG codons, *Proc Natl Acad Sci U S A 104*, 3141–3146.

28. Polycarpo, C. R., Herring, S., Berube, A., Wood, J. L., Soll, D., and Ambrogelly, A. (2006). Pyrrolysine analogues as substrates for pyrrolysyl-tRNA synthetase, *FEBS Lett 580*, 6695–6700.

29. Neumann, H., Hancock, S. M., Buning, R., Routh, A., Chapman, L., Somers, J., Owen-Hughes, T., van Noort, J., Rhodes, D., and Chin, J. W. (2009). A method for genetically installing site-specific acetylation in recombinant histones defines the effects of H3 K56 acetylation, *Mol Cell 36*, 153–163.

30. Tropberger, P., Pott, S., Keller, C., Kamieniarz-Gdula, K., Caron, M., Richter, F., Li, G., Mittler, G., Liu, E. T., Buhler, M., Margueron, R., and Schneider, R. (2013). Regulation of transcription through acetylation of H3K122 on the lateral surface of the histone octamer, *Cell 152*, 859–872.

31. Lammers, M., Neumann, H., Chin, J. W., and James, L. C. (2010). Acetylation regulates cyclophilin A catalysis, immunosuppression and HIV isomerization, *Nat Chem Biol 6*, 331–337.

32. Hsieh, Y. L., Kuo, H. Y., Chang, C. C., Naik, M. T., Liao, P. H., Ho, C. C., Huang, T. C., Jeng, J. C., Hsu, P. H., Tsai, M. D., Huang, T. H., and Shih, H. M. (2013). Ubc9 acetylation modulates distinct SUMO target modification and hypoxia response, *EMBO J 32*, 791–804.

33. Ullmann, R., Chien, C. D., Avantaggiati, M. L., and Muller, S. (2012). An acetylation switch regulates SUMO-dependent protein interaction networks, *Mol Cell 46*, 759–770.

34. Nguyen, D. P., Garcia Alai, M. M., Kapadnis, P. B., Neumann, H., and Chin, J. W. (2009). Genetically encoding N(epsilon)-methyl-L-lysine in recombinant histones, *J Am Chem Soc 131*, 14194–14195.

35. Nguyen, D. P., Garcia Alai, M. M., Virdee, S., and Chin, J. W. (2010). Genetically directing epsilon-N,N-dimethyl-L-lysine in recombinant histones, *Chem Biol 17*, 1072–1076.

36. Virdee, S., Ye, Y., Nguyen, D. P., Komander, D., and Chin, J. W. (2010). Engineered diubiquitin synthesis reveals Lys29-isopeptide specificity of an OTU deubiquitinase, *Nat Chem Biol 6*, 750–757.

37. Li, X., Fekner, T., Ottesen, J. J., and Chan, M. K. (2009). A pyrrolysine analogue for site-specific protein ubiquitination, *Angew Chem Int Ed Engl 48*, 9184–9187.

38. Virdee, S., Kapadnis, P. B., Elliott, T., Lang, K., Madrzak, J., Nguyen, D. P., Riechmann, L., and Chin, J. W. (2011). Traceless and site-specific ubiquitination of recombinant proteins, *J Am Chem Soc 133*, 10708–10711.

39. Xie, J., Supekova, L., and Schultz, P. G. (2007). A genetically encoded metabolically stable analogue of phosphotyrosine in *Escherichia coli*, *ACS Chem Biol 2*, 474–478.

40. Park, H. S., Hohn, M. J., Umehara, T., Guo, L. T., Osborne, E. M., Benner, J., Noren, C. J., Rinehart, J., and Soll, D. (2011). Expanding the genetic code of *Escherichia coli* with phosphoserine, *Science 333*, 1151–1154.

41. Lee, S., Oh, S., Yang, A., Kim, J., Soll, D., Lee, D., and Park, H. S. (2013). A facile strategy for selective incorporation of phosphoserine into histones, *Angew Chem Int Ed Engl 52*, 5771–5775.

42. Cooley, R. B., Feldman, J. L., Driggers, C. M., Bundy, T. A., Stokes, A. L., Karplus, P. A., and Mehl, R. A. (2014). Structural basis of improved second-generation 3-nitro-tyrosine tRNA synthetases, *Biochemistry 53*, 1916–1924.

43. Neumann, H., Hazen, J. L., Weinstein, J., Mehl, R. A., and Chin, J. W. (2008). Genetically encoding protein oxidative damage, *J Am Chem Soc 130*, 4028–4033.

44. Chin, J. W., Martin, A. B., King, D. S., Wang, L., and Schultz, P. G. (2002). Addition of a photocrosslinking amino acid to the genetic code of *Escherichia coli*, *Proc Natl Acad Sci U S A 99*, 11020–11024.

45. Mori, H., and Ito, K. (2006). Different modes of SecY-SecA interactions revealed by site-directed *in vivo* photo-cross-linking, *Proc Natl Acad Sci U S A 103*, 16159–16164.

46. Zimmer, J., Nam, Y., and Rapoport, T. A. (2008). Structure of a complex of the ATPase SecA and the protein-translocation channel, *Nature 455*, 936–943.

47. Schlieker, C., Weibezahn, J., Patzelt, H., Tessarz, P., Strub, C., Zeth, K., Erbse, A., Schneider-Mergener, J., Chin, J. W., Schultz, P. G., Bukau, B., and Mogk, A. (2004). Substrate recognition by the AAA+ chaperone ClpB, *Nat Struct Mol Biol 11*, 607–615.

48. Kaiser, C. M., Chang, H. C., Agashe, V. R., Lakshmipathy, S. K., Etchells, S. A., Hayer-Hartl, M., Hartl, F. U., and Barral, J. M. (2006). Real-time observation of trigger factor function on translating ribosomes, *Nature 444*, 455–460.

49. Merz, F., Boehringer, D., Schaffitzel, C., Preissler, S., Hoffmann, A., Maier, T., Rutkowska, A., Lozza, J., Ban, N., Bukau, B., and Deuerling, E. (2008). Molecular mechanism and structure of Trigger Factor bound to the translating ribosome, *EMBO J 27*, 1622–1632.

50. Shiota, T., Mabuchi, H., Tanaka-Yamano, S., Yamano, K., and Endo, T. (2011). *In vivo* protein-interaction mapping of a mitochondrial translocator protein Tom22 at work, *Proc Natl Acad Sci U S A 108*, 15179–15183.

51. Tamura, Y., Harada, Y., Shiota, T., Yamano, K., Watanabe, K., Yokota, M., Yamamoto, H., Sesaki, H., and Endo, T. (2009). Tim23-Tim50 pair coordinates functions of translocators and motor proteins in mitochondrial protein import, *J Cell Biol 184*, 129–141.

52. Yamano, K., Tanaka-Yamano, S., and Endo, T. (2010). Tom7 regulates Mdm10-mediated assembly of the mitochondrial import channel protein Tom40, *J Biol Chem 285*, 41222–41231.

53. Mohibullah, N., and Hahn, S. (2008). Site-specific cross-linking of TBP *in vivo* and *in vitro* reveals a direct functional interaction with the SAGA subunit Spt3, *Genes Dev 22*, 2994–3006.

54. Wu, C. C., Lin, Y. C., and Chen, H. T. (2011). The TFIIF-like Rpc37/53 dimer lies at the center of a protein network to connect TFIIIC, Bdp1, and the RNA polymerase III active center, *Mol Cell Biol 31*, 2715–2728.

55. Stanley, A. M., Carvalho, P., and Rapoport, T. (2011). Recognition of an ERAD-L substrate analyzed by site-specific *in vivo* photocrosslinking, *FEBS Lett 585*, 1281–1286.

56. Wilkins, B. J., Daggett, K. A., and Cropp, T. A. (2008). Peptide mass fingerprinting using isotopically encoded photo-crosslinking amino acids, *Mol Biosyst 4*, 934–936.

57. Forne, I., Ludwigsen, J., Imhof, A., Becker, P. B., and Mueller-Planitz, F. (2012). Probing the conformation of the ISWI ATPase domain with genetically encoded photoreactive crosslinkers and mass spectrometry, *Mol Cell Proteomics 11*, M111 012088.

58. Okuda, S., Freinkman, E., and Kahne, D. (2012). Cytoplasmic ATP hydrolysis powers transport of lipopolysaccharide across the periplasm in E. coli, *Science 338*, 1214–1217.

59. Wilkins, B. J., Rall, N. A., Ostwal, Y., Kruitwagen, T., Hiragami-Hamada, K., Winkler, M., Barral, Y., Fischle, W., and Neumann, H. (2014). A cascade of histone modifications induces chromatin condensation in mitosis, *Science 343*, 77–80.

60. Grunbeck, A., Huber, T., Abrol, R., Trzaskowski, B., Goddard, W. A., 3rd, and Sakmar, T. P. (2012). Genetically encoded photo-cross-linkers map the binding site of an allosteric drug on a G protein-coupled receptor, *ACS Chem Biol 7*, 967–972.

61. Tippmann, E. M., Liu, W., Summerer, D., Mack, A. V., and Schultz, P. G. (2007). A genetically encoded diazirine photocrosslinker in *Escherichia coli*, *Chembiochem 8*, 2210–2214.

62. Ai, H. W., Shen, W., Sagi, A., Chen, P. R., and Schultz, P. G. (2011). Probing protein-protein interactions with a genetically encoded photo-crosslinking amino acid, *Chembiochem 12*, 1854–1857.

63. Chou, C. J., Uprety, R., Davis, L., Chin, J. W., and Deiters, A. (2011). Genetically encoding an aliphatic diazirine for protein photocrosslinking, *Chem Sci 2*, 480–483.

64. Zhang, M., Lin, S., Song, X., Liu, J., Fu, Y., Ge, X., Fu, X., Chang, Z., and Chen, P. R. (2011). A genetically incorporated crosslinker reveals chaperone cooperation in acid resistance, *Nat Chem Biol 7*, 671–677.

65. Lin, S. X., Zhang, Z. R., Xu, H., Li, L., Chen, S., Li, J., Hao, Z. Y., and Chen, P. R. (2011). Site-specific incorporation of photo-cross-linker and bioorthogonal amino acids into enteric bacterial pathogens, *J Am Chem Soc 133*, 20581–20587.

66. Schmidt, M. J., and Summerer, D. (2013). Red-light-controlled protein-RNA crosslinking with a genetically encoded furan, *Angew Chem Int Edit 52*, 4690–4693.

67. Wu, Y. I., Frey, D., Lungu, O. I., Jaehrig, A., Schlichting, I., Kuhlman, B., and Hahn, K. M. (2009). A genetically encoded photoactivatable Rac controls the motility of living cells, *Nature 461*, 104–108.

68. Deiters, A., Groff, D., Ryu, Y., Xie, J., and Schultz, P. G. (2006). A genetically encoded photocaged tyrosine, *Angew Chem Int Ed Engl 45*, 2728–2731.

69. Luo, J., Uprety, R., Naro, Y., Chou, C., Nguyen, D. P., Chin, J. W., and Deiters, A. (2014). Genetically encoded optochemical probes for simultaneous fluorescence reporting and light activation of protein function with two-photon excitation, *J Am Chem Soc 136*, 15551–15558.

70. Gautier, A., Deiters, A., and Chin, J. W. (2011). Light-activated kinases enable temporal dissection of signaling networks in living cells, *J Am Chem Soc 133*, 2124–2127.

71. Kang, J. Y., Kawaguchi, D., Coin, I., Xiang, Z., O'Leary, D. D., Slesinger, P. A., and Wang, L. (2013). *In vivo* expression of a light-activatable potassium channel using unnatural amino acids, *Neuron 80*, 358–370.

72. Arbely, E., Torres-Kolbus, J., Deiters, A., and Chin, J. W. (2012). Photocontrol of tyrosine phosphorylation in mammalian cells via genetic encoding of photocaged tyrosine, *J Am Chem Soc. 134*, 11912–11915.

73. Szymanski, W., Beierle, J. M., Kistemaker, H. A. V., Velema, W. A., and Feringa, B. L. (2013). Reversible photocontrol of biological systems by the incorporation of molecular photoswitches, *Chem Rev 113*, 6114–6178.

74. Bose, M., Groff, D., Xie, J. M., Brustad, E., and Schultz, P. G. (2006). The incorporation of a photoisomerizable amino acid into proteins in E-coli, *J Am Chem Soc 128*, 388–389.

75. Kolb, H. C., Finn, M. G., and Sharpless, K. B. (2001). Click chemistry: Diverse chemical function from a few good reactions, *Angew Chem Int Ed Engl 40*, 2004–2021.

76. Chin, J. W., Santoro, S. W., Martin, A. B., King, D. S., Wang, L., and Schultz, P. G. (2002). Addition of p-azido-L-phenylalanine to the genetic code of *Escherichia coli*, *J Am Chem Soc 124*, 9026–9027.

77. Deiters, A., and Schultz, P. G. (2005). *In vivo* incorporation of an alkyne into proteins in *Escherichia coli*, *Bioorg Med Chem Lett 15*, 1521–1524.

78. Fekner, T., Li, X., Lee, M. M., and Chan, M. K. (2009). A pyrrolysine analogue for protein click chemistry, *Angew Chem Int Ed Engl 48*, 1633–1635.

79. Nguyen, D. P., Lusic, H., Neumann, H., Kapadnis, P. B., Deiters, A., and Chin, J. W. (2009). Genetic encoding and labeling of aliphatic azides and alkynes in recombinant proteins via a pyrrolysyl-tRNA Synthetase/tRNA(CUA) pair and click chemistry, *J Am Chem Soc 131*, 8720–8721.

80. Agard, N. J., Prescher, J. A., and Bertozzi, C. R. (2004). A strain-promoted [3 + 2] azide-alkyne cycloaddition for covalent modification of biomolecules in living systems, *J Am Chem Soc 126*, 15046–15047.

81. Baskin, J. M., Prescher, J. A., Laughlin, S. T., Agard, N. J., Chang, P. V., Miller, I. A., Lo, A., Codelli, J. A., and Bertozzi, C. R. (2007). Copper-free click chemistry for dynamic *in vivo* imaging, *Proc Natl Acad Sci U S A 104*, 16793–16797.

82. Plass, T., Milles, S., Koehler, C., Schultz, C., and Lemke, E. A. (2011). Genetically encoded copper-free click chemistry, *Angew Chem Int Ed Engl 50*, 3878–3881.

83. Lang, K., Davis, L., Torres-Kolbus, J., Chou, C., Deiters, A., and Chin, J. W. (2012). Genetically encoded norbornene directs site-specific cellular protein labelling via a rapid bioorthogonal reaction, *Nat Chem 4*, 298–304.

84. Lang, K., Davis, L., Wallace, S., Mahesh, M., Cox, D. J., Blackman, M. L., Fox, J. M., and Chin, J. W. (2012). Genetic encoding of bicyclononynes and trans-cyclooctenes for site-specific protein labeling *in vitro* and in live mammalian cells via rapid fluorogenic Diels-Alder reactions, *J Am Chem Soc 134*, 10317–10320.

85. Milles, S., Tyagi, S., Banterle, N., Koehler, C., VanDelinder, V., Plass, T., Neal, A. P., and Lemke, E. A. (2012). Click strategies for single-molecule protein fluorescence, *J Am Chem Soc 134*, 5187–5195.

86. Plass, T., Milles, S., Koehler, C., Szymanski, J., Mueller, R., Wiessler, M., Schultz, C., and Lemke, E. A. (2012). Amino acids for Diels-Alder reactions in living cells, *Angew Chem Int Ed Engl 51*, 4166–4170.

87. Brustad, E. M., Lemke, E. A., Schultz, P. G., and Deniz, A. A. (2008). A general and efficient method for the site-specific dual-labeling of proteins for single molecule fluorescence resonance energy transfer, *J Am Chem Soc 130*, 17664–17665.

88. Huang, Y., Wan, W., Russell, W. K., Pai, P. J., Wang, Z. Y., Russell, D. H., and Liu, W. S. (2010). Genetic incorporation of an aliphatic keto-containing amino acid into proteins for their site-specific modifications, *Bioorg Med Chem Lett 20*, 878–880.

89. Wang, L., Zhang, Z., Brock, A., and Schultz, P. G. (2003). Addition of the keto functional group to the genetic code of *Escherichia coli*, *Proc Natl Acad Sci U S A 100*, 56–61.

90. Nguyen, D. P., Elliott, T., Holt, M., Muir, T. W., and Chin, J. W. (2011). Genetically encoded 1,2-aminothiols facilitate rapid and site-specific protein labeling via a bio-orthogonal cyanobenzothiazole condensation, *J Am Chem Soc 133*, 11418–11421.

91. Cellitti, S. E., Ou, W., Chiu, H. P., Grunewald, J., Jones, D. H., Hao, X., Fan, Q., Quinn, L. L., Ng, K., Anfora, A. T., Lesley, S. A., Uno, T., Brock, A., and Geierstanger, B. H. (2011). D-Ornithine coopts pyrrolysine biosynthesis to make and insert pyrroline-carboxy-lysine, *Nat Chem Biol 7*, 528–530.

92. Ou, W., Uno, T., Chiu, H. P., Grunewald, J., Cellitti, S. E., Crossgrove, T., Hao, X., Fan, Q., Quinn, L. L., Patterson, P., Okach, L., Jones, D. H., Lesley, S. A., Brock, A., and Geierstanger, B. H. (2011). Site-specific protein modifications through pyrroline-carboxy-lysine residues, *Proc Natl Acad Sci U S A 108*, 10437–10442.

93. Milles, S., and Lemke, E. A. (2011). Single molecule study of the intrinsically disordered FG-repeat nucleoporin 153, *Biophys J 101*, 1710–1719.

94. Chakraborty, A., Wang, D., Ebright, Y. W., Korlann, Y., Kortkhonjia, E., Kim, T., Chowdhury, S., Wigneshweraraj, S., Irschik, H., Jansen, R., Nixon, B. T., Knight, J., Weiss, S., and Ebright, R. H. (2012). Opening and closing of the bacterial RNA polymerase clamp, *Science 337*, 591–595.

95. Rackham, O., and Chin, J. W. (2005). A network of orthogonal ribosome x mRNA pairs, *Nat Chem Biol 1*, 159–166.

96. Li, G. W., Oh, E., and Weissman, J. S. (2012). The anti-Shine-Dalgarno sequence drives translational pausing and codon choice in bacteria, *Nature 484*, 538–541.

97. Barrett, O. P., and Chin, J. W. (2010). Evolved orthogonal ribosome purification for *in vitro* characterization, *Nucleic Acids Res 38*, 2682–2691.

98. Wang, K., Neumann, H., Peak-Chew, S. Y., and Chin, J. W. (2007). Evolved orthogonal ribosomes enhance the efficiency of synthetic genetic code expansion, *Nat Biotechnol 25*, 770–777.

99. Neumann, H., Wang, K., Davis, L., Garcia-Alai, M., and Chin, J. W. (2010). Encoding multiple unnatural amino acids via evolution of a quadruplet-decoding ribosome, *Nature 464*, 441–444.

100. Lammers, C., Hahn, L. E., and Neumann, H. (2014). Optimized plasmid systems for the incorporation of multiple different unnatural amino acids by evolved orthogonal ribosomes, *Chembiochem 15*, 1800–1804.

101. Sachdeva, A., Wang, K., Elliott, T., and Chin, J. W. (2014). Concerted, rapid, quantitative, and site-specific dual labeling of proteins, *J Am Chem Soc 136*, 7785–7788.

102. Wang, K., Sachdeva, A., Cox, D. J., Wilf, N. W., Lang, K., Wallace, S., Mehl, R. A., and Chin, J. W. (2014). Optimized orthogonal translation of unnatural amino acids enables spontaneous protein double-labelling and FRET, *Nat Chem 6*, 393–403.

103. Neumann, H., Slusarczyk, A. L., and Chin, J. W. (2010). De novo generation of mutually orthogonal aminoacyl-tRNA synthetase/tRNA pairs, *J Am Chem Soc 132*, 2142–2144.

104. Farrell, I. S., Toroney, R., Hazen, J. L., Mehl, R. A., and Chin, J. W. (2005). Photo-cross-linking interacting proteins with a genetically encoded benzophenone, *Nat Methods 2*, 377–384.

105. Ryu, Y., and Schultz, P. G. (2006). Efficient incorporation of unnatural amino acids into proteins in *Escherichia coli*, *Nat Methods 3*, 263–265.

106. Young, T. S., Ahmad, I., Yin, J. A., and Schultz, P. G. (2010). An enhanced system for unnatural amino acid mutagenesis in *E. coli*, *J Mol Biol 395*, 361–374.

107. Ryden, S. M., and Isaksson, L. A. (1984). A temperature-sensitive mutant of *Escherichia coli* that shows enhanced misreading of UAG/A and increased efficiency for some tRNA nonsense suppressors, *Mol Gen Genet 193*, 38–45.

108. Wu, I. L., Patterson, M. A., Carpenter Desai, H. E., Mehl, R. A., Giorgi, G., and Conticello, V. P. (2013). Multiple site-selective insertions of noncanonical amino acids into sequence-repetitive polypeptides, *Chembiochem 14*, 968–978.

109. Huang, Y., Russell, W. K., Wan, W., Pai, P. J., Russell, D. H., and Liu, W. (2010). A convenient method for genetic incorporation of multiple noncanonical amino acids into one protein in *Escherichia coli*, *Mol Biosyst 6*, 683–686.

110. Mukai, T., Hayashi, A., Iraha, F., Sato, A., Ohtake, K., Yokoyama, S., and Sakamoto, K. (2010). Codon reassignment in the *Escherichia coli* genetic code, *Nucleic Acids Res 38*, 8188–8195.

111. Johnson, D. B., Xu, J., Shen, Z., Takimoto, J. K., Schultz, M. D., Schmitz, R. J., Xiang, Z., Ecker, J. R., Briggs, S. P., and Wang, L. (2011). RF1 knockout allows ribosomal incorporation of unnatural amino acids at multiple sites, *Nat Chem Biol 7*, 779–786.

112. Lajoie, M. J., Kosuri, S., Mosberg, J. A., Gregg, C. J., Zhang, D., and Church, G. M. (2013). Probing the limits of genetic recoding in essential genes, *Science 342*, 361–363.

113. Lajoie, M. J., Rovner, A. J., Goodman, D. B., Aerni, H. R., Haimovich, A. D., Kuznetsov, G., Mercer, J. A., Wang, H. H., Carr, P. A., Mosberg, J. A., Rohland, N., Schultz, P. G., Jacobson, J. M., Rinehart, J., Church, G. M., and Isaacs, F. J. (2013). Genomically recoded organisms expand biological functions, *Science 342*, 357–360.

114. Wang, H. H., Isaacs, F. J., Carr, P. A., Sun, Z. Z., Xu, G., Forest, C. R., and Church, G. M. (2009). Programming cells by multiplex genome engineering and accelerated evolution, *Nature 460*, 894–898.

Index

Printed and bound by CPI Group (UK) Ltd, Croydon, CR0 4YY

01/11/2024

01782601-0004